Palgrave Hate Studies

Series Editors

This series builds on recent developments in the broad and interdisciplinary field of hate studies. Palgrave Hate Studies aims to bring together in one series the very best scholars who are conducting hate studies research around the world. Reflecting the range and depth of research and scholarship in this burgeoning area, the series welcomes contributions from established hate studies researchers who have helped to shape the field, as well as new scholars who are building on this tradition and breaking new ground within and outside the existing canon of hate studies research.

More information about this series at
http://www.palgrave.com/series/14695

Andrew Jakubowicz • Kevin Dunn
Gail Mason • Yin Paradies
Ana-Maria Bliuc • Nasya Bahfen
Andre Oboler • Rosalie Atie
Karen Connelly

Cyber Racism and Community Resilience

Strategies for Combating Online Race Hate

palgrave
macmillan

Andrew Jakubowicz
School of Communication
University of Technology Sydney
Ultimo, New South Wales, Australia

Gail Mason
Law School
University of Sydney
Sydney, New South Wales, Australia

Ana-Maria Bliuc
Western Sydney University
Penrith, New South Wales, Australia

Andre Oboler
La Trobe University
Bundoora, Victoria, Australia

Karen Connelly
University of Technology Sydney
Ultimo, New South Wales, Australia

Kevin Dunn
Social Sciences and Psychology
Western Sydney University
Penrith, New South Wales, Australia

Yin Paradies
Deakin University
Burwood, Victoria, Australia

Nasya Bahfen
La Trobe University
Bundoora, Victoria, Australia

Rosalie Atie
Social Sciences and Psychology
Western Sydney University
Penrith, New South Wales, Australia

Palgrave Hate Studies
ISBN 978-3-319-64387-8 ISBN 978-3-319-64388-5 (eBook)
DOI 10.1007/978-3-319-64388-5

Library of Congress Control Number: 2017954206

Cover image provided by Simon Wiesenthal Center, Rabbi Abraham Cooper Associate Dean, Director Global Social Action Agenda

Printed on acid-free paper

This Palgrave Macmillan imprint is published by Springer Nature
The registered company is Springer International Publishing AG
The registered company address is: Gewerbestrasse 11, 6330 Cham, Switzerland

Preface

Background

Cyber racism, the spread of race hate speech through the Internet using the World Wide Web, has moved into centre stage in public debates across the world (Daniels 2010). Once considered a minor, if unfortunate, consequence of the freedom built into the Web, public concern has grown as those freedoms have magnified the impact of hate (The Guardian 2017). Targets include refugees and ethnic minorities in Europe, Muslim Blacks and Jews in the United States, Indigenous Australians and refugees and a multitude of ethnic and sub-national groups across Africa and Asia. Critical changes in both technology and politics have driven the growth of hate speech, as the multiplication of opportunities in social media has been incorporated into the growth of religious, ethnic and racial radicalisms. The Web enables the almost instantaneous creation of networks of association, while the isolation and potential anonymity of perpetrators protect them from accountability (Jakubowicz 2012). By early 2017, over 3.5 billion people, half the world's population, were connected to the Internet, a network that barely existed in 1997.

In the early 1990s, a team of researchers at the University of Technology Sydney undertook a study of Racism Ethnicity and the Media (Jakubowicz et al. 1994). The research took place just before the advent of the Internet and well before the appearance of the World Wide Web. At the conclusion of the study, the authors commented, "While the fact that the media are part of the power structure of Australian society may help explain the pattern of media representation, it does not excuse it…" (p. 192). That book argued that industry self-regulation would ensure the protection of corporate self-interest, that little of substance would change, while the politics of post-colonial nations would contribute to how the media interpreted and responded to growing cultural diversity and political inequalities.

This book takes the story on a generation, into the heart of Web 2.0 and into the shadow of the Internet of Things, where social media have transformed the nature of communication, even though the power of race to determine life chances and opportunity appears hardly to have changed. A generation ago, overt racism in the media was still apparent, though the Australian media often denied serious racism existed in Australia and urged governments not to succumb to claims by the Human Rights Commission in its Inquiry into Racist Violence (Moss and Castan 1991) that racism was widespread or structural. (Jakubowicz et al. 1994).

The project from which this book is drawn grew out of a sense of frustration among the project initiators that the spread of racism in the real world, fuelled by the spread of hate online, seemed to be resistant to traditional forms of social policy. Andrew Jakubowicz (sociologist), Gail Mason (criminologist) and Kevin Dunn (social geographer) had all been invited speakers at a 2010 Forum organised by the Australian Human Rights Commission (AHRC) on race hate speech online.

The Commission had held a previous forum in 2002, though its recommendations at that time that the government action to sign up to a European hate crime protocol were essentially ignored. Unlike many countries in Europe, Australia did not have national legislation on racial vilification at all until 1996, despite having signed the International Convention on the Eradication of All Forms of Racial Discrimination in 1966. Australia had acceded to the Convention in 1975, with the passing of the Racial Discrimination Act, but its reservation to the key Article 4 on vilification (similarly to the USA) meant that it had little in the way of rights-based legislation to direct against the spread of racism online. The trigger for the 2002 forum was the Australian government's consideration, as part of its accession to the European Convention on Cyber

Crime, to also take on the additional protocol on Cyber Racism. It had been a decade since the Human Rights Commission had proposed that Australia move ahead on legislation to combat racial vilification (Moss and Castan 1991). The Forum supported Australia signing on to the additional protocol, a position strongly opposed by industry representatives. Ultimately, despite the consensus among human rights groups of support, the industry (well before Facebook, Google or YouTube) successfully lobbied the government, and the additional protocol was dropped.

The 2010 event, occurring as social media appeared on the scene, was triggered by a request from the then new attorney general that the Commission investigate what might be done to deal with community complaints about antisemitic outbursts online posted by students at an exclusive Sydney private school, and the rise in hate speech directed against Muslims who as a members of a religious faith were not covered by Australian racial vilification laws. The second forum also produced a stalemate of an outcome—the Commission was reluctant to suggest legislative changes, the Internet industry was reluctant to accept any further regulation or responsibility, the media regulator was reluctant to extend its brief to detailed coverage of social media and the law enforcement authorities complained that there were no laws that might allow them to proactively intervene.

The three researchers then collaborated to initiate a research project that could explore the problem further. The scope would identify the extent and impact of hate speech associated with race and religion. Further, the project would explore with partners what sorts of strategies might prove to be effective in building resilience among targeted communities, support among bystander communities and innovative cooperation between civil society, government and industry.

The initial group was extended to include Yin Paradies (race relations researcher), a long-term research collaborator with Kevin Dunn, and then recruited two early career researchers, Ana-Maria Bliuc (social psychologist) and Nasya Bahfen (media studies), while Rosalie Atie joined from the Challenging Racism project. Coincidentally, the team thus comprised members with Indigenous, Jewish, Muslim, Indonesian/Arab and Eastern European backgrounds, as well as Anglo-Australian.

Half were male, and half were female. Institutional partners included the Victorian Community Health (VicHealth) foundation, the AHRC Discrimination commissioner and the Federation of Ethnic Communities Councils of Australia (FECCA). We joined with the Online Hate Prevention Institute, whose CEO Andre Oboler, with a background in both information technology and law, became part of our writing team. Finally, we recruited a doctoral student, anthropologist Karen Connelly. We were successful in securing an Australian Research Council Linkage Grant (LP120200115) (2012–2016) with the three community/industry research partners, the AHRC, VicHealth and the FECCA.

Our methodology needed to tie the empirical research into the extent and nature of the experiences of online racism, together with an exploration of how the phenomenon might be addressed through strategies ranging from legislative interventions to online engagement to wider social action. In the Australian context, it was quite fortuitous that our project began at a time when pressure was mounting from the conservative side of Australian politics for the withdrawal of the critical section on racial vilification from the Australian Racial Discrimination Act (Part IIA—especially Section 18c). This meant that there was ongoing and robust public debate about what might be encompassed by the idea of racial vilification, and what would be the range of legitimate responses. Major inquiries in 2014 and 2016 were launched by the government, to both of which our group made research-based submissions. In the second and public inquiry (the first was never released), our research was directly quoted in key deliberative sections of the Parliamentary committee report (Parliament of Australia 2017).

The public discussion of how to respond to race hate speech (both online and offline) provided invaluable insights into an evolving discourse around race and freedom of speech, particularly as the environment increasingly focused on threats to the social order occasioned by the spread of terrorism. One of our case studies, the online criticisms by Andrew Bolt, a conservative media commentator, of an Indigenous footballer, Adam Goodes, gave us two aspects of the complex situation. The Federal Court in 2011 found Bolt to have breached RDA Section 18c without an acceptable defence, and he was thus an "objectively" identifiable practitioner of race hate speech. His critique of Goodes in 2014 generated a rich, if poisoned, reservoir of commentary from his readers, generally abusing the prominent Indigenous celebrity (Australian of the Year 2014). Moreover, Bolt became a *cause célèbre* for the conservative cause, not only because he was employed by *News Limited*, which was closely associated through its newspapers with the conservative political interest, but also because many of the conservative and what became the alt-right pressure groups adopted his situation as a signal martyrdom for their cause (Gelber and McNamara 2013).

While the research was undertaken in a politically heightened context, at the same time we faced many of the same ethical issues that have been identified by other researchers in the field. People who pursue racist agendas often use the responses of their targets and the wider society as a means to propagate their views. They count on commentary in the media and the reactions of the community, governments and judicial bodies to amplify their messages. We may inadvertently alert propagators of racism to how they might become more effective. Our work will bring information about race hate propagandists and their "homes" to people who may not have previously been aware of them. It will add, if only a tiny amount, to the notoriety they crave. However, we believe it is crucial to lay out the narratives and arguments of racists, as they cannot be addressed nor "called out" if the wider society knows nothing of their existence. Moreover, a systematic codification of what should count as unacceptable race hate speech on the Internet contributes significantly to the recognition of racism for those who encounter it unawares. Our research shows how the capacity to recognise and identify racism online correlates highly with resistance to race hate speech online, while also being associated with anti-racist values in the offline world.

Our view could best be summed up in these words: active intervention that helps people recognise racism, call it out, fend off its hurt and join together with others to move forward, should be the focus for public policy. An aware, resourced and empathetic public sphere backed by effective legal resorts can play a major role in building resilient communities, undermining the corrosive effects of online racism, while sustaining democratic and liberal values.

Outline of the Book

Chapter 1 lays out the political economy of the Internet in the age of social media, proposing that the combination of technology and economy provides a powerful driver towards unimpeded opportunities to express and circulate racist ideas. Moreover, while regulation of the technology recognises the problems of hate speech, the barriers to effective intervention remain high. Regulation of content faces even greater challenges from a world divided over what should be controlled. The chapter then introduces some of the conflicting ideas about race and racism, including the impact of racism on those who are its targets and victims. The chapter concludes by canvassing the multiple meanings of "resilience," pointing to some of the implications of the concept as a realised program for social well-being.

Chapter 2 lays out an argument for the value of interdisciplinarity in research. Drawing on a range of social science and humanities disciplines, the review of approaches points to how this book utilised the diversity of perspectives in the research team.

Chapter 3 reports on the online survey that collected data from over 2000 Australian regular Internet users. We asked them about their encounters with racism, what they thought racism meant and what action they would like to see emerge to limit the pervasiveness and impact of racism online. The survey over-sampled some groups known to be the targets of racism, to ensure a depth of experience could be tapped. Targets of, bystanders to and perpetrators of racism online were all examined to understand the differences between them and the issues that they prioritised.

Chapter 4 drills down into what attracts racists to the Internet, and the different ways in which racism is "performed" online. Three cases—associated with Jewish, Muslim and Indigenous communities—are explored, which serve to link psychological, sociological and political insights, and the implications for resilience down the track.

Chapter 5 then explores the experiences of targets of cyber racism, both in terms of institutional experiences and responses, and through the eyes of groups who rarely have their voices heard. Six groups and two community organisations provided detailed information—covering

Muslims, Jews, Africans, people from the Middle East, Chinese and Australian Indigenous online users. The targets share among themselves a sense of the growing impact of hate speech, both among their ethnicities of origin and through their responses. These responses have been characterised by withdrawal from arenas where bridging social capital might have blossomed, to locations where they focus on building intra-communal bonding social capital locked away from a wider and more threatening world. However, different communities experience the dangers and threats in very different ways and with highly varied impact.

Chapter 6 examines the discourses of racism, and how narratives are formulated to carry racist messages and sustain the dominance of some groups over others. Narratives about Australian nationalism provide a framework for the examination of specific discursive elements in racism online. These include discourses that seek to legitimise and delegitimise acceptable identities, the use of rhetoric and the forms of language that appear in racist discourse. Particular emphasis is placed on how communities are built in cyberspace, and how they are threatened by racist attacks—indeed how such attacks are used to attract and hold counter-communities.

Chapter 7 looks to how communities of resistance to racism might be built and their solidarity be achieved and sustained. A range of online communities are described, and then six Australian groups are analysed in terms of their capacity to counteract specific forms of racism online. Reflections on the more successful strategies provide elements for a model of resilience.

Chapter 8 focuses on the regulatory regimes that have been constructed at national and transnational levels, identifying the issues in different countries or groups of countries in relation to philosophies of law. Specific attention is paid to self-regulatory approaches by the Internet industry, criminal law and its limitations, and civil and administrative law remedies.

Chapter 9 reflects on what the findings of the empirical research mean for the current approaches to pushing back against race hate online. Drawing on actions in Europe, North America and Australia, the difficulties with current approaches are summarised and the outcomes of current initiatives are crystallised. Drawing on our research we create an analytical matrix that can help shape specific strategies for the different fields we have discovered underpin the growth of Internet racism. Finally, the chapter offers a portfolio of approaches that together might build community resilience in a way that ensures the collaboration of civil society, the Internet industries and the government at the local, national and transnational levels.

Acknowledgements

A project like this always represents collaboration among many minds. The primary authors acknowledge the researchers who provided much of the background material. These included Mark Bahnisch, Jonathan Westenberg, Hazel Maxwell, Matty Johnson and James Maccallum at the UTS, Natalie Czapski and Carolyn McKay at University of Sydney and Amanuel Elias at Deakin University. Our community allies at the Online Hate Prevention Institute (Andre Oboler and Chetna Prakash), All Together Now (Priscilla Brice) and the Federation of Ethnic Communities Councils of Australia (Tanya von Ahlefeldt, Dini Liyanarachchi, Pino Migliorino and Joe Caputo) have ensured our research remains rooted in the real world. Our contract research teams at CIRCA (led by Victoria Smith) played a key role in ensuring the qualitative data we needed were collected efficiently and processed with sensitivity and accuracy. Penny Jones has been a stalwart editorial eye and voice of reason in the final rush to the line.

We also acknowledge the funding and in-kind support from the Australian Research Council (under Grant LP120200115), the Victorian Health Promotion Foundation (VicHealth), the Federation of Ethnic Communities Councils of Australia (FECCA), and the Australian Human Rights Commission (AHRC), and the support from our universities especially Melbourne University, Deakin University and University of Technology Sydney for the Higher Degree Research student stipend attached to the project. We particularly want to thank the various "wranglers" at our partner researchers, especially Rivka Nissim, Tim Soutphommasane and Helen Szoke at the AHRC, Irene Verins and her team at VicHealth, Pino Migliorino and Joe Caputo, chairs of FECCA, and Andre Oboler who joined the writing team from his deep experience as CEO of the Online Hate Prevention Institute (OHPI).

Special thanks to colleagues who have commended this book to our readers, Prof David Theo Goldberg, Director and Professor University of California Humanities Research Institute|, Executive Director Digital Media and Learning Research Hub, Prof Karen Ross, Karen Ross, Professor of Gender and Media, School of Arts and Cultures Newcastle University UK, and Prof James Curran Goldsmiths College, University of London. We also recognise the foundation work in the field of cyber racism and human rights by Istanbul Bilgi University Professor of Law Dr Yaman Akdeniz, whose insights on cyber racism have been critical in the development of the research on which this book is based, and whose public stand on freedom of the Internet has been inspiring.

Finally our families in their diverse and sometimes changing forms have played an important part in sustaining our sense of human contact and ensuring we realise that there is a life outside the research pond. Ultimately of course this book is our work and we accept full responsibility for what we have made of our data and what we feel empowered to say about its implications.

Sydney and Melbourne, August 2017
The CRaCR research group

Daniels, J. (2010). Cyber Racism & The Future of Free Speech (Updated). *Racism Review: Scholarship and Activism Towards Racial Justice*, 2016(13 November). http://www.racismreview.com/blog/2010/11/16/cyber-racism-future-of-free-speech/. Accessed 16 Dec 2016.

Gelber, K. & McNamara, L. (2013). Freedom of Speech and Racial Vilification in Australia: 'The Bolt Case' in Public Discourse. *Australian Journal of Political Science,* 48(4), pp. 470–484.

Jakubowicz, A. (2012). 'Cyber Racism', in *More or Less: Democracy and New Media* [Online]. Version. http://www.futureleaders.com.au/book_chapters/pdf/More-or-Less/Andrew_Jakubowicz.pdf. Accessed 16 July 2016.

Jakubowicz, A., Goodall, H., Martin, J., Mitchell, T., Randall, L. & Seneviratne, K. (1994). *Racism, Ethnicity and the Media.* Sydney: Allen and Unwin.

Moss, I. & Castan, R. (1991). *Racist Violence: Report of the National Inquiry into Racist Violence.* Canberra: Human Rights and Equal Opportunity Commission. https://view.officeapps.live.com/op/view.aspx?src=https://www.humanrights.gov.au/sites/default/files/document/publication/NIRV.doc. Accessed 6 Sept 2016.

Parliament of Australia. (2017). *Freedom of Speech in Australia: Inquiry into the Operation of Part IIA of the Racial Discrimination Act 1975 (Cth) and Related Procedures Under the Australian Human Rights Commission Act 1986 (Cth).* Canberra: Parliament of Australia. http://www.aph.gov.au/Parliamentary_Business/Committees/Joint/Human_Rights_inquiries/FreedomspeechAustralia/Report. Accessed 3 Apr 2017.

The Guardian. (2017). *Facebook Files: The Guardian* (Online). https://www.theguardian.com/news/series/facebook-files. Accessed 24 May 2017.

Contents

CRaCR Author Biographies

Rosalie Atie is a PhD candidate in the school of Social Sciences and Psychology at Western Sydney University. Her PhD project looks at slam poetry as a space for alternative storytelling. Rosalie also works with the Challenging Racism Project, and her other areas of interest include racism and anti-racism and Islam in Australia.

Nasya Bahfen is a Senior Lecturer in Media and Communication at La Trobe University and an Australian Football League multicultural ambassador. Her research interests include the intersection of media, sport and diversity.

Ana-Maria Bliuc is a Lecturer in Social Psychology at the Western Sydney University. Broadly, her research explores intergroup relations in contexts of socio-political conflict using diverse methodological approaches. As such, most of her expertise is around potentially divisive issues such as racism, White supremacy and multiculturalism.

Karen Connelly is a PhD candidate at the University of Technology Sydney whose research is part of the Cyber-Racism and Community Resilience Project. As an anthropologist, Karen focuses on cultural exclusion narratives and counter-narratives on Facebook and Twitter.

Kevin Dunn is Dean of Social Sciences and Psychology, Professor in Human Geography and Urban Studies, Western Sydney University. His areas of research include the geographies of racism and anti-racism, and Islam in Australia. He leads the national Challenging Racism Project.

Andrew Jakubowicz is Professor of Sociology at the University of Technology Sydney. He holds an honours degree in Government, and a PhD in Sociology. His research covers media, social policy and cultural diversity.

Gail Mason is Professor of Criminology at the University of Sydney. Her research interests include crime and social justice, hate crime and sexual assault. She is the co-author of "Policing Hate Crime: Understanding Communities and Prejudice'" (2017: Routledge).

Andre Oboler is a Lecturer in Law at La Trobe University and CEO of the Online Hate Prevention Institute. He holds a PhD in Computer Science from Lancaster University and a Juris Doctor from Monash University.

Yin Paradies is Professor in Race Relations at the Alfred Deakin Research Institute for Citizenship and Globalisation, Deakin University. He conducts interdisciplinary research on health, social and economic effects of racism as well as anti-racism theory, policy and practice.

List of Figures

List of Tables

1

Context: "Cyberspace," "Race" and Community Resilience

Word and Deed

The cyber world and the world of action are not distinct. On January 29, 2017, a Canadian student attacked a mosque in Quebec City, murdering six worshippers. Alexandre Bissonnette (Narayan 2017), reputedly a strong nationalist, joined Norwegian killer Anders Breivik (Jakubowicz 2011) as an outstanding global example of the murderous capacity of White power. Bissonnette had apparently been turned towards White power by the visit to Canada of Marine Le Pen, head of France's Front National, and Breivik had fondly quoted Australian anti-multiculturalists in his online manifesto. Each had long histories of participating in White power online communities. On the same day, the Australian media (Olding 2017) reported the case of a local "White supremacist" Michael Holt who had pleaded guilty to stockpiling weapons in preparation for a mass rampage in a local shopping mall. He was not known to have been directly connected with any White power group, but had been actively involved in online hate speech. He had also searched the Web, constantly in pursuit of information about what he should believe and how he should act. There is no doubt that the opportunities afforded by the online world have expanded the reach and impact of race hate across the planet.

A. Jakubowicz et al., *Cyber Racism and Community Resilience*, Palgrave Hate Studies,
DOI 10.1007/978-3-319-64388-5_1

From the earliest appearance of the Internet as a publicly accessible network of communication, individuals and organisations with radical racist agendas have seen it as their natural hunting ground (Richardson 1998). American White power groups, Islamist crusaders, anti-Islamic crusaders, European neo-Nazis, alt-right anti-Semites, Australian ultra-nationalists and revanchist separatist groups using race or ethnicity as their imaginary nirvana have all discovered the almost endless possibilities of community, networking, secret communication and anonymous targeting of enemies. The Internet has too often attracted users who want to access it as a place to pursue the targets of their hate, described by one of our research subjects as people who want "to twist in the knife".

This book reports the results of a five-year study of the evolution of racism on the Internet, mainly in Australia, but always within the borderless reality of global cyberspace. It describes the way in which the Internet has afforded the opportunities for racism to grow, spurred on by the intimate relationship between technology, economy and power that underpins this transformation of global realities. It explores how cyber racism erodes trust and the underpinning of social cohesion in multicultural societies. It concludes that this erosion can only be met by strategies that build community resilience.

In the first chapter, we lay out the broad political economy of race and the Internet. We explore the structure of technology, economy and power that has come to be realised in the emergence of the transnational super-corporations within whose structures and through whose products and services racism occurs. We then drill down to the next level, where processes of regulation form, are resisted and transform. We look at how things have changed since the major studies undertaken in the first decade of the century have been overtaken by new technologies, new questions of regulation and new environments of racialised conflict and racial empowerment.

How the Internet Began to Grow Opportunities for Racism

As a networked transnational and in some ways post-national global society, cyberspace has shown the presence of racism as one of its most easily recognisable features. It is scarcely possible to enter any sector of the contemporary social media world without tripping over an image, meme, video, Instagram post, Facebook page or Tweet that uses racialised difference to demean or intimidate somebody or some group. It may be done with humour, introducing a joking relationship of power between the perpetrator, the bystanders and the targets or victims. It may ooze unconsciously out of the preferences expressed and the messages left on dating sites. It may stridently announce itself in the self-aggrandising websites, blogs and Facebook pages of individuals whose pride in their communal heritage has transformed into the systematic abuse of people who differ from them. It may exude from thousands of trolling tweets on Twitter, where racist and sexist abuse have become some of the more problematic and unattended consequences of un-bordered communication. But wherever it flows, its intent is hurt, and its aspiration is humiliation.

Racism is a relationship between people who are defined in some sense by their genotypes, phenotypes or cultures—who their parents were, what they look like and how they live their lives (Lentin 2008). Racism is always a relationship of power, where either the practices of discrimination and oppression or the language of those relationships illuminates deeper histories and continually contested realities (Roseman 2014). Racism can be embedded in structures of societies (Williams 2009), in the benefits that flow and the disadvantages that inhibit. It can affect peoples' sense of agency, corrupting their awareness of common humanity and permitting deep and callous pain to be inflicted (Emirbayer and Desmond 2012). Racism erodes the perpetrator and corrodes the victim; it distorts economies and limits opportunities; it lessens social development and shortens lives. It can produce in all parties to the relationship an increasingly destructive process of aggression, leading, in some cases, to violence and murder.

Racism today cannot be understood without understanding the Internet; the Internet today cannot be understood without understanding racism (Daniels 2013). Over the past 15 years, a number of insightful studies have shown us how the development of the Internet has been so inflected by racism. In North America, the analysis of global White power and its local manifestations, or, perhaps better, the spread and infilling of White power into the interstices of the spreading Internet, has developed as a key point of intersection for committed scholars concerned about future tendencies in intergroup conflict.

The concept of "cyber racism" has its origins in research by Les Back in the UK early in the period of the Internet (Back 2002). Drawing on Walter Benjamin's analysis of Nazism as a philosophy of terror built on an aesthetic of communitarianism, Back explores how fascism developed its own popular culture that could carry, through emotion, large masses of people into the fold of regimented authoritarianism through a type of "moral indolence." "The celluloid enhancement of racial narcissism prompted a widespread indifference towards its victims," Back argues (Back 2002, p. 628), thereby embedding a normalised barbarism as part of the everyday. Moreover, the means of mechanical reproduction—photography, cinema and radio—were critical to the rise and spread of Nazism (Ware and Back 2002). Eighty years later, "the Net has provided a means for people to sense, listen, feel and be involved intimately in racist culture from a distance," both publicly through the Internet and privately in front of their own screens (Back 2002, p. 629). Back set out to develop an analysis that could "make critical theory speak to political realities and vice versa" (Back 2002, p. 631) in order to move beyond what he saw as sterile conversations about Internet censorship, and technologically unattached explorations of cultures of fascism.

He deployed "the notion of 'cyber racism'" to encompass a range of sub-cultural movements, which commonly contain rhetorics of racial, national or sectarian uniqueness and common destiny, leavened by ideas of racial supremacy and separation, fed by a repertoire of racial otherness and sustained usually by a utopian (or dystopian) worldview (Back 2002, p. 632). Back focused on documenting the emergent culture of White power, in particular its globalisation, through the sharing of discourses of whiteness. White was defined against the Jew, the Black and the mongrelised races, reinvigorating the once-thought dead stereotypes and imagery of Nazism's targets, and circulating them widely to new audiences and across national borders. Moreover, Back was able to show how, importantly, the Internet had provided culturally supportive spaces for racists, such that when one of Back's subjects abandoned White power, the activist was bereft. Intense, competitive but short-lived seemed to be the cultural attraction for some people of life in online White power communities.

Back also raised the role that the technology of the Internet played in squeezing time and space, and thereby intensifying schismatic tendencies within and between racist groups. Moreover, he was extremely prescient in his predictions, albeit he did not imagine the explosion generated by Web 2.0 and the growth of social media. "The real danger is perhaps that in the 'informational age' isolated acts of racist terrorism may become commonplace" (Back 2002, p. 642). Breivik, Bissonnette and Holt are just three of those who trod the road that Back fearfully foretold.

Drawing inspiration from Back, Jesse Daniels focused more centrally on the USA to explore how White supremacists used the Internet to attract young people, identifying the key dynamic in "the persuasive story telling of hate" (Daniels 2009). In doing so, she pointed to the links between masculinities and racism, and how the technologies of gender intertwined with those of race. In imaginative research, she sought to identify why young people were attracted to the ideologies of race, and how their own senses of self were drawn into the narratives they consumed and then reproduced. One of her more challenging insights identified their sense of empowerment, of increased agency, that becoming part of a racist community provided to its participants. As was clear in the cases of Bissonnette, Breivik and Holt, and indeed as has been shown with the radicalisation of young Muslims (Young et al. 2016), the online community can also provide a place for people, who are otherwise isolated in their lives, to create a sort of meaning and enter into relationships that bolster identity.

Cyber racism research thus has its roots in studies of the political culture of White power in the digital age. Don Black, the founder of Stormfront, a US-based, but now global, White power site (an Australian example of which we discuss in Chap. 6), proudly posted a cutting in the mid-2000s, dating back to 1998 at the birth of the Internet. Black, a former Ku Klux Klan grand dragon, had been interviewed by the Montgomery, Alabama, *Advertiser* (Richardson 1998) journalist Sandee Richardson for a story canvassing the growth of hate sites. Montgomery is also the home town of the Southern Poverty Law Center (SPLC), a key organisation that has tracked race hate online since its first appearance. The article captures the moment well—Black, proudly proclaiming how the Internet would be the making of the global White power movement,

Joe Roy of SPLC decrying the expansion and David Goldman of SPLC Hatewatch foretelling that "We're seeing just the beginning phase of the potential for a racist coalition made possible by the Net."

Understanding the Multilayered Networking Processes That Enable the Internet

The history of the Internet has been well-rehearsed, beginning with technologies that allowed the digitisation of information in different localities to be connected through technologies of communication (Cohen-Almagor 2011). Commencing in the world of the military, where protected communication increasingly became the underpinning of modern warfare (Hafner and Lyon 1998; Kleinrock 2008), key scientific centres supported by government funds extended the links between them. Soon the networks designed to fortify the state entered into resource deals with commerce, while the innovation communities that were generated by these hubs spun off into new private corporations, now some of the largest in the world (Hanson 2015). At the heart of the Internet, a new military industrial complex with many players expands continually. Two broad systems emerged, one technological, the other economic, facilitated by governments, especially that of the USA, at the outset. Each system had its own ecological hierarchy, the imperatives of which both complemented and challenged each other.

In his "Internet-map," Ruslan Enikeev has described cyberspace as "a huge quantity of utterly unstructured information" (Enikeev 2011). However, once he colour-coded the 350,000 sites he selected and the 2 million interlinkages he collected and visualised as a universe, it became clear it is centred on Google and Facebook, which are surrounded by country networks—the USA the most dominant, under challenge from China, with Russia and Japan closely following behind. These central pivots—a search engine and a transit station—summarise the Internet in the period of the mature Web 2.0. Increasingly, they frame and delimit what can be "known" about Internet relationships.

The Internet continues to grow dramatically, with users increasing from just over one billion in 2005 to 3.5 billion in 2016, almost half the population of the planet, (Statista 2017). Statista has shown the spread of the Internet, and the demographic reach—from Europe at 79% to Africa at 25% (Statista 2016)—which we would suggest reflects patterns of power reaching back hundreds of years into the spread of European imperialism. Regional penetration of the Internet has also intensified—by 2017, North America stood at 88%, Europe 84%, Oceania (dominated by Australia) 68%, Eastern Europe 67%, South Asia 33% and Africa 29%. That is, the access to information has become structured in ways and follows the patterns that reflect broader economic and political forces.

Access to and use of the Internet have not been evenly distributed—the developed Western world and particularly its younger populations are the ones most likely to have access to and make use of the Internet as the backbone to their economic and communal lives. Older people, with lower levels of literacy, from less economically developed societies are far less likely to use the Internet. This digital divide remains both a global and a societal phenomenon, though the expansion of the Internet has made this far less dramatic a phenomenon, especially in the West, than it was in the first Internet decade (Daniels 2013). For example, in the Australian context, this would be evident in the differences between the access and use among recent older African refugees and younger, tertiary educated Indian immigrants.

The political economy of the Internet requires us to determine how and where value is created. The Internet depends on and contributes to globalisation, which can be defined as the acceleration and expansion of the circulation of capital, culture and populations, as barriers that might otherwise inhibit them are reduced (Fuchs 2009). However, this circulation does not occur randomly or in an unstructured way, but rather through attractors that seek on the one hand to speed up the circulation of economic capital, and on the other selectively extract from populations the value that they produce. Culture carries much of the potential to create value, which is born both in the transfer of technologies as finished goods, and in the cultural capital embedded in people. Various structures of regulation exist to manage this system, or at least keep it somewhat monitored. Meanwhile, national governments or their corporate proxies and partners use similar processes to continually

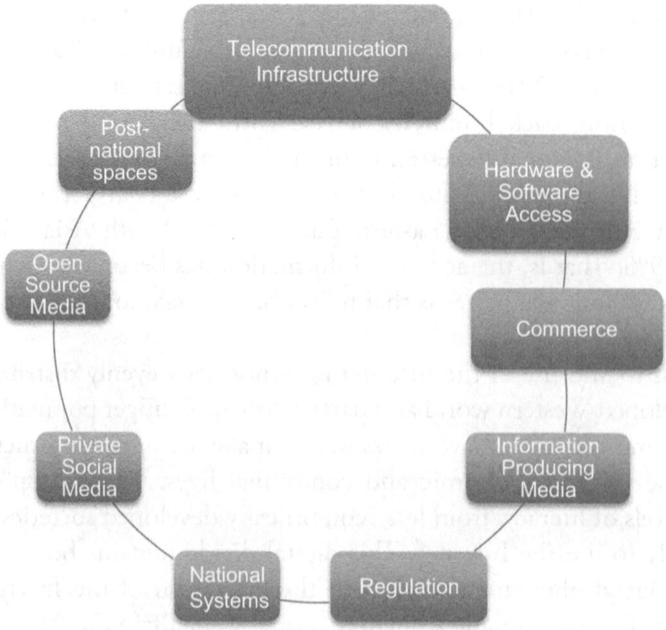

Fig. 1.1 Thematic representation of Internet

reposition themselves in relation to the wealth potential associated with control over the globalisation flows.

Figure 1.1 shows a simple "cut down" display of the key elements in the political economy of the Internet, operating across cyberspace. Each element can exist only when the others do, so the interrelationship is symbiotic, but also continually morphing, as new capital, culture and demand from populations enter the scene. The telecommunications infrastructure corporations provide the link between local and global populations and commerce.

When we start to add names to these labels, as in Fig. 1.2, the diagrammatic outline takes shape as a scene of purposeful and competing actors. Over a decade or so since Fuch's 2009 study, the relative place of different corporations (and their national backers) has changed dramatically. In 2007, the top four telecoms were ATT, Vodafone, Verizon and Deutsche Telekom (Fuchs 2009). By 2015, China Mobile had become the "leading

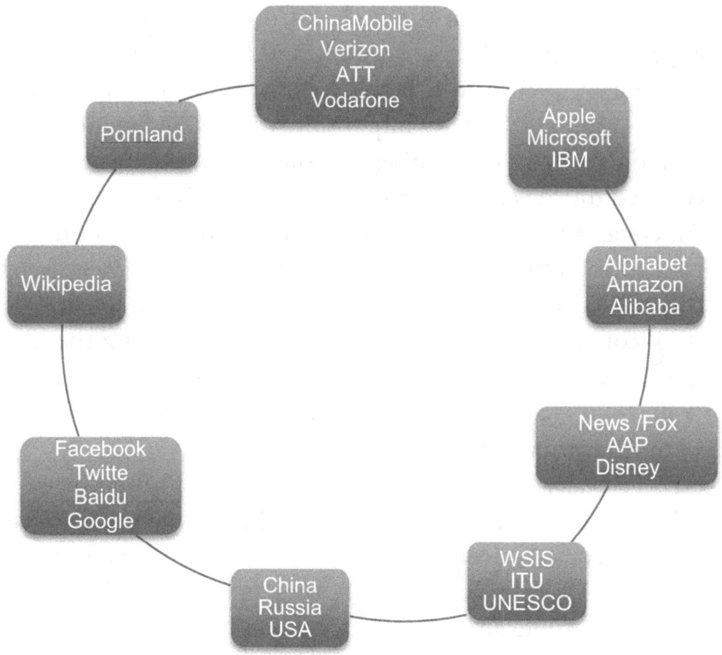

Fig 1.2 Representation of the Internet with selected key players

provider of telecommunications services by subscribers," with a market value of US$280 billion (Chen 2015). It was well ahead of Verizon and ATT (USA) and Vodafone (UK), who were leaders just a decade earlier. More recently, Fuchs has argued that it remains crucial to understand the Internet as primarily driven by capitalist dynamics of accumulation, without which the current energy for expansion would soon dissipate. This remains centrally the case, especially in apparently hybrid economies such as China, where the process of capitalist accumulation remains dominant even where the state is centrally involved in managing overall economic development. (Fuchs 2017).

The production of hardware and software that links people to the telecommunications companies remains a global process—design, production and assembly can happen in different countries where the costs of labour vary, usually as a consequence of historic stages of development (and under-development). Hierarchies of race and patterns of exploitation have long been embedded in the nodes of the supply chains underpinning companies such as Microsoft, Apple and IBM (Kosoff 2016a, b). In the contemporary period, the mobility of some sectors of intellectual labour has become increasingly important, with skilled engineering and software graduates from developing economies either seeking to enter or being recruited for the expanding software and hardware corporations in North America. The process became a concern for the ultra-nationalist right in the USA, where Steve Bannon, formerly of *Breitbart News* and later political adviser to President Donald Trump, attacked the dependence of Silicon Valley on imported foreign labour during the US election in 2016 (Kosoff 2016b). Bannon suggested an ethnic cleansing for Silicon Valley—it was not just about commerce for him but was also a matter of defending American culture.

In early February 2017, when President Trump announced a US immigration hold on entry from seven "Muslim-majority" countries, many senior executives from key global US-based companies declared their opposition (Wattles et al. 2017). These included technology firms General Electric, Apple and Microsoft, communication firms Facebook, 21st Century Fox, Google and Twitter, and commerce firms Amazon, Chobani, Ford, Coca-Cola, Netflix, Uber, Starbucks, Tesla and Expedia. A number of these companies identified immigrant workers as their first concern, with Starbucks vowing to employ 10,000 refugees worldwide in its stores. Many with immigrant (or children thereof) CEOs argued that the free flow of populations was crucial for the migration of labour in its search for opportunity, bringing with them the global cultural capital that contributed to American wealth and innovation.

The flow of capital on the Internet reveals a huge commercial space in which users can buy and sell almost anything. Importantly though, for the largest firms, such as those that voiced opposition to the US block on arrivals from some Muslim-majority states, the advent of Web 2.0 with its capacity for user-generated content has added to the ways in which value can be produced and appropriated. In the pre-social media period, value was created in the processes of production by the application of labour power to the non-labour factors of production (Fuchs 2009, 2017). The relationships of the market at that time more clearly separated producer/worker, owner of the means of production, and consumers.

Value was created at the point of production, but realised when sold. Now, however, much of the value is produced by social media or search engine users. Their presence *en masse* on these sites gives site owners a valuable new commodity: user data. This information can be onsold to advertisers and corporations looking for data about potential consumers. Thus, the real value of these sites to their owners depends on the unimpeded access for users. These users can be onsold to various corporates as customers, clients and marketing targets. Much of the value is created by people whose labour is free to the owner of the site, albeit potentially costly to attract. It is realised, in terms of the pre-Internet critique, through delivering audiences to advertisers (Ross and Nightingale 2003). Thus, the underlying economic impetus comes from freedom of access so that the political economy of the Internet has emerged as a space to resist regulation that might limit that freedom; the centrality of that resistance has lain in the two mega-constellations of users on Google and Facebook.

Nothing could have better tested our argument about the structure of the Internet and its affordances for racism than the eruption in March 2017 of the Google/YouTube advertiser boycott. The boycott began when media reports exposed the posting of advertisements for major national and global brands on offensive YouTube pages, drawn by the algorithms that attached advertisements to valued sites and their followers. While we did not predict the crisis that occurred, the logic of our position spoke to the inevitable manifestation of racism at the heart of the Internet business model perfected by Google and its associated Alphabet companies. The algorithms used to serve advertising content on popular sites, or search results inevitably deliver advertising from global companies to sites produced by people with racist, homophobic, sexist and nationalistic agendas. Writing on the discovery of this issue by thousands of advertisers and their cessation of advertising on Alphabet sites as a consequence, online journal *Wired*'s Davey Alba argued "To succeed as a content producer, it [YouTube] needs to let its creators push boundaries. To succeed as a platform that won't alienate users or advertisers, it needs to exert quality control" (Alba 2017a). It appeared furthermore that Alphabet was caught by a third factor—its commitment to freedom of speech meant that it would not remove offensive material, even where the offensiveness of the material might have significantly impacted on the standing and reputation of the advertisers (Alba 2017b). The intersection of freedom to speak as a value underpinning the marketisation of opinion, and the desire for maximising profit (and indeed paying offensive sites for the traffic they generated to advertisers), had the inevitable outcome—an almost impossible to modify push of cash towards racist, homophobic and sexist sites.

Regulating the Internet as an Infrastructure

This of course is not to say that there is no Internet regulation at the global level. The International Telecommunications Union (ITU), an international association of national bodies established in an earlier age of telegraph and the earliest telephony, and now an agency of the United Nations (UN), regulates the broad physical infrastructure. Its main role has been to ensure that the primary technology of the global communications system remains one that facilitates the flow of information and minimises idiosyncratic interference by:

> defining the frameworks that will ensure optimum functioning of all services, both existing and future… by connecting the world and fulfilling everyone's fundamental right to communicate, we strive to make the world a better and safer place. (International Telecommunications Union 2017)

The ITU was heavily involved during the 2000s in bridging the digital divide in less developed countries through the development of information and communication technologies.

It was during the early to mid-2000s, with the ebbing away of the former Soviet bloc, that the international community created the World Summit on the Information Society (WSIS) (International Telecommunications Union 2014). It followed earlier debates over what has been called the New World Information and Communication Order (NWICO), an at-the-time contentious reconceptualisation by a UNESCO panel of the cultural inequalities in the world. The panel argued these inequalities derived from issues of neo-colonialism and domination of knowledge and information by advanced capitalist societies (Fuchs 2015). Embedded in this analysis, racism was presented as a continuing consequence of earlier imperialism by the Western nations. The rationales that had enabled the colonial powers to invade, seize and exploit much of the world were supposedly still extant in the period of post-colonial liberation. They still constrained the opportunities of less developed countries, which had become dependent on their former colonial controllers for networks of finance and markets. Many of these former metropolitan powers were the destination for millions of their migrating former colonial subjects seeking the possibilities of employment and education not available in their own societies.

The UN Declaration on Human Rights played a central role within the 67 Principles of the WSIS adopted in Geneva in 2003. The WSIS would advance Article 29 of the UN charter, namely, that:

> everyone has the right to freedom of opinion and expression… Communication is a fundamental social process, a basic human need and the foundation of all social organization" (at Principle 4). Moreover "building an inclusive Information Society requires new forms of solidarity, partnership and cooperation among governments and other stakeholders… bridging the digital divide and ensuring harmonious, fair and equitable development for all… we call for digital solidarity, both at national and international levels" (at Principle 17). Under paragraph B8, the WSIS asserted cultural diversity as the common heritage of humankind, in which the new Information Society "should be founded on and stimulate respect for cultural identity, cultural and linguistic diversity, tradition and religions, and foster dialogue among cultures and civilizations.

A decade after Geneva, the WSIS+10 review concluded that the evolution of the Information Society was continuing to contribute to "the development of knowledge societies around the world that are based on …respect for cultural and linguistic diversity and cultural heritage" (International Telecommunications Union 2014). Yet challenges remained in "the need to protect and reinforce all human rights, and to recognise their importance to realise economic and social development" (p. 16). The most challenging human rights questions related to the empowerment of women for full participation in society. However, the Summit also reiterated the:

> "need to respect human diversity in all its forms, in particular, cultural and linguistic diversity as well as diversity of tradition, religious beliefs and convictions" and adopt policies that "safeguard endangered languages and preserve cultural and linguistic heritage, including support for multilingualism." (p. 19)

By the 2015 Summit, the regulatory and program bodies involved in addressing these issues included the ITU, UNESCO, the UN Development program (UNDP), UNCTAD, UNDESA, FAO, UNEP, WHO, ILO, WMO, ITC, UPU and the Regional Commission.[1] There was also a growing concern for cyber security, including privacy protections (p. 28), that stressed the critical importance of the affordability of Information and Communications Technology (ICTs) and the role such technologies would play in overcoming global inequality through empowering marginalised and vulnerable people. In the midst of these human rights type of reflections, the Summit recognised that "in spite of recent progress, there remains an important and growing digital divide between developed and developing countries … [which needs to be reversed to] contribute to narrowing existing and emerging socio-economic inequalities" (pp. 34–5). In commenting on the human rights issues, the Summit affirmed that "the same rights people have offline must also be protected online, and that this is applicable to media on all its platforms" (p. 46).

The spread of ICTs, the process which essentially builds the Internet, requires ongoing coordination and collaboration, nowhere more so than on how the Internet's spread affects women's empowerment and participation, cultural linguistic and religious diversity and people with disabilities. The regulatory bodies might well have managed the synchronisation of technical connections and the standardisation of high-technology devices. However, issues of human rights and the associated implications as populations moved across the globe have been left to the scattering of UN conventions and their implementing agencies to try to embed in societies affected by the changes.

These principles of human development were launched in a world where the corporations that were building themselves as the new generation of cyberspace moguls were constantly seeking to valorise their technologies and assets (Deane et al. 2003). These highly competitive organisations had little interest initially in human rights issues, despite the proclivities of some of their founders towards empathy with disadvantaged and abused populations. Where the global priorities of WSIS might have affected their orientations were the situation otherwise, the national marketplaces and forms of government in fact established the initial rules under which they operated. These national systems adopted their own regulatory environments, covering everything from technology, spectrum and infrastructure, to the legal status and responsibilities of the major intermediaries such as the service providers and the platforms. The national systems also "ported" the existing power relations of the society into cyberspace, including those reflecting gender and racial inequalities.

Globalisation has been portrayed as an unstoppable process of progress in which national borders fade in importance. Global markets emerge in which comparative advantage rules the game (Kellner 2001). Companies seek out regions where the costs of production are lowest (usually among racialised and exploited populations). In the process of developing productive capacity, intellectual capital follows the financial capital into those locations and the quality of local production grows, very much the development model advanced by the WSIS (Srnicek 2016). In 2006, China moved to free itself from the US-licenced Internet Corporation for Assigned Names and Numbers (ICANN), sparking fears that it would in effect destroy the Internet by creating new top-level domains that would not be compatible with the ICANN network.

While this total separation did not occur, China did create a domestic system with its own enumeration procedures, installing a "Great Firewall" to limit Chinese-user access to the outside world (and the world's access to them) (Liu 2012). In 2016, China took another step towards reinstating control over its domestic Internet, by potentially cutting off Chinese access to the global Internet (Bradshaw and DeNardis 2016). The US National Telecommunication and Information Administration argued that intensified local registration rules would in fact effectively separate

China from the ICANN system, or at least give it total control over standards that might soon ensure incompatibility, actions that "infringe upon internationally recognized commitments on free expression and trade …. China is threatening to fragment the Internet, which would limit the Internet's ability to operate as a global platform for human communication, commerce, and creativity" (Sepulveda & Strickling 2016).

Such global moves to defend national states at the potential expense of Internet freedoms have rapidly taken on major political overtones, affecting the way in which the political economy of the Internet has been shaped. During 2016, these dynamics intensified, with allegations of cyber hacks, cyber penetration of nation-state institutions (from power stations to elections) and commercial espionage (Mehzer et al. 2016). Meanwhile the expansion of user-generated content sites, ranging from Wikipedia through dozens of messaging and meeting systems, contributed to opportunities for agency for the so-called prosumers of cyberspace (Miller 2016; Fuchs 2017). While the most significant arena for such interactions may have continued to be what Enikeev labelled as "pornland," a free-floating constellation of often-associated sites (Enikeev 2011), the most prominent (its space censored from pornography) continued to be Facebook. In May 2017, Facebook had a market capitalisation of US$435 billion,[2] growing from US$67 billion in 2013 to US$165 billion in 2014, US$220 billion in 2015, and US$290 billion in 2016. This compares with the Alphabet group (owner of Google) at US$657 billion, up from US$258 billion in 2013, and Twitter at US$13 billion, down from a 2013 high of US$40 billion. In China, during 2016, Ali Baba, the online global marketing company, passed China Mobile in terms of market capitalisation, though the search-engine company Baidu, protected by Chinese government bans on Google, had the highest calculated brand value of the three Internet-related companies.[3]

A political economy analysis of the Internet demonstrates the way in which national identities remain important dimensions of key actors—from the governments of countries like China, the USA and Russia, through to the search engines that enable the systems to work by connecting users and desires. Issues of ethno-cultural conflicts, carried through international competition, generate awareness among policy actors that open, free and respectful communication has never been more valuable in facilitating opportunities and supporting development. Yet, it is the very rapidity of the Internet and the massive complexity of its pathways that make it such a powerful refuge for racism, and so difficult a region to monitor, let alone regulate.

Reflecting on the political economy of the Internet and the myths that had grown and then dissolved in the wake of social media, Curran noted that far from being an arena of freedom, the Internet was set in "world… divided by bitter conflicts of value, belief and interest. These can find expression in websites that foment—rather than assuage—animosity" (Curran 2012, p. 10).

Regulating Racism on the Internet

The main global framework for the regulation of race hate can be found in the 1966 International Convention on the Elimination of All Forms of Racial Discrimination (ICERD). It was adopted on the sixth anniversary of the 1960 Sharpeville massacre in South Africa, a date commemorated annually as the World Day for the elimination of all forms of racial discrimination. In Australia, it has been known as Harmony Day since 1998.

The USA ratified the ICERD only in 1994, while reserving any action under Articles 4 or 7 to the agreement of the Senate (Harris 2008). Australia had filed a significant and continuing reservation on the issue of hate speech under Article 4 on signing in 1967, effectively refusing to criminalise any hate speech that incited hatred short of violence. The broad international community has, however, accepted the Convention and approved of Article 4.

Even so, a universally agreed definition of "racist hate speech" remains unresolved, not the least because the idea of race itself is contested. The characteristics of racism are vigorously debated and dispute remains as to whether hate is best regarded as an extreme emotion or motivation of the perpetrator, or a perception of extreme hurt by the victim. The ICERD has been described as an "outlier" among UN Conventions, requiring signatories to pass laws that make certain expressions punishable by law—that is criminal offences—rather than prohibited by law, which may allow civil constraints to apply. Discussing the relationship between freedom of speech and freedom from hate, Tarlach McGonagle has suggested that "hate speech" "is less a term of art than a term of convenience" (McGonagle 2012). The contentious Article 4 begins: "States Parties condemn all propaganda and all organizations which are based on ideas or theories of superiority of one race or group of persons of one colour or ethnic origin, or which attempt to justify or promote racial hatred and discrimination in any form." Neither the USA nor Australia was willing to accede to Article 4, rationalising their refusal by reference to constitutional and legal issues associated with the more important dimensions of freedom of speech.

The ICERD was in place and widely ratified well over two decades before the Internet emerged as a global phenomenon. The creation of the Internet can really be identified at the point in 1982 when the Internet protocol suite TCP/IP (Transmission Control Protocol/Internet Protocol) was deemed robust enough to be widely used. The first commercial providers of Internet services emerged in the USA in the late 1980s, while the last limitation on commercial provision was withdrawn in 1995, with the closure of the US National Science Foundation Network. The invention of the hypertext protocols necessary to permit the World Wide Web to operate occurred around 1990 (Berners-Lee 2017)[4].

From that point on it took only a few years for the UN to become aware that the Internet was emerging as a critical location for the advancement of racial discrimination and racist propaganda. By 1997, the UN ICERD Committee was reflecting on how it might respond, focusing on urging its signatories to use Article 4 and Article 7 (educational priorities) to block hate speech and build capacity in their societies to resist its spread. In some cases, the advocacy was effective. While the USA and Australia avoided the request, Europe began to take the issues more seriously (Swiss Institute of Comparative Law 2000/2009).

Yaman Akdeniz reported for the UN at the end of the Internet's first public decade that governmental responses had had variable impact, even where quite tough laws were in place such as in France and Germany (Akdeniz 2006). In the wake of the Durban Declaration with its focus on spreading human rights and combatting racism, Akdeniz concluded that states could not agree on a global strategy, torn between concerns with social cohesion and security on the one hand, and the priority for free speech on the other. These ongoing disagreements over fundamental strategy were, however, being overshadowed by a shared awareness of the role the Internet was playing in the growth of terrorism. He suggested that the incapacity of governments to act in concert might shift the focus towards "governance" rather than on "government," where private corporations began to take responsibility (Akdeniz 2006).

The European Commission did move as a group on cyber racism, by including an additional protocol to its Convention on Cybercrime in 2003. In the Explanatory Report, the Commission noted that:

> the emergence of international communication networks like the Internet provide certain persons with modern and powerful means to support racism and xenophobia and enables them to disseminate easily and widely expressions containing such ideas. In order to investigate and prosecute such persons, international co-operation is vital. (Council of Europe 2003)

The development of the additional protocol, deemed too divisive to be included in the main Convention, was passed to the *Committee of experts on the criminalisation of acts of a racist and xenophobic nature committed through Computer Systems,* a group reporting to the *European Committee on Crime Problems.*

In developing the Protocol, the Commission took into account that the right to freedom of speech could be restricted if it violated the rights of others. The concern was not for the content of the materials, but for the potential those materials had to affect the conduct of people exposed to them. Religion could be included if it was used as a pretext for "race, colour, descent or national or ethnic origin" (Council of Europe 2003). The only non-European nations who signed onto the Protocol were Canada and South Africa. Australia had included the additional Protocol in the exposure draft of the Treaty, but then removed it without explanation, apparently after pressure from free speech lobbyists (Jakubowicz 2012). The main advocate against Australia's involvement with the Protocol was Electronic Frontiers Australia (EFA), which in 2002 had strongly opposed any limitation on online speech, arguing that "when encountering racist or hateful speech, the best remedy ...is generally more speech, not enforced silence" (Electronic Frontiers Australia 2002).

A decade later, Akdeniz's comment from 2006, that governance would increasingly have to replace government in this arena, was appearing to be accurate. For Europe, the Internet had become an increasingly critical space for the rise of conservative social movements, providing the avenues through which anti-immigrant, anti-Europe and anti-foreigner movements like Brexit; anti-immigrant and anti-Islamic political parties in all European countries; and global White power alliances could bring their followers together, deluging their supporters with propaganda and harassing their opponents and targets. In a historic meeting in Berlin in January 2016, Mark Zuckerberg of Facebook told young Germans that Facebook would work to stop itself being used as a home for hatred (Zuckerberg 2016). The Online Civil Courage Initiative launched in Berlin was founded by Facebook, and designed to "promote civil courage displayed by organisations and grassroots activists carrying out valuable counter-speech work online." However, Facebook continued to come under attack from German activists who were concerned that the site was too slow to take down reported hate posts (Maier 2016a).

In mid-2016, New York Times editor Jonathan Weisman quit his Twitter account with its 35,000 followers, exasperated that Twitter was slow to take down reported hate tweets, leaving many in place (Maier 2016b). The argument that the responsibility to police hate should be with the service providers may well have enabled governments such as Australia to avoid initiatives that might increase protection for the targets of hate. There has, however, been rising pressure on providers to accept responsibility, even where they endeavour to avoid doing so (Simon Wiesenthal Center 2016).

By 2017, the World Wide Web Foundation created by Tim Berners-Lee was well aware that the dreams of equality and freedom that had driven the "free gifting" of its protocols to the world in a digital revolution were being undermined by "new patterns of privilege and discrimination." Furthermore, it was "amplifying voices of fear and hate just as much as voices for tolerance and rationality" (World Wide Web Foundation 2017; Berners-Lee 2017).

Moreover, as the heat continued to increase in global debate about the responsibility of platforms for the dissemination of hate speech, Zuckerberg offered an extensive analysis and program for how Facebook would build accessible, more safe and civil communities. He indicated that Facebook would no longer abandon users to the predators who, he admitted, had found opportunities among the billions of Facebook members. On the other hand, he proposed that the best way forward to ensure a safer Internet would be to crowdsource standards, allowing Facebook communities region by region to set the range for acceptable material, while individuals could protect themselves from the material they found unacceptable through their individual settings (Zuckerberg 2017).

Understanding Racism and Its Place on the Internet

Racism cannot be taken as a given, not even in terms of its relationship to the concept of "race." Race means different things in different parts of the world, especially in the wake of biological science declaring its lack of scientific value. There is now broad agreement among both natural and social scientists, let alone humanists, that race remains a powerful socio-political framework, but it has no genetic value in predicting behaviour.

For instance, Yudell et al. (2016) have summarised their position as geneticists thus—"We believe the use of biological concepts of race in human genetic research—so disputed and so mired in confusion—is problematic at best and harmful at worst." Psychologists, while finding the social value of race continues to indicate important disparities, have specified that race should be seen as "dynamic, malleable, and socially constructed, shifting across time, place, perceiver, and target" (Richeson and Sommers 2016).

Race is quintessentially a social and cultural construct, drawing on local and global historically significant markers of difference to justify unequal relations of power. Once race is used to describe difference, it becomes a broad definitional edifice that separates and classifies. Race is deeply embedded in the psychological menus people draw on to manage their lives. Race operates at many levels in the social ecology of modernity, from the broad panorama of global difference and power, to the internal dynamics of cognition and identity in the individual. We examine next how this all works and how the multiple levels of race interact.

Following that elaboration, we identify how the points of tension and conflict, which are often described as racism, relate to each other.

At its global level, the concept of race in European thought first emerged within the context of imperialism undertaken by Western metropolitan nations (Ransom 2013). As explorers and adventurers encountered ever more diverse populations through either trade or conquest, the distinctive characteristics of these peoples were recorded and codified. While differentiating concepts had been widely used in Western religious, secular and scientific literature for centuries, it was really not until the nineteenth century, when the full press for imperial domination erupted, that "race" took on its particularly modern qualities. That is, race became powerful as a concept when it became embedded in ideologies of racism, which stressed hierarchies of races and in particular the place of White Europeans at the apex of the human pyramid. Racism is perhaps captured as ideology most sharply by the phrase "survival of the fittest" coined by sociologist Herbert Spencer (Paul 1988), who used it to "explain" the success of Western European martial societies, their economic and scientific advances and their apparently overarching capacity to rule and lead "lesser breeds." Spencer's social Darwinism became so intertwined with the biological narratives of Charles Darwin and other natural scientists that race as a model of global social power was soon embedded in public and political discourse across Europe, the Americas and their imperial possessions. The political philosophy of racism mobilised the scientistic concept of race to assert the power of the dominant political order in multiracial social situations.

The link that was made was crucial—that biological differences were inevitably reflected in social differences, and that the quality of civilisation could be deduced from the physiognomy of a people. Moreover, different races were thought to fill different ecological niches, their physiology and behavioural practices determined by their reaction to specific local circumstances. More recently though, the wide range of behavioural diversity and the smaller range of genetic diversity have led biologists towards abandonment of a simple Darwinian logic in relation to humans (Smith 2011).

Yet the impact of Spencerian thinking on ideologies of imperial expansion was not as simply applied as might have been thought. For instance,

Eric Love (Love 2004) in his analysis of the expansion of the US empire into the Caribbean and the Pacific after the Civil War argued that no serious imperialist would justify the invasion and possession of Cuba, Hawaii and the Philippines by reference to the lesser breeds beyond the law, manifest destiny or anything else that placed "non-Whites" at the centre of imperial policies. Love pointed to the campaigns in the nineteenth century to suppress non-Whites (especially Chinese) and exclude them from the USA, expropriate the Native Americans and prevent African Americans from gaining political power, which were at the forefront of US racist politics. He pointed as well to the pressure from White Americans to expedite the "return to Africa" of freed slaves, and the increasing closing down of non-White immigration (including, after 1921, by Jews from Eastern Europe, Italians, Greeks and other eastern and southern Europeans). Interestingly, it was to Australia's great benefit in the early 1920s that the 1924 US Immigration Act, with its associated National Origins Act and Asian Exclusion Act, was passed "to preserve the ideal of American homogeneity" (Office of the Historian n.d.).

Australia had passed its own Immigration Restriction Act in 1902 designed to keep out non-Whites (Huttenback 1976; Fitzgerald 2007), establishing the basis for what has been described as an ethnocracy (Jakubowicz 2016). Even so, it became a destination after 1921 for Southern Europeans blocked from passage to the USA. At the time, they were anxiously sought by the tropical agricultural industries whose traditional South Sea Islander labour had been expelled in the years after 1902 (Rivett 1962).

The ideology of racism that pervaded Europe, the Americas, much of Africa, Asia and the Pacific during the nineteenth century firmly embedded racialised worldviews into the everyday realities of metropolitan societies, as well as their colonies and ex-colonies. Racist ideology can be defined as a system of ideas and values that operate to promote the interests of one racialised group over others by shaping perceptions, expectations and responses to diversity of appearance and culture. Racial hierarchy was created through systems of classification, imposed through the exercise or threat of force and reproduced in the societal institutions that benefited from its organisational imperatives. For example, the first US Naturalization Act of 1790 had restricted citizenship to a "free White person."

Antisemitism increasingly took on racial overtones, where once the antagonism had been fed from religious prejudice. Anti-Chinese feelings flourished in the Americas and Australia, driven by anger over competition in exploitative economic industries such as mining (Price 1974). The vicious treatment of Africans in the colonies seized by nearly all the main European powers was poignantly captured in Joseph Conrad's *Heart of Darkness* (Conrad 1899/2011). Even so Conrad wrote about the humanity of Black Africans but not about their equality with Whites.

Racial science became an increasingly central part of theories of modernity. Racism was also one of the bolsters for the edifice of nationalism, especially volatile and violent during the economic crises in the 1930s. In a 1934 book *Whither Away?* dedicated to "the workers of Australia with or without collars," John Bostock and L. Jarvis Nye wrote: "We have approached our problems and expressed our opinions in the light of a scientific investigation ['of race Psychology and the factors leading to Australia's national decline']. …Human societies] are on many planes of development, intellect and culture; some we call high, others low" (Bostock and Nye 1934). Strongly influenced by fascist ideologue Oswald Spengler, they used as their epigraph his quote "World history… always it has sacrificed truth and justice to might and race." Channelling Spencer, they declared: "race safety, in our present state of knowledge, must be grounded on the experience of history and the immutable laws of evolution (p. 1)…in order to ensure race safety there must be maintained in continuity, a national spirit impelling the race towards a definite objective" (p. 69)…. "Race safety depends on race discipline" (p. 84). Such narratives of nation and race remain within the portfolio of racist ideologies today, even if the scientific value of race has long been rejected.

One of the consequential problems for social analysis remains the very local history of the term Racism. A 2004 report of the US National Research Council of the National Academies Measuring Racial Discrimination (Blank et al. 2004) offered an uncomfortable consensus on the meaning of race. Noting that there was no single concept of race, the panel accepted that race referred to "a subjective social construct based on observed or ascribed characteristics that have acquired socially significant meaning" (p. 2). Reflecting that race and ethnicity blur into each other, especially with Hispanics of colour, the report opted to utilise the US Government categorisation of races—Black or African American, American Indian or Alaskan Native, Asian, native Hawaiian or Pacific Islander, and White; they also included one ethnic group, Hispanics "who may be of any race."

For the Australian authors of this book, these American categories are conceptually curious and politically difficult. From an Australian perspective, the same underlying categorisation criteria seem not to apply. A Black, a Negro, or an African American does not appear to share the same criteria of selectivity as does a "White": the first grouping speaks of a political history embedded in slavery, and in the second, concerning whiteness, no specific slave history may be relevant. Even so, White privilege may be so deeply embedded that the residue of slavery continues to affect people whose personal histories were unconnected. Former president Barak Obama is "Black" because his father was African, but he had no history of slavery, and his mother was "White." He recognised though that the widespread racism against Blacks in the USA had grown through hundreds of years of slavery and post-slavery discrimination and exploitation. It appears to us that these categories of distinctiveness reflect narratives of exclusion and structures of oppression in the US history. Asia, for instance, is a Western concept that assigns to people from the edge of Europe to the Sea of Japan a common category. In the USA, it usually applies to people from the east of Asia—Chinese, Japanese, Korean, Vietnamese, Indonesian, and what might once have been described as members of the Mongol race (Bushmakin 1928). For the 2020 Census, the USA has been reconsidering what to call the things to which people are asked to assign themselves—perhaps opting for "Which categories describe the person?" (Cohn 2015).

Australia essentially abandoned "racial" categories based on skin colour in the 1980s, replacing the concept with four indicators of difference—language spoken at home, country of birth, self-identity as an Indigenous or Torres Strait Islander person and ethnic heritage (Australian Bureau of Statistics 2011a, b). This can be elaborated by reference to other languages spoken more generally, and in some cases deepened by association with religion. In Australia, Asia gets divided into sub-regions so that geographic origin rather than similar physiognomy drives assignment to the category—west (from Afghanistan to Israel), south (India and Sri Lanka), north (China, Korea and Japan), and southeast (the ASEAN or Association of Southeast Asian Nations). By way of another example, Australia's largest African population is mainly "Whites" from South Africa, though Africa also includes Black Africans (cultural heritage and national or tribal language spoken at home) and Arabs (transnational language) from the north. However, Australians tend to use "race" to refer to any non-Whites who are culturally non-European, and ethnicity to refer to the remainder—though ethnic heritage in the Census covers them all. In Australia, Jews, no matter from where they come, are regarded as a "race" because of their common culture, rather than their religion, and irrespective of their "colour." In fact, it reflects both negative drivers—as in early twentieth century understandings by antisemites of Jews as a racial group, the same categorisation used by the Nazis to plan their extermination—and positive drivers as an affirmation of connectivities through history, origins, social practices and myths. Declarations of faith are not required; the opposite is the case for Muslims, for whom declaration of faith is crucial for membership of the global *ummah*. In the Australian state of New South Wales (NSW), ethno-religious groups are covered by anti-discrimination laws, but not at the Commonwealth level, that is, Jews and Sikhs are covered, Muslims and Christians are not.

Compare both of these countries with the UK. The Office of National Statistics reported on the 2001 and 2011 Censuses, using the following categories: White (divided into Irish, Gypsy and Other White), Asian/British, Black/African/Caribbean/British, Mixed and Other (in 2011 separating Arab). The White category (pre-Brexit) encompassed the Europeans (e.g., Poles), Irish and Gypsies—none of which had the same parameters for inclusion—as well as the White British. Arabs and Gypsies were missing from the 2001 enumeration. The label attributed to all these groups by the Institute for Race Relations was that of "ethnicities." However, these are not necessarily identities, nor are they "race" in any common traditional or modern sense.

If so many of these categories and criteria have at one time or another been deemed to be about "race," while others have not, what counts as racism can be highly controversial. For instance, someone in the state of NSW in Australia of Iranian Shi'a ethno-religious heritage may or may not be found to have experienced racial discrimination, depending on how much their ethnic heritage contributed to their plight. Yet, in

Victoria, the adjacent state, their religion can be accepted as an independent factor in their complaint of discrimination.

What Does "Race" Do to Peoples' Life Experiences?

The scientific literature on the health impact of racialised social environments on non-majority communities has proposed a number of conditions in any use of the idea of race. Race has been identified as a social construct, rather than a scientifically valid term, one "premised on the erroneous belief of meaningful genetic or biological differences of groups in the population based on arbitrary physical markers, such as skin color" (Spence et al. 2016). However, even though "race" does not provide a useful model for predicting individual or group behaviour, its social meaning does offer a way of identifying fault lines for intergroup conflicts and social differences. Spence et al. (2016) make distinctions between "race," which they see as a descriptive and categorical term only with no scientific value, "racism" as an ideology of racial difference implying hierarchy and power and "racial discrimination," which they see as a series of practices linked to the justifying ideologies.

Whatever the parameters of the categories described as "races," racisms operate in fairly similar ways. Racism can only be a social question where there are categories differentiated by race. Once that differentiation occurs, and evidence suggests the historical trajectories that produce it are today the consequence of the expansion of empires and their overlay of race on earlier forms of difference (for instance, tribal distinctions), a socio-political reality emerges and settles like a political sediment, in layered bands of distinctiveness. These distinctions may find expression at the societal level—in what we have already alluded to as an ethnocracy—where key political and economic roles are sequestered by dominant groups, and other groups are marginalised, exploited or harassed. Such an ecological expression of the individual within the multiple terrains of race from the individual and psychological to the global (Fig 1.3) demonstrates the layers and criss-crossing, and sometimes reinforcing, influences at work on individual freedom and opportunity.

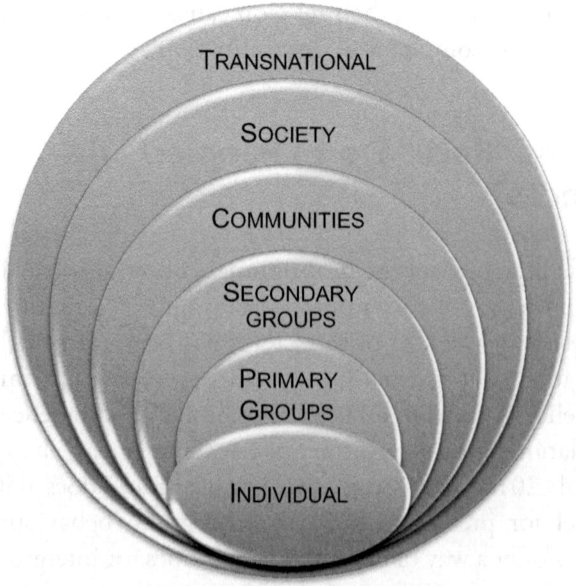

Fig. 1.3 Ecological rings of influences on attitudes to and experiences of racially inflected difference

Such differentiation will soon generate cultural rationales (as in so-called race science), infiltrating the worldviews of all social groups, though shaped by the particular politics of race that each separately experiences. The more racialised the striations of inequality, the more the culture of each group and the shared cultural mapping of the society will impose race as a principle of social order. At that point, access to resources, societal outcomes and group interaction will have become inseparable from narratives of racial difference. The culture will be embedded in and sustained through narratives used to explain the *status quo* and its points of tension and conflict. Identities will have become heavily inflected by concepts of race, ethnic culture, religiosity and possibly victimhood. This narrative dynamic has become the "stuff" of cyber racism, to the point where the very perceptions of difference have burrowed deep into the worldviews of all society's members.

However, this discursive terrain does not emerge mechanically and roll on without meeting resistance, being changed or redirected. Spence et al. (2016) have noted in their studies of Canadian First Nations people that experiences of racial discrimination increased stress among their research subjects. Their summary proposed that such discrimination (essentially discrimination that restricts opportunity and limits the potential for individual growth and flourishing) has been linked to "manifold out-comes, such as poor self-rated health, immune dysregulation, hypertension, obesity, sleep difficulty, preterm birth, frequent colds, mental health, and substance use disorders" (Spence et al. 2016). Their extensive review of the research findings in many different environments reinforces the sense that racial hierarchy continues to have destructive consequences, both due to poorer institutional supports for health, education and so forth, and due to damaged and undermined cultural supports for identity and sense of self-worth. Crucially they refer to the erosion of the individual's support within community.

In the Australian context, Yin Paradies (a co-author of this book) has written extensively about the health and personal impacts of racism, noting it has only been in the past generation that an understanding of the health impacts of racism has come under sustained scrutiny.

> An individual's social locations then become the backdrop upon which interpersonal and internalized oppression are perpetrated. Interpersonal oppression is the interaction between individuals that serve to increase power differentials between social groups. Internalized oppression can be defined as the incorporation of attitudes, beliefs, or ideologies within an individual's worldview that result in the unequal distribution of power among social groups. (Paradies 2017)

Paradies proposes that exposure to racism can generate internalised domination among those privileged by the system and internalised oppression for those deemed inferior in the system. The decomposition of self that results from sustained oppressive exposure to racism, without support, has become arguably the most corrosive aspect of racism on the Internet.

People respond to racism in many different ways. Some succumb to the onslaught, inetrnalising self-hate and at times self-harming or suiciding (Campbell et al. 2016). Others may internalise racial distinctions and manage hurt through aggressive and potentially racist counter-actions.

In his analysis of how racially excluded or suppressed groups behave in mixed environments, Mark Orbe has offered a behavioural matrix, which seeks to capture potential standpoints that protagonists can adopt (Orbe 1998). He preferred the use of the term "co-cultural," rather than minority or sub-culture, for its value in resetting the assumptions about hierarchy that permeate racially differentiated societies such as the United States.

> A co-cultural approach to communication is situated in the belief that the United States is a country of many cultures, each of which exists simultaneously within, as well as apart from, other cultures. The term co-culture is embraced over other terminology to signify the notion that no one culture in our society is inherently superior (but may be dominant) over other co-existing cultures. (Orbe 1998, p. 2)

The Orbe matrix contains two axes, each with three levels of engagement, producing nine cells in which Orbe identifies some 28 different tactical positions. Along the vertical axis, the cells move from non-assertive, through assertive to aggressive. Along the horizontal axis, the cells move from separation, through accommodation to assimilation (Orbe 1998). Thus, the least engaged responses (separated/non-assertive) have individuals keeping themselves separate and maintaining high interpersonal barriers. Aggressive responses may include attempts to sabotage others, confront them and try to gain advantage; or when seeking assimilation, ridiculing themselves or simply mirroring the behaviour of the dominant others. Together these tactics, identified in conjunction with members of co-cultural communities, create a powerful landscape of how subordinated groups orient themselves when confronted with oppressive situations.

The categories labeled as "race" when activated in situations powered by racist ideologies through practices of discrimination generate a field of power. This power, as has been argued earlier, affects the relationships of those subject to the field, often destructively. The field of power can be brought to life within and through the Internet, where the many tactical positions that Orbe (1998) describes can be discerned. Moreover, as Spence et al. (2016) and Campbell et al. (2016) conclude, building resilience among those subject to these fields of power has emerged as the most critical element in resisting and overcoming the destructive forces of racism. In the final section of this chapter, we mark out some of the key ideas associated with resilience and racism.

Practical Ways to Build Resilience in the Face of Threat

When Spence and his colleagues were exploring what factors appeared to have helped Canadian First Nations members to recover from the destructive effects of racism, they identified reinvigorating a sense of self through cultural affirmation (2016). Resilience theory has become increasingly important, as societies face not only physical threats, but also social challenges from forces that undermine social cohesion and break down social capital. Racism can be identified as such a force, given that the effects of racism are to separate and isolate minority groups, force their disengagement from wider social agendas and reduce their influence on the development of the broader social landscape. Moreover, racism also fragments the groups themselves, producing individuals isolated from their peers, and likely to act in destructive ways—either against themselves or against those they see as their enemies. At the same time, racism seeks to build communities of activists who will collaborate to advance their agenda, silence opponents and intimidate their victims. In so far as racism produces public health effects, and the evidence is strong for exactly this impact, the spread of racism can be considered a public health crisis.

When the USA began to consider its National Public Health Strategy after 2008, the RAND Corporation assessed how it might build community resilience into planning, especially for more under-resourced communities likely to bear heavier impacts from threats (Chandra et al. 2010). It defined "community resilience" as the "sustained ability of a community to utilise available resources to respond to, withstand, and recover from adverse situations." Moreover, a resilient community would be able to determine what it needs to reduce damage, and then continue to strengthen health, social and economic systems. The community would learn from the past and improve preparedness, primarily through effective social capital development—"engaged in community and connected to each other" (RAND Corporation 2013).

In the Australian context, community resilience has been developed in two significant contexts—the first in relation to natural disasters (fires and floods), the second in relation to counter-radicalisation of Muslim communities where young people might be drawn to violence. The terminology is shared, and even confused, though some of the basic ideas are rather different. However, the resilience that has been referred to in relation to racism does not mean "putting up with it," though it can mean "bouncing forward" (Houston 2015).

Community resilience requires at its heart an effective and sustained communication strategy. Without communication, there can be no community and therefore no resilience. Racism seeks to isolate people and wear them down with repeated humiliation. Resistance to racism requires the building of community and the reinforcement of identity through positive narratives and shared responses to threats. As we will see in later chapters, building these communities requires care and sustained innovation. A resilience strategy identifies common interests, communal resources and modes of cooperation.

Research by Pfefferbaum and colleagues (2015b) has identified five key dimensions that support or enhance that ability, most of which have their roots in "strengthening social infrastructure at the community level" (Aldrich and Meyer 2015). These are primarily communication-focused, as it is the capacity to own the processes through which meaning is communicated that plays a critical role in the capacity to bounce forward. Thus, communities experiencing trauma require the capacity to:

- Connect with and care for each other;
- Understand the transformative potential of the trauma;
- Manage the actual event in an effective and intelligent manner;
- Identify and access the resources that are needed and
- Access and utilise information and communication in a timely and accurate manner.

A critical part of this process lies in the capacity of communities to reknit through a shared process of visioning a future post the trauma.

In doing so, governments and communities need to work together in ways that:

(a) use a multi hazard approach relevant to the local context;
(b) utilise community assessment;
(c) focus on community engagement;
(d) adhere to bioethical principles;
(e) emphasize both assets and needs; and
(f) encourage skill development.
(Pfefferbaum et al. 2015a)

The focus on arming with resilience communities that are confronting traumatising events, such as sustained racism and exclusion, helps them to re-assert their own agency within their own frames of reference, without entering long-term conflictual relations with the wider society.

One model, developed by Houston's Missouri team, illuminates the social capital issues that contribute to the validation of the perspectives and deeper values of communities under threat. Houston et al. (2015) demonstrate how the interrelationship of media and communication perspectives and strategies that underpin resilience offers a robust and self-supporting way forward. The importance of building bonding social capital inside the communities to overcome isolation and fragmentation can be recognised quite clearly, but they also illuminate the crucial role of bridging social capital that "knits" individuals and groups into wider networks through layers of "weak ties" (Granovetter 1973; Roithmayr 2014). By focusing on the sphere of communication and demonstrating the interrelationships between the elements, the model points to the range of check-off issues that an effective community resilience program would need (Houston et al. 2015).

One source of empirical data available to understand how building community resilience in relation to cyber racism might be fashioned exists in the now closed program of Community Resilience, operated by the Australian Department of Attorney General. An analytical policy review of the concept has been undertaken by Rivka Nissim of the Australian Human Rights Commission (Nissim 2014).

Commenced by the Federal Labor government in 2010, and closed down in June 2013 by the incoming Coalition, the Community Resilience program funded 59 projects. We were able to examine 11 in some detail. Drawing on work that looked at how to build resilience in Indigenous communities against racism (Bodkin-Andrews et al. 2013), which was then adapted and extended by Rivkah Nissim (2014), four sets of key activities can be identified. They reflect similar insights to the work that the Pfefferbaum network has undertaken in the USA (Pfefferbaum et al. 2015b).

The 59 Australian project summaries (Jakubowicz 2015) highlighted the following key activities:

1. **Knowledge and awareness through education**—the development of multifaith educational resources, learning about radicalisation, understanding the counter-narrative, challenging myths and racism and cultural tolerance;
2. **Skill building**—peace building activities, social media training, website creation, leadership skills, sporting skills, employment and business skills, multimedia skills and team work activities;
3. **Mentoring**—training for mentors, working with community leaders, mentoring for support and role modelling for others and
4. **Networks and civic participation**—community forums, community events, volunteering, interactive websites, social activities and sporting activities.

These are crucial features for effective community development in the face of potentially demobilising trauma, where recognising the way the event can produce trauma and having access to models of response and recovery fit closely into the "bouncing forward" approach. The building of collectives for survival involves mutual sharing of trauma responses, thereby ensuring that the networks that both bond communities together and allow them to bridge to other communities can be more fully realised so that the individuals within them are strengthened.

Resilience allows us to consider what might be involved in developing pathways forward. As we extend our analysis, we will explore how cyber racism can be confronted and undermined by community resilience. Firstly though, in the following chapter, we identify how we went about our research, demonstrating the range of methods and the relationship between the different disciplinary orientations we brought together. For, as the Internet affords many different ways of knowing, so cyber racism demands a wide-ranging set of perspectives that can collaborate to promote understanding, insight and an informed practice.

Notes

1. ITU: International Telecommunications Union; UNESCO: United Nations Educational Social and Cultural Organisation; UNCTAD: Conference on Trade and Development; UNDESA: Department of Economic and Social Affairs; FAO: Food and Agricultural Organisation; UNEP: Environment Programme; WHO: World Health Organisation; ILO: International Labor Organisation; WMO: World Meteorological Organisation; ITC: International Trade Center (part of WTO: World Trade Organisation); UPU: Universal Postal Union.
2. https://ycharts.com/companies/FB/market_cap
3. http://www.insidermonkey.com/blog/10-most-valuable-brands-in-china-525227/2/
4. http://webfoundation.org/about/vision/history-of-the-web/

References

Akdeniz, Y. (2006, January). *Stocktaking on Efforts to Combat Racism on the Internet, Commission on Human Rights,* Fourth Session, Geneva, 16–27, Sixty-Second session.

Alba, D. (2017a, February 14). Pewdiepie's Anti-Semitic Videos Put YouTube's Business on a Bind. *Wired.* https://www.wired.com/2017/02/pewdiepies-anti-semitic-videos-trigger-business-nightmare-youtube/. Accessed 23 Mar 2017.

Alba, D. (2017b, March 25). YouTube's Ad Problems Finally Blow Up in Google's Face. *Wired.* https://www.wired.com/2017/03/youtubes-ad-problems-finally-blow-googles-face/. Accessed 13 Apr 2017.

Aldrich, D. P., & Meyer, M. A. (2015). Social Capital and Community Resilience. *American Behavioral Scientist, 59*(2), 254–269.

Australian Bureau of Statistics. (2011a). *1249.0—Australian Standard Classification of Cultural and Ethnic Groups (ASCCEG), 2011.* Australian Bureau of Statistics. http://www.abs.gov.au/ausstats/abs@.nsf/Lookup/1249.0main+features22011. Accessed 14 Sep 2016.

Australian Bureau of Statistics. (2011b). Indigenous Status (INGP). *2901.0—Census Dictionary, 2011.* Australian Bureau of Statistics. http://www.abs.gov.au/ausstats/abs@.nsf/Lookup/2901.0Chapter37202011. Accessed 30 Apr 2016.

Back, L. (2002). Aryans Reading Adorno: Cyber-Culture and Twenty-First-Century Racism. *Ethnic and Racial Studies, 25*(4), 628–651.

Berners-Lee, T. (2017). Tim Berners-Lee: I Invented the Web. Here Are Three Things We Need to Change to Save It. *The Guardian.* https://www.theguardian.com/technology/2017/mar/11/tim-berners-lee-web-inventor-save-internet. Accessed 15 Mar.

Blank, R., Dabady, M., & Citro, C. (Eds.). (2004). *Measuring Racial Discrimination.* Washington, DC: Panel on Methods for Assessing Discrimination, National Academies Press.

Bodkin-Andrews, G., Newey, K., O'Rourke, V., & Craven, R. (2013). Promoting Resiliency to Counter Racism: The Lived Wisdom Within Aboriginal Voices. *InPsych: The Bulletin of the Australian Psychological Society Ltd, 34*(4), 14–15.

Bostock, J., & Nye, L. J. (1934). *Whither Away? A Study of Race Psychology and the Factors Leading to Australia's National Decline by a Psychologist and a Physician.* Sydney: Angus and Robertson.

Bradshaw, S., & DeNardis, L. (2016). The Politicization of the Internet's Domain Name System: Implications for Internet Security, Universality, and Freedom. *New Media and Society,* 1–19. http://journals.sagepub.com/doi/pdf/10.1177/1461444816662932. Accessed 7 Mar 2017.

Bushmakin, N. (1928). Characteristics of the Brain of the Mongol Race. *American Journal of Physical Anthropology, 12*(2), 221–243.

Campbell, A., Balaratnasingam, S., McHugh, C., Janca, A., & Chapman, M. (2016). Alarming Increase of Suicide in a Remote Indigenous Australian Population: An Audit of Data from 2005 to 2014. *World Psychiatry, 15,* 296–297. Accessed 27 Feb 2017.

Chandra, A., Acosta, J., Meredith, L. S., Sanches, K., Stern, S., Uscher-Pines, L., Williams, M. V., & Yeung, D. (2010). *Understanding Community Resilience in the Context of National Health Security: A Literature Review.* RAND Corporation. http://www.rand.org/pubs/working_papers/WR737. Accessed 12 Sep 2015.

Chen, L. (2015, June 1). The World's Largest Telecom Companies: China Mobile Beats Verizon, AT&T Again. *Forbes.* https://www.forbes.com/forbes/welcome/?toURL=https://www.forbes.com/sites/liyanchen/2015/06/01/the-worlds-largest-telecom-companies-china-mobile-beats-verizon-att-again/&refURL=https://www.google.com.au/&referrer=https://www.google.com.au/. Accessed 23 Aug 2016.

Cohen-Almagor, R. (2011). Internet History. *International Journal of Technoethics, 2*(2): 45–64. http://www.hull.ac.uk/rca/docs/articles/internet-history.pdf

Cohn, D. V. (2015, June 18). *Census Considers New Approach to Asking About Race—by Not Using the Term at All.* http://www.pewresearch.org/fact-tank/2015/06/18/census-considers-new-approach-to-asking-about-race-by-not-using-the-term-at-all/. Accessed 14 Feb 2017.

Conrad, J. (1899/2011). *Heart of Darkness.* London: Penguin.

Council of Europe. (2003). *Additional Protocol to the Convention on Cybercrime, Concerning the Criminalisation of Acts of a Racist and Xenophobic Nature Committed Through Computer Systems* (ETS No. 189). http://www.coe.int/en/web/conventions/full-list/-/conventions/treaty/189. Accessed 23 July 2015.

Curran, J. (2012). Reinterpreting the Internet. In J. Curran, N. Fenton, & D. Freedman (Eds.), *Misunderstanding the Internet* (pp. 3–33). London: Routledge.

Daniels, J. (2009). *Cyber Racism: White Supremacy Online and the New Attack on Civil Rights Perspectives on a Multiracial America.* Lanham: Rowan and Littlefield.

Daniels, J. (2013). Race and Racism in Internet Studies: A Review and Critique. *New Media and Society, 15*(5): 695–719. http://journals.sagepub.com/doi/abs/10.1177/1461444812462849. Accessed 17 June 2014.

Deane, J., Dixit, K., Mue, N., Banda, F., & Waisbord, S. (2003). The Other Information Revolution: Media and Empowerment in Developing Countries. In *Communicating in the Information Society* (pp. 65–100). [Online]. Version. http://www.unrisd.org/80256B3C005BCCF9/(httpAuxPages)/5DCA28E932BB8CFDC1256E240029A075/$file/InfoSoc.pdf. Accessed 9 Nov 2016.

Electronic Frontiers Australia. (2002). Online Hate Speech and Section 18C of the Australian Racial Discrimination Act 1975. In *Submission to HREOC's Background Paper for the Cyber-Racism Symposium 2002.* Electronic Frontiers Australia. https://www.efa.org.au/Publish/hreocsubm2002.html

Emirbayer, M., & Desmond, M. (2012). Race and Reflexivity. *Ethnic and Racial Studies, 35*(4), 574–599.

Enikeev, R. (2011). *Internet Map.* http://internet-map.net. Accessed 21 Nov 2016.

Fitzgerald, J. (2007). *Big White Lie : Chinese Australians in White Australia.* Sydney: University of New South Wales Press.

Fuchs, C. (2009). Information and Communication Technologies and Society: A Contribution to the Critique of the Political Economy of the Internet. *European Journal of Communication, 24*(1), 69–87. http://journals.sagepub.com/doi/abs/10.1177/0267323108098947. Accessed 23 Aug 2015.

Fuchs, C. (2015). The Macbride Report in Twenty-First Century Capitalism, the Age of Social Media, and the BRICS. *Javnost—The Public: Journal of the European Institute for Communication and Culture, 22*(3.) http://www.tandfonline.com/doi/abs/10.1080/13183222.2015.1059626. Accessed 23 Sep 2016.

Fuchs, C. (2017). *Social Media: A Critical Introduction* (2nd ed.). London: Sage.

Granovetter, M. (1973). The Strength of Weak Ties. *Amercian Journal of Sociology, 78*(6), 1360–1380.

Hafner, A., & Lyon, M. (1998). *Where Wizards Stay Up Late: The Origins of the Internet*. New York: Simon and Schuster.

Hanson, F. (2015). *Internet Wars : The Struggle for Power in the Twenty-First Century*. Haberfield: Longueville Media.

Harris, H. (2008). Race Across Borders: The U.S. and ICERD. *Harvard BlackLetter Law Journal, 24*, 61–67. http://hjrej.com/wp-content/uploads/2012/11/vol24/Harris.pdf. Accessed 20 Apr 2016.

Houston, J. B. (2015). Bouncing Forward: Assessing Advances in Community Resilience Assessment, Intervention, and Theory to Guide Future Work. *American Behavioral Scientist, 59*(2), 175–180.

Houston, J. B., Spialek, M. L., Cox, J., Greenwood, M. M., & First, J. (2015). The Centrality of Communication and Media in Fostering Community Resilience: A Framework for Assessment and Intervention. *American Behavioral Scientist, 59*(2), 270–283. http://abs.sagepub.com/content/59/2/270.abstract. Accessed 1 Feb 2015.

Huttenback, R. A. (1976). *Racism and Empire: White Settlers and Coloured Immigrants in the British Self-Governing Colonies 1830–1910*. New York: Cornell University Press.

International Telecommunications Union. (2014). *WSIS +10*. http://www.itu.int/net/wsis/implementation/2014/forum/inc/doc/outcome/362828V2E.pdf. Accessed 30 Sep 2016.

International Telecommunications Union. (2017). *What We Do*. http://www.itu.int/en/about/Pages/whatwedo.aspx. Accessed 21 Jan 2017.

Jakubowicz, A. (2011). Anders Breivik, Australian Anti-Multiculturalists and the Threat to Social Cohesion. *The Conversation*. https://theconversation.com/anders-breivik-australian-anti-multiculturalists-and-the-threat-to-social-cohesion-2542. Accessed 12 Sep 2016.

Jakubowicz, A. (2012). Cyber Racism. In *More or Less: Democracy and New Media*. [Online]. Version. http://www.futureleaders.com.au/book_chapters/pdf/More-or-Less/Andrew_Jakubowicz.pdf. Accessed 16 July 2016.

Jakubowicz, A. (2015). Hunting for the Snark and Finding the Boojum—Building Community Resilience Against Race Hate Cyber Swarms. In *Perspectives on the Racial Discrimination Act: Papers from the 40 Years of the Racial Discrimination Act Conference* (pp. 109–120). [Online]. Version. https://www.humanrights.gov.au/sites/default/files/document/publication/Conference Papers_Publication layout Final_0.pdf. Accessed 17 Oct 2016.

Jakubowicz, A. (2016). Once upon a Time in … Ethnocratic Australia: Migration, Refugees, Diversity and Contested Discourses of Inclusion and Exclusion. *Cosmopolitan Civil Societies: An Interdisciplinary Journal, 8*(3), 5239.

Kellner, D. (2001). Globalisation, Technopolitics and Revolution. *Theoria: A Journal of Social and Political Theory, 48*(4), 14–34. https://www.jstor.org/stable/41802171?seq=1-page_scan_tab_contents. Accessed 23 Mar 2016.

Kleinrock, L. (2008). History of the Internet and Its Flexible Future. *IEEE Wireless Communications, 15*(1), 8–18. http://ieeexplore.ieee.org/document/4454699/. Accessed 6 June 2016.

Kosoff, M. (2016a). Does Silicon Valley Have a Racism Problem? *Vanity Fair HIVE*. http://www.vanityfair.com/news/2016/06/does-silicon-valley-have-a-racism-problem. Accessed 23 Sep 2016.

Kosoff, M. (2016b). Steve Bannon's Racist Comments About Silicon Valley Are Also Wildly Inaccurate. *Vanity Fair HIVE*. http://www.vanityfair.com/news/2016/11/steve-bannon-racist-comments-silicon-valley-inaccurate. Accessed 14 Jan 2017.

Lentin, A. (2008). *Racism: A Beginner's Guide*. Oxford: Oneworld Publications.

Liu, Y. (2012). The Rise of China and Global Internet Governance. *China Media Research, 8*(2), 46–55.

Love, E. (2004). *Race Over Empire: Racism and U.S. Imperialism, 1865–1900*. Chapel Hill: University of North Carolina Press.

Maier, L. (2016a). Germany Investigates Mark Zuckerberg and Facebook Over Slow Removal of Hate Speech. *Forward*. http://forward.com/news/world/353625/germany-investigates-mark-zuckerberg-and-facebook-over-slow-removal-of-hate/. Accessed 23 Mar 2017.

Maier, L. (2016b). Twitter Gives Online Hate Speech Its Biggest Platform—Why? *Forward*. http://forward.com/news/352133/twitter-gives-online-hate-speech-its-biggest-platform-why/?attribution=articles-article-related-1-headline. Accessed 23 Jan 2017.

McGonagle, T. (2012). The Troubled Relationship Between Free Speech and Racist Hate Speech: The Ambiguous Roles of the Media and Internet. *Day of Thematic Discussion "Racist Hate Speech"*. UN Committee on the Elimination of Racial Discrimination. https://www.ivir.nl/publicaties/download/Expert_Paper_Racist_Hate_Speech_UN.pdf. Accessed 24 May 2016.

Mehzer, T., El Khatib, S., & Maduwanthi, S. (2016). Cyberattacks on Critical Infrastructure and Potential Sustainable Development Impacts. *Civil and Environmental Engineering: Concepts, Methodologies, Tools, and Applications*. http://www.igi-global.com/chapter/cyberattacks-on-critical-infrastructure-and-potential-sustainable-development-impacts/144514. Accessed 3 Feb 2017.

Miller, T. (2016). Cybertarian Flexibility—When Prosumers Join the Cognitariat, All That Is Scholarship Melts into Air. In M. Curtin & K. Sanson (Eds.), *Precarious Creativity: Global Media, Local Labor*. Oakland: University of California Press.

Narayan, C. (2017). Quebec Mosque Suspect Known to People Who Monitor Far-Right Groups. *CNN*. http://edition.cnn.com/2017/01/31/americas/quebec-mosque-shooting-suspect/. Accessed 11 Feb 2017.

Nissim, R. (2014). Building Resilience in the Face of Racism: Options for Anti-Racism Strategies. *Australian Policy Online and Australian Human Rights Commission*. Australian Human Rights Commission. http://apo.org.au/files/Resource/rnissim_buildingresilienceinthefaceofracism_oct_2014.pdf. Accessed 23 July 2015.

Office of the Historian. (n.d.). The Immigration Act of 1924 (The Johnson-Reed Act). *Bureau of Public Affairs*. United States Department of State. https://history.state.gov/milestones/1921-1936/immigration-act. Accessed 27 Nov 2016.

Olding, R. (2017). White Supremacist Threatened to Shoot Up Central Coast Shopping Centre. *Sydney Morning Herald*. http://www.smh.com.au/nsw/white-supremacist-threatened-to-shoot-up-sydney-shopping-centre-20170120-gtvod0.html. Accessed 28 Jan 2017.

Orbe, M. (1998). From The Standpoint(s) of Traditionally Muted Groups: Explicating A Co-Cultural Communication Theoretical Model. *Communication Theory, 8*(1), 1–26. https://wmich.pure.elsevier.com/en/publications/from-the-standpoints-of-traditionally-muted-groups-explicating-a--3

Paradies, Y. (2017). Racism and Health. In *International Encyclopedia of Public Health* (pp. 249–259). [Online]. Version. http://www.sciencedirect.com/science/article/pii/B9780128036785003726. Accessed 2 Mar 2017.

Paul, D. (1988). The Selection of the "Survival of the Fittest". *Journal of the History of Biology, 21*(3), 411–424. http://www.jstor.org/stable/4331067. Accessed 2 Mar 2017.

Pfefferbaum, B., Pfefferbaum, R. L., & Van Horn, R. L. (2015a). Community Resilience Interventions: Participatory, Assessment-Based, Action-Oriented Processes. *American Behavioral Scientist, 59*(2), 238–253. doi:10.1177/0002764214550298. Accessed 1 Feb 2016.

Pfefferbaum, R. L., Pfefferbaum, B., Nitiéma, P., Houston, J. B., & Van Horn, R. L. (2015b). Assessing Community Resilience: An Application of the Expanded CART Survey Instrument with Affiliated Volunteer Responders. *American Behavioral Scientist, 59*(2), 181–199. http://abs.sagepub.com/content/59/2/181.abstract. Accessed 1 Feb 2016.

Price, C. (1974). *The Great White Walls Are Built: Immigration to North America and Australasia, 1836–1888*. Canberra: ANU Press.

RAND Corporation. (2013). *Community Resilience Learn and Tell Toolkit*. RAND Corporation. http://www.rand.org/pubs/tools/TL163.html. Accessed 23 Nov 2016.

Ransom, J. (2013). *Race and Empire*. London: Routledge.

Richardson, S. (1998). Net Spreads Message, Draws in Recruits. *Montgomery Advertiser*. https://www.stormfront.org/dblack/press062898.htm. Accessed 28 June 1998.

Richeson, J., & Sommers, S. (2016). Toward a Social Psychology of Race and Race Relations for the Twenty-First Century. *Annual Review of Psychology, 67*, 439–463. http://annualreviews.org/doi/abs/10.1146/annurev-psych-010213-115115. Accessed 30 Jan 2017.

Rivett, K. (Ed.) (1962). *Immigration: Control or Colour Bar?: The Background to "White Australia" and a Proposal for Change, with a Historical Chapter by David Johanson* (Revised and Enlarged ed.). Melbourne: Melbourne University Press for Immigration Reform Group.

Roithmayr, D. (2014). *Reproducing Racism: How Everyday Choices Lock in White Advantage*. New York: New York University Press.

Roseman, C. (2014). Troublesome Reflection: Racism as the Blind Spot in the Scientific Critique of Race. *Human Biology Open Access Pre-Prints*, Paper 67. http://digitalcommons.wayne.edu/humbiol_preprints/67. Accessed 27 Aug 2016.

Ross, K., & Nightingale, V. (2003). *Media and Audience: New Perspectives*. Maidenhead: Open University Press.

Sepulveda, D., & Strickling, L. (2016, May 26). China's Internet Domain Name Measures and the Digital Economy. National Telecommunications and Information Administration, Blog. https://www.ntia.doc.gov/blog/2016/china-s-internet-domain-name-measures-and-digital-economy. Accessed 23 June 2016.

Simon Wiesenthal Center. (2016). Wiesenthal Center Digital Terrorism & Hate Report 2016: Terrorists' Online Encryption A Threat to Us All. *Simon Wiesenthal Center*. http://www.wiesenthal.com/site/apps/nlnet/content.aspx?c=lsKWLbPJLnF&b=8776547&ct=14827597. Accessed 15 Dec 2016.

Smith, E. A. (2011). Endless Forms: Human Behavioural Diversity and Evolved Universals. *Philosophical Transactions of the Royal Society B (Biological Sciences), 366*(1563). The Royal Society. doi:10.1098/rstb.2010.0233. Accessed 16 Nov 2016.

Spence, N., Wells, S., Graham, K., & George, J. (2016). Racial Discrimination, Cultural Resilience, and Stress. *The Canadian Journal of Psychiatry, 61*(5), 298–307. https://www.ncbi.nlm.nih.gov/pubmed/27254805. Accessed 2 Feb 2017.

Srnicek, N. (2016). *Platform Capitalism*. London: Polity.

Statista. (2016). *The Not So World Wide Web*. https://www.statista.com/chart/3512/internet-adoption-in-2015. Accessed 15 Feb 2017.

Statista. (2017). *Number of Internet Users World Wide*. https://www.statista.com/statistics/273018/number-of-internet-users-worldwide/. Accessed 10 Feb 2017.

Swiss Institute of Comparative Law. (2000/2009). *Legal Instruments for Combatting Racism on the Internet*. Strasbourg: Council of Europe. https://book.coe.int/eur/en/human-rights-and-democracy/4177-legal-instruments-for-combating-racism-on-the-internet.html. Accessed 14 Oct 2015.

Ware, V., & Back, L. (2002). *Out of Whiteness: Color, Politics and Culture*. Chicago: University of Chicago Press.

Wattles, J., Smith, A., & Gupta, S. (2017). Trump's Travel Ban: Companies and Executives Speak Out. *CNN Money*. http://money.cnn.com/2017/01/30/news/companies/travel-ban-executives-business-reactions/. Accessed 14 Feb 2017.

Williams, G. (2009, April 7). Racist Premise of Our Constitution Remains. *Sydney Morning Herald*. http://www.smh.com.au/federal-politics/racist-premise-of-our-constitution-remains-20090406-9utm.html. Accessed 5 Jan 2017.

World Wide Web Foundation. (2017). Delivering Digital Equality: The Web Foundation's 2017–2022 Strategy. *Web Foundation*. http://webfoundation.org/2017/02/delivering-digital-equality-the-web-foundations-2017-2022-strategy/. Accessed 12 Mar 2017.

Young, H., Rooze, M., Russell, J., Ebner, J., & Schulten, N. (2016). *Evidence-Based Policy Advice: Final Report. Prevention, De-radicalisation and Citizenship.* TerRa: Terrorism and Radicalisation. http://terratoolkit.eu/wp-content/uploads/2016/12/TERRA-Evidence-based-Policy-Advice_English_Final-Report.pdf. Accessed 4 Mar 2017.

Yudell, M., Roberts, D., DeSalle, R., & Tishkoff, S. (2016). Taking Race Out of Human Genetics. *Science, 351*(6273), 564–565. http://science.sciencemag.org/content/351/6273/564. Accessed 13 Jan 2017.

Zuckerberg, M. (2016). Mark Zuckerberg to Hold Facebook Q&A Session in Berlin. *Zuckerberg Transcripts 189.* http://dc.uwm.edu/zuckerberg_files_transcripts/189. https://www.youtube.com/watch?v=5ZAMx-1mDmI. Accessed 23 Feb 2017.

Zuckerberg, M. (2017). Building Global Community. *Facebook.* https://www.facebook.com/notes/mark-zuckerberg/building-global-community/10154544292806634/. Accessed 2 May 2017.

2

Researching Cyber Racism: Methodologies and Methods Across Disciplines

Introduction

Racism refers to a range of power relationships between human groups, reflecting social beliefs and institutional practices about phenotypical, genotypical and cultural differences, and their psychological, social and cultural consequences. Cyber racism remains a complex phenomenon, demonstrated in the layered elements discussed in the previous chapter, that together produce the dynamics experienced by online users. Our approach recognises that these layers operate at global, national, local, communal, group and individual levels. Cyber racism as a process of intergroup and interpersonal relationships calls for research approaches that both draw on the specific strengths of the applied disciplines and seek new synergies and integrated approaches between them.

Our intention as researchers lies in presenting how we have come to understand these layered complexities from multiple perspectives so that we can propose initiatives that might reduce the impact of the phenomenon on cyber safety, societal harmony, community resilience and individual well-being. In Chap. 1, we provided an ecological model of the individual located in the concentric planes of association that affect identity (Fig. 1.3).

A. Jakubowicz et al., *Cyber Racism and Community Resilience*, Palgrave Hate Studies,
DOI 10.1007/978-3-319-64388-5_2

Table 2.1 Actors in the "chain of cyber racism": data generated by each, and methods used in CRaCR (Cyber Racism and Community Resilience)

Actors	Data	Methods
• Victims • Bystanders • Perpetrators • Legal context	• Online survey data • Online rhetoric (from online sites, Facebook, forums, etc.) • Focus groups • Australian rules and regulations	• Statistic analyses • Thematic analysis • Narrative analysis • Content analysis

In Table 2.1, we show how the relationships between actors, data and methods can be summarised. Each of these clusters can be approached from different methodological perspectives, drawing on different and integrated methods.

This integration of approaches has allowed us to collect rich and diverse data (e.g., archival as well as newly generated, both qualitative and quantitative) from all locations along the "chain." We are able to examine racist texts online, in their forms as propaganda, commentary posts and websites. We can peruse the conversations and interactions of perpetrators in places such as racist forums and Facebook pages. Their survey responses allow us to explore how perpetrators see the world, looking at both their fears and aspirations. The targets of racism can recount their experiences through the survey and through intensive focus groups. We can also explore how people who feel they are just bystanders are indeed implicated in the dynamics of the Web and its interflows of racism. We soon understood that the experience of racism and the opportunities it presents reflect the broader socio-legal context, nation by nation, even where the Internet has already redefined the meaning of transnational to make it more integrated and all-embracing.

Racism has global, transnational, national, regional, local and personal dimensions. This makes it open to analysis from the full range of scholarly disciplines. Some aspects of racism are specific, local and heavily shaped by long histories of intergroup contact and conflict. Other aspects are more universal, apparently reflecting many anthropological and psychological patterns derived from evolutionary survival strategies. Now that racism has been so facilitated by the ease of online communication, the challenges for scholars to understand, analyse and interpret have intensified. The growth of cyber racism and its capacity to destabilise social relations intensify the need for scholars to theorise and research the complexities presented by new technologies and burgeoning contact between once-separated populations.

From theology (Taylor 2003) to physics (Jego and Roehner 2007), and through the full range of humanities and social sciences, scholars have sought to define racism (as separate from the debate about whether race exists in a scientific sense) and research its dynamics, sources, impact and possible societal responses. In this chapter, we set out to provide a broad description of how the pathway from disciplinary framework to research question, to method, to research outcome operates in the study of cyber racism. That is, we are interested in how scholars have sought to define the particular characteristics of the Internet and social media in relation to race-targeting hate speech. We then want to explain and explore how we have built on that research to define our research questions in relation to the approaches we have taken to understand the phenomena. These many points of tension between the power and interpersonal relations of racism have been ever-more complicated by the vast range of interactions afforded by the development of the Internet and its many contexts.

Disciplinary approaches to racism affect the questions asked and how they might be pursued. However, a growing press towards interdisciplinarity marks many of the approaches, such as those discussed in recent reviews of key disciplines. For example, a discussion of research in social psychology has identified a range of approaches including personality theories, social identity theory, critical discursive approaches and racism as an embodied experience. The authors note that even though it has been posited that these approaches are inconsistent, "recent attempts at integration are providing richer accounts" (Augoustinos and Every 2015).

In the *Sociology of Racism*, the authors of a 2015 review define the topic area as studying the relationship "between racism, racial discrimination and racial inequality." The three terms refer to attitudes and values, practices and outcomes, thereby blending into social psychology, political science and economics and health studies. Moreover, they reject criticisms of this "conceptual broadening." They claim that most social scientists accept such a multivalent approach, especially where contemporary sociology understands racism in its more diffuse and subtle forms, thereby pointing to factors that sustain racism in "purportedly 'postracial' society" (Clair and Denis 2015).

History has also been changing, from initially considering racism as "a theory of difference," to now seeing it as "an ideology with an integral relationship to domination or exclusion" (Fredrickson 2015). Such broad interdisciplinarity underpins the creation of our research team and the integration of our disciplinary methodologies.

Given the increased relevance of cyber racism in the current socio-political context, research in cyber racism has accelerated as the social issues it reveals become more pressing, and the awareness of the ways in which the Internet magnifies and alters racism has become more evident. Research from a range of disciplines has been focusing on addressing important questions regarding how racist groups and individuals use the Internet to:

- disseminate their messages in order to validate and endorse views (in the case of individuals) and
- widen their support base and strengthen the commitment of existing members (in the case of groups and organisations).

The Internet is now widely used by these groups to build transnational communities united under the common goal of spreading racial hatred.

Transdisciplinarity and Approaches in Research on Cyber Racism

Cyber racism is a topic of research in many disciplines in humanities, social sciences, public health and general science. The diversity of disciplines that have approached the study of cyber racism is reflected in multifaceted, complex but often fragmented conceptualisations. The relative lack of cohesion in defining cyber racism can lead to difficulties in integrating the research findings from those various disciplines. This makes it difficult for new research to systematically build on the existing work (and progress in the field to be identified and quantified).

The way in which a particular construct is conceptualised determines the approach that is adopted in its researching. These conceptualisations are not just attempts to deconstruct and define the world. They reflect and reveal the epistemological and philosophical leanings of those conducting the research. In an attempt to provide a more structured mode of approaching research in this area, we review a number of definitions of cyber racism proposed in different disciplinary fields and analyse the ways in which these definitions shape the methodologies used in each of the research fields (see Table 2.2). Then, based on this analysis, we propose an integrative conceptualisation of cyber racism and discuss a number of valid ways in which different methodologies, both traditional and newly emerging, can be applied to the study of cyber racism.

Table 2.2 Cyber racism research questions and methods across disciplines

Discipline	Research questions examples	Methods
Sociology	How do racist groups find and recruit new members? How are racist identities (collectively) constructed online? What are the rhetoric strategies used by racists online? How do racist groups connect as part of a broader white supremacist movement? What are the goals of online messages communicated by racist groups? How is humour used to mainstream racism?	Discourse analysis Critical discourse analysis Thematic analysis Social network analysis Grounded theory
Anthropology	How exclusionary racist narratives are constructed online?	Narrative analysis
Political science	How do racist groups recruit online? How (transnational) are racist identities constructed and mobilised to engage in collective action?	Content analysis Discourse analysis Social network analysis
Criminology	How is the Internet used to victimise targets from racially and culturally diverse backgrounds? What is the extent to which video games advocate racist violence?	Discourse analysis Content analysis Machine learning applications to "big data" analysis

(continued)

Table 2.2 (continued)

Discipline	Research questions examples	Methods
Social psychology	What is the role of humour in spreading racism online? What is the role of racism in explaining online reactions to news events? What strategies do racist groups use online to promote themselves? What is the extent to which racists advocate online violence in response to economic and cultural threats?	Discourse analysis Content analysis Conversation analysis Statistical analysis (inferential statistics)
Communication studies	How does online discourse promote racist ideologies? How are fantasy themes used as discursive strategies by racist groups? What are the characteristics of hate speech online? What is the role of online communication in creating and exploiting moral panic?	Critical discourse analysis Thematic analysis Content analysis Virtual ethnography
Cultural studies	What discursive strategies are used in racist groups (in particular in relation to women)? What are the online experiences of gamers from minority ethnic backgrounds?	Content analysis Virtual ethnography
Information science	What are the characteristics of racist groups in blogs? How do racist groups connect online?	Semi-automated Web mining Social network analysis Data visualisation Content analysis

The Study of Cyber Racism in Different Disciplinary Fields

Sociology

In sociology, cyber racism is studied as an extension of offline racism in the virtual domain while at the same time considering the affordances of living in a highly connected world from the perspective of perpetrators (i.e., how racist groups use these affordances in their favour). The primary sources of data used by sociologists are participant interviews, online conversation archives (Burke and Goodman 2012) and websites of, and materials produced by, racist or white supremacist groups (Adam and Roscigno 2005; Atton 2006). This preference can be at least partially explained by the overarching aims that sociological research in the domain of cyber racism is set to achieve. That is, sociological studies often seek to uncover the specific strategies used by racist groups online to exert influence and achieve their goals of societal dominance. For instance, it was proposed that humour is often used as part of a strategy to introduce racist messages to an audience in a way that would make that audience believe racism is a benign response to structural and economic changes occurring at a given time in society. In terms of the most effective methodologies to use so that subtle strategies and nuances (such as the use of humour) can be captured, researchers in sociology often use *discourse analysis* (DA) and in particular *critical discourse analysis* (CDA). CDA is a preferred method in sociology because it is specifically designed to capture those societal "ripples" that are the result of power struggles between various socio-political actors. It is defined by Teun van Dijk (1993a) as an approach that seeks to identify those "(…) structures, strategies and other properties of the text, talk, verbal interaction or communicative events" (p. 250) that play a role in the production, reproduction and challenge of dominance as the social power exercised by different groups and organisations with the aim of undermining social equality. It is different from other types of DA in that in CDA the researchers' stance is not only explicitly stated, but represents an integral part of the method. In the words of Teun van Dijk:

> Unlike other discourse analysts, critical discourse analysts (should) take an explicit sociopolitical stance: they spell out their point of view perspective, principles and aims, both within their discipline and within society at large. Although not in each stage of theory formation and analysis, their work is admittedly and ultimately political. Their hope, if occasionally illusory, is change through critical understanding. Their perspective, if

> possible, that of those who suffer most from dominance and inequality. Their critical targets are the power elites that enact, sustain, legitimate, condone or ignore social inequality and injustice. That is, one of the criteria of their work is solidarity with those who need it most. (p. 252)

Applied to the study of cyber racism, CDA is used, for instance, to understand and elucidate how racism is dynamically constructed online through the interplay between discourses and identities formed around concepts of power, culture and oppression with the aim of legitimising oppression of the "Other" (Atton 2006), or, in other instances, to capture the use of humour as clever persuasion devices (Weaver 2010, 2011, 2013; Grigg and Manderson, 2015).

Also, specific to sociological approaches to understanding cyber racism is *social network analysis* (SNA). SNA is a particularly useful methodological technique in relation to understanding cyber racism as it is based on the idea that social networks are key to understanding interactions, relationships and transfer of social influence between different "agents" or "nodes" in the network. And what is cyber racism if not a massive, global online network of ideologically driven actors that connect and communicate to spread race hate. The SNA approach is driven by the reality of interconnectedness (and interinfluence) of social actors, which in a virtual context is particularly relevant. An added benefit of the approach is that it can be used to explore both intragroup interaction (by looking at linkages between members of the same groups; e.g., direct communication on an online forum) and intergroup dynamics (by looking at linkages between different organisations and their facilitating influence on particular aspects of a group's actions; see Burris et al. 2000).

Political Science

In political science, many theoretical constructs are derived from sociological theory, so it is not surprising that their methodological approaches in researching cyber racism overlap to some degree. However, a method that is often used by political scientists and less by sociologists is content analysis (which also seems to be a predominant methodology in studies of cyber racism conducted by social psychologists and communication studies researchers). For instance, Caiani and Parenti (2010) use content analysis across Italian racist websites to examine how racist groups disseminate their ideologies online. Their data set is made up of "classical" Nazi texts as well as biographies, collections of fascist stories and works by fascist and neo-fascist intellectuals, hate symbols, banners and photos. They analyse the content of all these materials to identify the various uses of particular types of communication (e.g., how visual symbols are used to endorse group identity for existing supporters, while multimedia tools are predominantly used for recruitment purposes).

Criminology

Studies of cyber racism in criminology have primarily focused on understanding cyber racism in terms of the harm perpetrated against victims (including consequences for victims) in an online environment as opposed to physical attacks. The research (in both its design and interpretation of findings) is often embedded in discussions of legislation and the legal implications of cyber racism (Burnap and Williams 2016). As such, in a criminological context, cyber racism has been approached under the broader umbrella of "virtual victimisation" (Awan 2014; Awan and Zempi 2015) and "cyber hate" (as hateful and antagonistic content; see Williams and Burnap 2016). Both terms include a number of types of online victimising behaviours such as homophobia, antisemitism, racism and sexism. In terms of the methodologies used in criminology, they are particularly eclectic, ranging from different types of qualitative discursive analyses of participant interviews (Awan and Zempi 2015) and content analysis (Schafer 2002) to applications of machine learning approaches to analyse large data sets from the Internet and social media (Burnap and Williams 2016; Williams and Burnap 2016).

Surprisingly, analyses driven by grounded theory (Strauss and Corbin 1998)—a predominant methodological approach in sociology—are not as frequent as one would have thought. Based on a survey of the existing literature on cyber racism, we were only able to identify one study that used this approach to identify the specific aims of online messages in racist groups in the USA (McNamee et al. 2010).

Cultural Studies and Anthropology

In cultural studies, a new branch of "digital cultural studies" has emerged to deal with the complexities that online communication experiences entail. One iconic work that shaped this area of inquiry in relation to cyber racism is the book *Cyber Racism: White Supremacy Online and the New Attack on Civil Rights* by Jessie Daniels. In a recent research review, Daniels highlights the idea (previously discussed by Castells 1997 and others; e.g., Byrne 2007; Nakamura and Chow-White 2013) that "people seek out racial identity and community online. People use the Internet to both form and reaffirm individual racial identity and seek out communities based on race and racial understandings of the world" (Daniels 2013, p. 699). This is reflected in the emergence of not only various online diaspora communities, but also online communities based on shared views of racial discrimination and oppression of particular ethnic and religious groups. Those online communities that desire to enlarge racialised perspectives are in large part responsible for the current proliferation of racism in different locations in the cyberspace. However many other online communities are formed not with the intent to extend a racist agenda or intimidate others, but rather to sustain their own communities and re-engage with cultural heritage across the generations, or amongst the diaspora. Sometimes the line between these orientations can be blurred, where celebration of one ethnic heritage grows to depend on the denigration of others.

One important contribution from the intersection of digital cultural studies and anthropology extends the focus from well-known and (relatively) common virtual places of online interaction (such as forums, discussion sites and chat rooms) to more current and less researched areas where such interactions occur—gaming playgrounds. Researchers such as Kishiona Gray (2012) document the experiences of interacting in online gaming communities from the perspective of American African players using "*virtual ethnography.*" She explores how minority players are subjected to racial stereotyping and racism in the context of the gamers' community where whiteness and masculinity are dominant values. The method of virtual ethnography is derived from classic ethnography and, as such, uses the same three primary sources of information (i.e., observation, participant observation and interviews), but in an online context. In the case of Gray's research, she used Xbox Live as the location of her study with games such as Halo Reach, Gears of War and Call of Duty as field sites. The semi-structured interviews were conducted online in the private chat rooms of Xbox Live space or via Skype and ooVoo.

In anthropology, *narrative approaches* were used to examine Facebook materials that undermine cosmopolitanism by promoting messages of cultural exclusion (Connelly 2016). Narrative analysis (also a key methodology in historical sociology; Gotham and Staples 1996) was used by Teun van Dijk (1993b) in the study of reproduction of discursive racism by looking at the role of everyday storytelling about ethnic or racial minorities.

Communication Science

From the perspective of communication science, Klein argues that:

> (…) the World Wide Web has brought its own unique properties of communication into the information age, such as search engines, blogs, and social networks. Just as these properties have affected not only the flow, but also, subsequently, the form of traditional media content (i.e., journalism), so too have they begun to reshape the appearance and profile of hate speech in cyberspace. When thought of as a rhetorical strategy, hate speech should be understood as the tactical employment of words, images, and symbols, as well as links, downloads, news threads, conspiracy theories, politics, and even pop culture, all of which have become the complex machinery of effective inflammatory rhetoric. (Klein 2012, p. 428)

Some of the methods that communication science researchers use are:

- **CDA** (for an analysis of online white supremacist ideology, see Brown 2009; and for characteristics of hate speech, see Erjavec and Kovacic 2012);

- **Thematic analysis** (as fantasy theme analysis to identify discursive strategies such as the use of fantasy themes in research, see Duffy 2003); and
- **Online or virtual ethnography** (to explore the role of the Internet in creating and exploiting moral panic, see Flores-Yeffal et al. 2011).

Psychology

In psychology (and particularly social psychology), in researching cyber racism, quantitative approaches specific to the discipline are supplemented by qualitative methodologies including content analysis, DA and CDA. Social psychologists are primarily interested in understanding how the online medium can facilitate (or hinder) the expression and formation of particular "social identities" that are framed around intergroup relations and processes (Bliuc et al. 2012). Socio-psychological dimensions that are captured in written language (e.g., on racist websites) are identified and quantified in terms of their prevalence—as in Douglas et al. (2005) where dimensions such as intergroup hostility and social creativity were identified and coded (to assess their levels) in white supremacist online speech. Conversation analysis has been used (in addition to DA) to examine the role of race and racism in online social interactions about black protesters in South Africa (Durrheim et al. 2015) and in explaining online reactions to news (Creswell et al. 2014). Other psychological research has employed online surveys to capture the effects of cyber racism on the psychological adjustment of young people (Tynes et al. 2008), and on perceptions of racial climate in a university campus (Tynes et al. 2013).

To understand how racist arguments are constructed online using moral disengagement—as a strategy of making morally problematic positions seem more justified—Faulkner and Bliuc (2016) use thematic analysis when analysing online rhetoric around racist incidents in Australia (Braun and Clark 2006). Thematic analysis as a methodological approach is particularly useful in psychology where the researchers' approach is usually based on a clear assumption (derived from theory or existing research) that needs to be systematically tested, that is in confirmatory rather than exploratory research where a bottom-up approach based on grounded theory would be more suitable.

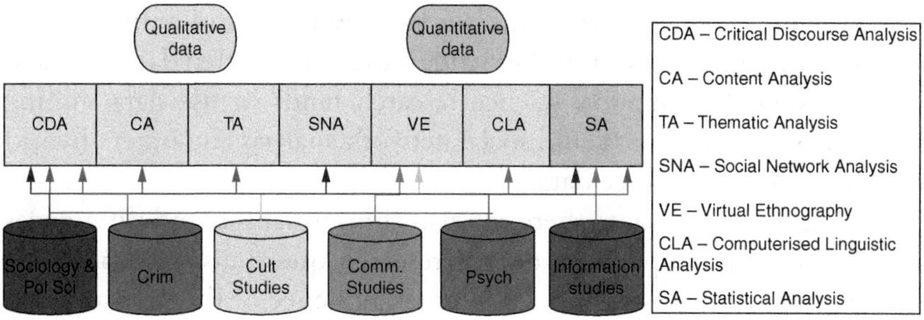

Fig. 2.1 Some of the key methodologies used in cyber racism research across disciplines

Information Science

Finally, research on cyber racism also comes from information science where the focus has been mostly on understanding how the Internet is used by racists (rather than the content of their communication or their strategies). From a methodological point of view, information scientists use sophisticated quantitative methodologies based on large data sets obtained through automated and semi-automated data mining. For example, to identify characteristics of racist groups and understand how these groups connect to each other, SNA complemented by data visualisation was used (Chau and Xu 2006) (Fig. 2.1).

An Integrated Understanding of Research Focus and Methodologies Used Across Disciplines

The cross-disciplinary overlap in research focus emerges naturally from shared research questions. For example, the question of how the Internet is used by racists has been addressed from the perspective of sociology, social psychology and cultural studies. However, methodologies employed to answer these questions are more discipline-specific. That is, disciplines such as sociology and criminology tend to use qualitative analyses of archival data, generally extracted from racist websites and online discussion forums. Psychological studies tend to use self-reported data collected using surveys. Information science research tends to use data mining techniques to extract data, and social network analysis and other "hybrid methods" to analyse these data.

In collecting their data there are also commonalities and differences across disciplines depending on the research questions that each particular study seeks to address. The target audience of the (racist) communication has implications for what data might be available for analysis, and therefore for the methodologies employed. According to how they are obtained, the data can be classified into participant-generated data (spontaneously generated by participants in discussion forums, news commentary websites or racist groups), or researcher-driven data (data that is driven by the questions that are asked by researchers). In a sense, participant-generated data can be used to conduct more exploratory research, which can provide naturalistic representations of different aspects of cyber racism. On the other hand, researcher-driven data can be used for explanatory and confirmatory research where issues identified through analyses of participant-generated data can be further researched, that is, specific questions emerging from the data can be formulated and tested as hypotheses (see Table 2.3).

Table 2.3 Mapping methodological approaches by types of data, context of research and target audience

Type	Self-generated data (e.g., website comments)			Researcher-driven data (e.g., survey completions)	
Sources	Groups		Individuals	Individuals	
Online context	Racist group websites		Forums, chat rooms, blogs and commentary websites	Questionnaires	Interviews
Target audience	In-group	Out-group	In-group	No target audience	

Research Methodologies in the CRaCR Project

Our Cyber Racism and Community Resilience (CRaCR) project was designed to examine the "chain of cyber racism" considering the socio-political and current legal context of cyber racism. To achieve this, we developed an integrated methodology, drawing on a number of approaches in several disciplinary fields. For example, we used DA of online content (see Chap. 6), survey-based data on individual encounters (see Chap. 3), textual and process analyses of regulatory regimes (see Chap. 8), case-study analysis with expert informants, government and non-government organisations (see Chap. 5) and qualitative focus groups with targeted users (see Chap. 5).

Drawing on approaches from (social) psychology and sociology, to analyse online discourses of race hate (and their counter-discourses) we used thematic analysis applied to racist and anti-racist posts in online news sites, debating two (relatively) recent notable instances of blatant racism in an Australian context (see Chap. 6 and also Faulkner and Bliuc 2016):

1. A racist incident on public transport in Sydney (anti-Asian racism)
2. Racial abuse suffered by Adam Goodes, an Indigenous Australian footballer during a football game (anti-Indigenous racism)

In addition, to analyse white supremacist discourse we extracted online interactions from the Australian branch of Stormfront—at the time the world's largest white supremacist online community—from 2001 to 2015. We examined key recurring themes using natural language processing (NLP) software (Bliuc et al. 2016). Text analysis software is particularly effective in cases where "big data" are available for analysis—an occurrence that is likely to become more frequent given the massive increase in online racist (and counter-racist) exchanges.

An online survey of attitudes and behaviours of regular Internet users forms the basis for the data used in Chap. 3 of this book. It draws a representative sample, compounded by additional participants who came from the communities most likely to be targeted by online racism.

From a criminological perspective, our project also incorporated a systematic *analysis of regulation* such as laws, regulatory and industry codes of practice, conditions of use, records and statements of key stakeholders, to model approaches and identify outcomes that enable assessments of efficacy (see Chap. 8).

A Case for an Interdisciplinary Approach to the Study of Contemporary Cyber Racism into the Future

With significant technological advances being made on an almost daily basis, proponents of racism and more broadly far-right ideologies have proved to be very adept in not only adapting to these changes, but also quickly capitalising on them. New research which seeks to understand these dynamics must be able to capture both ideological changes (with their implications on strategies and behaviours) and the ways in which these changes are enabled by the sophisticated use of new technologies. The methodology of new research needs to be current and adaptable, and by adopting an interdisciplinary approach (i.e., problem-driven rather than theory-driven) to capturing changes in ideology, we can reconcile aspects of research on how we ask the questions, what questions are we asking and where to look for the data. Emerging research approaches from linguistics and the psychology of language (enabled by developments in a range of psychologically relevant linguistic analysis software) can be integrated with discursive analysis techniques to enable the identification of relationships between changes in the socio-political context and changes in attitudes, beliefs, values and ideologies of racists on the Internet. Media commentators have already noted how effective far-right groups are in capitalising on the recent string of dramatic terrorist incidents in Europe (e.g., the case of the 2016 Christmas Berlin market attack; see Johnston 2016). Research can provide evidence and detail on how such events impact not only on the language they use in their communications but also on the internal states of members of these groups. For instance, existing software developed by social psychologists can be used specifically to identify changes in emotions and psychological dimensions such as social processes and cognitive complexity. By using such instruments, the currently lacking longitudinal research on cyber racism and its underpinning factors can be finally produced.

Conclusion

The integration of theoretical perspectives from complementary disciplines with a range of methodologies has served to produce a very rich evidence base for this study. Singular disciplines can of course offer a great deal, yet by testing propositions against different explanatory frameworks and integrating the results in comprehensive conclusions, the capacity for better explanation and the contribution to social resilience can be advanced. Research where there is strong alignment between both conceptual development and methodologies pushes outcomes in a more solution-focused direction. More collaboration between disciplines can enable the development of complementary approaches to finding solutions to the increasingly complex problems revealed by investigations of cyber racism.

References

Adams, J., & Roscigno, V. J. (2005). White Supremacists, Oppositional Culture and the World Wide Web. *Social Forces, 84*(2), 759–778.

Atton, C. (2006). Far-Right Media on the Internet: Culture, Discourse and Power. *New Media & Society, 8*(4), 573–587.

Augoustinos, M., & Every, D. (2015). Racism: Social Psychological Perspectives. In J. Wright (Ed.), *International Encyclopedia of the Social & Behavioral Sciences* (pp. 864–869). Amsterdam: Elsevier.

Awan, I. (2014). Islamophobia and Twitter: A Typology of Online Hate Against Muslims on Social Media. *Policy and Internet, 6*(2), 133–150.

Awan, I., & Zempi, I. (2015). 'I Will Blow Your Face off'—Virtual and Physical World Anti-Muslim Hate Crime. *British Journal of Criminology, 57*(2), 362–380.

Bliuc, A. M., McGarty, C., Hartley, L., & Muntele Hendres, D. (2012). Manipulating National Identity: The Strategic Use of Rhetoric by Supporters and Opponents of the 'Cronulla Riots' in Australia. *Ethnic and Racial Studies, 35*(12), 2174–2194.

Bliuc, A.-M., Faulkner, N., & Jakubowicz, A. (2016). *Online Networks of Hate: A Systematic Review of Research on Cyber-Racism* (Unpublished Manuscript).

Braun, V., & Clarke, V. (2006). Using Thematic Analysis in Psychology. *Qualitative Research in Psychology, 3*(2), 77–101.

Brown, C. (2009). WWW.HATE.COM: White Supremacist Discourse on the Internet and the Construction of Whiteness Ideology. *The Howard Journal of Communications, 20*, 189–208.

Burke, S., & Goodman, S. (2012). 'Bring Back Hitler's Gas Chambers': Asylum Seeking, Nazis and Facebook – A Discursive Analysis. *Discourse and Society, 23*(1), 19–33.

Burnap, P., & Williams, M. L. (2016). Us and Them: Identifying Cyber Hate on Twitter Across Multiple Protected Characteristics. *EPJ Data Science, 5*(1), 1.

Burris, V., Smith, E., & Strahm, A. (2000). White Supremacist Networks on the Internet. *Sociological Focus, 33*(2), 215–235.

Byrne, D. N. (2007). Public Discourse, Community Concerns, and Civic Engagement: Exploring Black Social Networking Traditions on BlackPlanet.com. *Journal of Computer-Mediated Communication, 13*(1), 319–340.

Caiani, M., & Parenti, L. (2010). The Dark Side of the Web: Italian Right-Wing Extremist Groups and the Internet. *Southern European Society and Politics, 14*(3), 273–294.

Castells, M. (1997). *The Power of Identity: The Information Age: Economy, Society and Culture* (Vol. II). Oxford: Blackwell.

Chau, M., & Xu, J. (2006). Mining Communities and Their Relationships in Blogs: A study of Online Hate Groups. *International Journal of Human-Computer Studies, 65*, 57–70.

Clair, M., & Denis, J. (2015). Sociology of Racism. In *International Encyclopedia of the Social and Behavioral Sciences* [Online]. Version. http://www.sciencedirect.com/science/article/pii/B9780080970868321225

Connelly, K. (2016). Undermining Cosmopolitanism: Cyber-Racism Narratives in Australia. *Sites: A Journal of Social Anthropology & Cultural Studies, 13*(1).

Cresswell, C., Whitehead, K. A., & Durrheim, K. (2014). The Anatomy of 'Race Trouble' in Online Interactions. *Ethnic and Racial Studies, 37*(14), 2512–2528.

Daniels, J. (2013). Race and Racism in Internet Studies: A Review and Critique. *New Media & Society, 15*(5), 695–719.

Douglas, K. M., McGarty, C., Bliuc, A.-M., & Lala, G. (2005). Understanding Cyberhate. *Social Science Computer Review, 23*(1), 68–76.

Duffy, M. E. (2003). Web of Hate: A Fantasy Theme Analysis of the Rhetorical Vision of Hate Groups Online. *Journal of Communication Inquiry, 27*(3), 291–312.

Durrheim, K., Greener, R., & Whitehead, K. A. (2015). Race Trouble: Attending to Race and Racism in Online Interaction. *British Journal of Social Psychology, 54*(1), 84–99.

Erjavec, K., & Kovacic, M. P. (2012). 'You Don't Understand, This is a New War!' Analysis of Hate Speech in News Websites' Comments. *Mass Communication and Society, 15*(6), 899–920.

Faulkner, N., & Bliuc, A. M. (2016). 'It's Okay to be Racist': Moral Disengagement in Online Discussions of Racist Incidents in Australia. *Ethnic and Racial Studies, 39*(14), 2545–2563.

Flores-Yeffal, N. Y., Vidales, G., & Plemons, A. (2011). The Latino Cyber-Moral Panic Processes in the United States. *Information, Communication and Society, 14*(4), 568–589.

Fredrickson, G. M. (2015). *Diverse Nations: Explorations in the History of Racial and Ethnic Pluralism*. London/New York: Routledge.

Gotham, K. F., & Staples, W. G. (1996). Narrative Analysis and the New Historical Sociology. *The Sociological Quarterly, 37*(3), 481–501.

Gray, K. L. (2012). Deviant Bodies, Stigmatized Identities, and Racist Acts: Examining Experiences of African-American Gamers in Xbox Live. *New Review of Hypermedia and Multimedia, 18*(4), 261–276.

Grigg, K., & Manderson, L. (2015). "Just a Joke": Young Australian Understandings of Racism. *International Journal of Intercultural Relations, 47*, 195–208. https://doi.org/10.1016/j.ijintrel.2015.06.006. Accessed 17 Aug 2016.

Jego, C., & Roehner, B. (2007). *A Physicist's View of the Notion of "Racism"*. e-Print arXiv.org/physics/0704.2883, Cornell University Library.

Johnston, I. (2016). Berlin Attack: German Far-Right Leader Frauke Petry Attacks Call for 'Humanity' as 'Political Correctness'. *The Independent*. http://www.independent.co.uk/news/world/europe/berlin-attack-germany-frauke-petry-afd-far-right-refugees-response-statement-horror-arrived-facebook-a7486171.html. Accessed 5 Nov 2016.

Klein, A. (2012). Slipping Racism into the Mainstream: A Theory of Information Laundering. *Communication Theory, 22*(4), 427–448.

McNamee, L. G., Peterson, B. L., & Pena, J. (2010). A call to Educate, Participate, Invoke and Indict: Understanding the Communication of Online Hate Groups. *Communication Monographs, 77*(2), 257–280.

Nakamura, L., & Chow-White, P. (Eds.). (2013). *Race After the Internet*. New York: Routledge.

Schafer, J. A. (2002). Spinning the Web of Hate: Web-Based Hate Propagation by Extremist Organizations. *Journal of Criminal Justice and Popular Culture, 9*(2), 69–88.

Strauss, A., & Corbin, J. (1998). *Basics of Qualitative Research: Techniques and Procedures for Developing Grounded Theory*. London: Sage Publications Inc.

Taylor, G. H. (2003). *Racism as the Nation's Crucial Sin: Theology and Derrick Bell. Michigan Journal of Race & Language, 9*, 269.

Tynes, B. M., Giang, M. T., Williams, D. R., & Thompson, G. N. (2008). Online Racial Discrimination and Psychological Adjustment Among Adolescents. *Journal of Adolescent Health, 43*, 565–569.

Tynes, B. M., Rose, C. A., & Markoe, S. L. (2013). Extending Campus Life to the Internet: Social Media, Discrimination, and Perceptions of Racial Climate. *Journal of Diversity in Higher Education, 6*(2), 102–114.

Van Dijk, T. A. (1993a). Principles of Critical Discourse Analysis. *Discourse & Society, 4*(2), 249–283.

Van Dijk, T. A. (1993b). Stories and Racism. In *Narrative and Social Control: Critical Perspectives* (pp. 121–142).

Weaver, S. (2010). Developing a Rhetorical Analysis of Racist Humour: Examining Anti-Black Jokes on the Internet. *Social Semiotics, 20*(5), 537–555.

Weaver, S. (2011). Jokes, Rhetoric and Embodied Racism: A Rhetorical Discourse Analysis of the Logics of Racist Jokes on the Internet. *Ethnicities, 11*(4), 413–435.

Weaver, S. (2013). A Rhetorical Discourse Analysis of Online Anti-Muslim and Antisemitic Jokes. *Ethnic and Racial Studies, 36*(3), 483–499.

Williams, M. L., & Burnap, P. (2016). Cyberhate on Social Media in the Aftermath of Woolwich: A Case Study in Computational Criminology and Big Data. *British Journal of Criminology, 56*(2), 211–238.

3

How Cyber Users Experience and Respond to Racism: Evidence from an Online Survey

Introduction

The proliferation of social media across the globe has transformed the ways people engage with each other and consume information online. The provision of information is increasingly complemented or even supplanted by opportunities to socialise, as users integrate social media into their daily lives. Yet, the ubiquity of social media has opened Internet users to a host of threats, ranging from criminal exploitation to cultural and individual harassment. Where people might once have sought the solidarity of an online community, they can now find themselves exposed to dangerous and destructive interactions. This not only threatens their self-image and personal integrity, but can often negatively impact their physical and mental health.

Online harassment can motivate serious modifications in online behaviour and significantly undermine personal well-being. A 2016 study of 3000 Americans (Lenhart et al. 2016) found that 87% of them had online lives, and 47% had been harassed, becoming victims of "unwanted contact that is used to create an intimidating, annoying, frightening, or even hostile environment." While the report covers harassment on grounds of sexuality and gender, as well as race, it demonstrates how prevalent threats

A. Jakubowicz et al., *Cyber Racism and Community Resilience*, Palgrave Hate Studies,
DOI 10.1007/978-3-319-64388-5_3

are, and how dangerous the Internet has become. Moreover, 25% had self-censored their online lives to avoid drawing attention from trolls or exposing themselves as targets. Even when not directly attacked, an overwhelming majority had witnessed others being harassed online. Younger people tended to be more likely to be harassed than older people, women more than men and gay people more than straight. The most likely to self-censor were younger women. In general, Black users were far more likely to have witnessed other people being harassed and abused online, and therefore describe online behaviour by others as "unkind." Together, these findings suggest a social media environment, which is gendered and affected by race and age, is much like the offline world.

Other international research suggests similar types of dangers. The German Association of the Internet Industry, *eco*, sponsored an online survey during 2015 of German users, which reported that over one-third of respondents had encountered racist hate speech online (eco 2016). As with the US research, young people were much more likely to encounter racism—up to 62% of 18–24-year-olds, of whom 7% had been the direct targets of racism. The *eco* Complaints Office saw a 150% increase in justified complaints about racism in 2015 compared to 2014. While 30% of the respondents had reported racist material and 14% had responded (counter posts), 45% simply ignored the encounters. In its December 2016 Annual Review, *eco* pointed to the rising concerns about hate speech, and tensions in the industry caused by Government proposals that would mandate the industry to proactively weed out hate speech (eco 2017).

The spread of racist hate speech has become one of the most threatening dimensions of the Internet, driven on the one hand by increasingly powerful technologies, and on the other by increasingly knowledgeable perpetrators. In addition, the normalisation of racist world views built around the explosion of right-wing nationalist political movements and parties feeds racism online. Today, racism is perpetrated on gay and straight dating sites, in games rooms, on the comments pages of the mass media and in the social media locales such as Facebook, Twitter and YouTube. As will be seen later in the chapter, Australia provides a valuable case study for understanding online interactions, where encounters with racism produce many consequences for victims, bystanders and perpetrators.

The Australian Online CRaCR Survey

Australia has arguably one of the highest proportional uses of social media in the world, with Australian Internet users spending a considerable amount of time visiting social networks and blogs (Nelson 2011a, b). As such, Australians are potentially more exposed to cyber racism than residents of other countries. Both peak NGOs (Paradies et al. 2009) and leading academics (Lewis et al. 2015) have identified cyber racism as a key priority because of its impact on health and well-being. With the Australian government's announcement of a National Anti-Racism Partnership and Strategy (Australian Human Rights Commission 2012), racism has been repositioned as a social policy priority (AMAC 2010). The Australian Human Rights Commission (AHRC) advised that close to one in ten racial complaints in 2015–16 were concerned with Internet-based racism (AHRC n.d.). However, the official complaints and reports represent only a small proportion of the encounters with racism, and, we assert, a small proportion of the responses to cyber racism. So, to what extent, and how, do people take action against the racism they encounter online?

Research, to date, on racist online content has focused on hate group poetics and politics, cultural privilege and stereotypes in online news and the reception of such messaging (Back 2002; Daniels 2009a, b, 2013; Douglas et al. 2005; Hughey and Daniels 2013; Locke 2012; Steinfeldt et al. 2010). Tynes and Markoe's (2010) study is one of the few that analysed how users respond to cyber racism, examining the link between colour-blind racial attitudes and online racism using photographs posted on Facebook and Myspace as prompts. They found that users were largely unaffected by racist content, and some even found it amusing. Outside of this research, there is surprisingly little research on the negotiation of racism online and how racism shapes Internet use at an everyday level (Daniels 2013). In general, there is an astounding lack of data that investigate the frequency of, and responses to, online racist content or the effectiveness of regulatory measures for dealing with it.

In this chapter, we detail the main findings of a survey conducted in December 2013 as a part of the Australian online CRaCR Project. Over

2,000 Internet users across Australia were surveyed regarding their encounters with cyber racism, the impact of these encounters and their responses to them. We examined to what extent gender, age or ethnicity affected the likelihood of experiencing racist content, and whether these categories affect peoples' reactions and responses to cyber racism or attitudes to racism. We also examined attitudes to debates about the regulation of racism, especially cyber racism, and the closely associated issue of free speech. As we show using survey data, a significant number of Internet users are in harm's way as a result of racism, as targets but also as witnesses. We also show that a small but prolific group are publishing racist content online and then broadcasting it to a wider audience than was ever possible before. There are similarities with offline rates of racism, but some interesting differences too, particularly in terms of the way people respond to the racist content they encounter online.

Method

The CRaCR survey used established prompts on encounters with different types of racism, including name-calling, exclusion and violent incitement, narrowed and adapted to experiences likely to occur across mainstay Internet platforms such as Facebook, Twitter, YouTube and, more broadly, Internet forums. The survey also examined the effects of such encounters on a respondent's sense of well-being and the actions they might have taken, such as making a formal complaint, ignoring it or engaging with the racist.

The sample was generated by a commercial online provider, MyOpinion (now SSI), who constructed two online panels: one mirroring the ethnic diversity of the Australian population aged 15–54 as at the 2011 Census; the other drawing from groups significantly at risk of racism, including Indigenous Australians ($n = 58$), Australians of North African and Middle Eastern background ($n = 34$), Australians of Southeast Asian background ($n = 192$), Australians from Northeast Asian background ($n = 266$) and Australians from Southern and Central Asian background ($n = 142$). The data were collected in December 2013 from a total of 2,141 respondents.

The age range of respondents was spread fairly evenly across most age brackets, and roughly half the respondents were male and half female.

The data collected and reported on here are compared to findings from the National Challenging Racism Project (CRP), which has collected substantial data on racism in Australia since 2001. The CRP survey of attitudes to cultural diversity, racism and the so-called out-groups was rolled out across the nation between 2001 and 2008, with a final collective sample of n = 12,512 representing participants from all Australian states and territories, compiled in a national dataset. The demographic profile of these participants is comparable to that of the CRaCR survey. Table 3.1 demonstrates this with regards to gender and age profile. A comparison of socio-demographic characteristics of the participants from

Table 3.1 Age, gender and level of education of CRaCR survey participants in 2013 (*n* = 2,141) compared to National CRP survey participants in 2001–2008 (*n* = 12,512)

	CRaCR		CRP	
	n	%	*n*	%
Gender				
Male	1095	51.2	6087	48.6
Female	1043	48.8	6425	51.4
Total	2138	100	12,512	100
Age (years)				
18–24	178	8.3	1431	11.4
25–34	316	14.8	2347	18.8
35–44	389	18.2	2393	19.1
45–54	403	18.8	2330	18.6
55–64	394	18.4	1793	14.3
65+	458	21.4	2218	17.8
Total	2138	100	12,512	100
Education				
Primary school or no formal qualifications	20	0.9	935	7.5
Some high school or Year 10 Certificate or equivalent	132	6.2	2877	23.1
High School Certificate	532	24.9	2609	21
Trade or Technical and Further Education (TAFE) qualifications or other tertiary qualifications	597	27.9	2598	20.9
University degree or higher	854	39.9	3420	27.5
Total	2135	100	12,440	100

Source: Cyber Racism and Community Resilience (CRaCR) Survey 2013 and Challenging Racism Project (CRP) National Findings 2011

both samples using level of education reveals that the CRaCR survey was a more highly educated sample, and this partly reflects the online delivery mode of that sample.

The attitudes profiles of respondents in both survey samples, on matters of diversity and racism, can be compared (see Table 3.2). There was more apparent uncertainty among the "CRaCR sample." These respon-

Table 3.2 Attitudes towards diversity and equality in 2013 (*n* = 2,141), compared to the national average, 2001–2008 (*n* = 12,512)

	Agree		Neither		Disagree		Total	
	CRaCR	CRP	CRaCR	CRP	CRaCR	CRP	CRaCR	CRP
	%	%	%	%	%	%	%	%
It is a good thing for a society to be made up of people from different cultures	80.7	87.3	16	6.2	3.2	6.5	100	100
I feel uncomfortable around people from other racial, ethnic or religious backgrounds	11.7	9.6	22.5	11.2	65.8	79.3	100	100
People from racial, ethnic or religious minority groups experience discrimination in Australia	60	85.2	30.1	6.8	9.9	8	100	100
I am prejudiced against other cultures	8.2	12.4	28.9	6.9	62.9	80.7	100	100
Australians from an Anglo background (i.e., of British descent) enjoy an advantaged position in our society	40.7	42.3	34.8	57.7	24.5	41.5	100	100

(continued)

Table 3.2 (continued)

	Agree		Neither		Disagree		Total	
	CRaCR	CRP	CRaCR	CRP	CRaCR	CRP	CRaCR	CRP
	%	%	%	%	%	%	%	%
The Australian way of life is weakened by people from minority racial, ethnic or religious backgrounds maintaining their cultural beliefs and values	33.7	41.9	30.1	15.3	36.2	42.8	100	100
There is no one racial, ethnic, cultural or religious group that is superior to others	82.1	85	12.9	4.1	5	10.9	100	100

Source: Cyber Racism and Community Resilience (CRaCR) Survey 2013 and Challenging Racism Project (CRP) National Findings 2011

dents were consistently more likely to select the "neither" option in response to propositions on diversity and racism. On some questions, the proportion of "neither" selections was reflected in a proportional reduction for both disagree and agree responses (e.g., stance on diversity, assimilationism, and racial supremacism). But for most questions, the higher "neither" responses reflected a reduced agreement (or disagreement) with the more "controversial side of the response options," such as less acknowledgement of racism and of Anglo privilege, and less self-identification as prejudiced.

The similarity of the demographic and attitudinal profiles of the respondents in both surveys provides a sense of the reliability of the perspectives offered from the CRaCR survey. This provides some assurance that the survey sample allows an insight into the Australian experience of cyber racism as either targets or witnesses. Furthermore, we can analyse these survey results to gain a glimpse into everyday Australian's attitudes about cyber racism and mechanisms for regulating civility, including our novel questions on free speech and racial vilification provisions.

Encountering Cyber Racism: Witnesses and Targets

The Australian survey bears out the findings of the reported studies in the USA and in Germany (Lenhart et al. 2016; eco 2016). One-third of respondents indicated that they had witnessed racist content online (34.8%; see Table 3.3). Most of these encounters had occurred quite recently, having taken place in the month before the survey (40.3%). A small, but significant, number of respondents indicated that they had personally been targets of cyber racism (4.8%). Among these respondents, when asked to think back to the most recent time they had been targeted, one-third indicated they were targeted within the last month and two-thirds within the last six months. This indicates that a considerable proportion of Australians are regularly exposed to cyber racism, and the literature suggests this would result in negative impacts for many of them (Tynes et al. 2016). Gender was not a consistent predictor of respondent encounter with cyber racism as a witness, target or author. Younger users of the platforms were more likely than their older peers to

Table 3.3 Most recent encounter as a witness to, or target of, cyber racism, 2013 (*n* = 2,129 and 2,132)

	Witness		Target	
	n	%	*n*	%
In the last day	30	4.0	80	7.8
In the last week	109	14.7	13	12.7
In the last month	160	21.6	12	11.8
In the last six months	242	32.7	31	30.4
In the last year	123	16.6	14	13.7
More than a year ago	77	10.4	24	23.5
Total who witnessed cyber racism	741	34.8	102	4.8
Total sample	2129	100	2,132	100

Source: Cyber Racism and Community Resilience (CRaCR) Survey 2013

report being a target and a witness of cyber racism. Our analysis of the attitudes of witnesses and targets of cyber racism presents both surprising and expected results. For example, 11% of those aged 18–25 had been targets, compared to only 3% of those aged over 55 ($\chi 2(5, N = 2132) = 23.066$; $p = 0.000$). Those in the younger age bracket were also twice as likely to be witnesses to cyber racism than the older group (50+% for the younger group, and 25% of less for the older categories: $\chi 2(10, N = 2138) = 151.226$; $p = 0.000$).

The different encounters with cyber racism are associated with different attitudinal profiles. Targets of racism were more likely than non-targets to acknowledge the cultural prejudice towards and difficulties faced by Aboriginal Australians (51% vs 39%; $\chi 2(2, N = 2{,}101) = 8.631$; $p = 0.013$). Targets were more likely to feel uncomfortable around people from other racial, ethnic or religious backgrounds (19% vs 11%; $\chi 2(2, N = 2{,}101) = 7.516$; $p = 0.023$) and more likely to admit their own prejudice (16% vs 8%; $\chi 2(2, N = 2{,}101) = 8.129$; $p = 0.017$). Targets shared some of this attitudinal profile with the authors of racist content, a point we return to later in the chapter. The 35% of respondents who had witnessed cyber racism had an attitudinal profile that was quite distinct from those that said they had not witnessed such content. Witnesses were consistently better disposed to diversity and difference, and more tolerant. Specifically, witnesses were more pro-diversity (87% vs 78%; $\chi 2(2, N = 2{,}101) = 24.237$; $p = 0.000$), less assimilatory (29% vs 36%; $\chi 2(2, N = 2{,}101) = 76.484$; $p = 0.000$), less separatist (10% vs 13%; $\chi 2(2, N = 2{,}101) = 22.693$; $p = 0.000$) and more likely to acknowledge racism (75% vs 52%; $\chi 2(2, N = 2{,}101) = 106.413$; $p = 0.000$) and Anglo privilege (52% vs 35%; $\chi 2(2, N = 2{,}101) = 62.567$; $p = 0.000$). They were also more sympathetic to Aboriginal Australians (48% vs 36%; $\chi 2(2, N = 2{,}101) = 31.724$; $p = 0.000$) and asylum seekers (46% vs 31%; $\chi 2(2, N = 2{,}101) = 44.797$; $p = 0.000$). Witnesses (one-third of users) were more pro-diversity and less racist group than non-witnesses of cyber racism. Being a witness may require some literacy about what constitutes racism, and perhaps a moral conviction that it is a notable issue. Witnessing, racist literacy and an anti-racist conviction are very important components of cyber anti-racism.

Racist content was most commonly encountered on Facebook (40.1%; see Table 3.4). Other platforms where racist content appeared to be more prevalent were in the commentaries of online news stories, at almost one-fifth (18.5%) of reports, and in YouTube (15.7%). Facebook was also the platform where respondents spent more time, on average, than the other platforms, at 69 minutes per day. This was more than double the other platforms and may help explain some of the higher rates of encounter with racism on Facebook. Almost half (44.6%) of targets indicated that their most recent encounter of cyber racism was on Facebook. Similar to the results for witnesses, targets said that racist content was most likely to take the form of racist talk (84.3%). However, racist exclusion and violent incitement were at higher rates for targets, than for witnesses of racism.

Table 3.4 Platform on which cyber racism was witnessed, time spent on each platform per day and type of racism witnessed, 2013 ($n = 2{,}141$)

Platform	Witness		Target		Average time spent on platform/day (minutes)
Platform	*n*	%	*n*	%	
Comments—online news	135	18.5	9	9.8	33.5
Email	42	5.7	4	4.3	34.2
Facebook	293	40.1	41	44.6	69
Forums	94	12.9	16	17.4	20.7
Other	38	5.2	11	12	
Twitter	14	1.9	1	1.1	22.1
YouTube	115	15.7	10	10.9	38.1
Total	731	100	92	100	
Form[a]	*n*	%	*n*	%	
Talk	667	89.7	86	84.3	
Exclusion	90	12.1	28	27.4	
Violent incitement	88	11.8	17	16.7	

Source: Cyber Racism and Community Resilience (CRaCR) Survey 2013

[a]What form did this racist content take? **Talk**, for example, verbal abuse, name-calling, offensive jokes/slang, stereotypes, ridicule based on cultural background including in the form of videos and photos; **Exclusion**, for example, being left out in an online group or discussion, being ignored, treated as inferior by others online, treated with suspicion because of your cultural background or told that your cultural group doesn't belong; **Violent incitement**, for example, you or your cultural group were the subject of an online campaign (message or online group or page) aimed at encouraging or supporting physical attacks on people or physical damage/defacement of property

The forms of racism that are likely to generate more acute impacts were therefore more prevalent for targets than for witnesses.

Table 3.5 compares ethnic groups as a proportion of the overall survey sample to the percentage of respondents within those ethnic groups who had been targets of racist content. Online anti-Aboriginal sentiment appears to be very strong. Aboriginal Australians were by far the most likely to indicate that they had been targets of cyber racism, with one-fifth indicating that this was the case (20.7%). This was followed by North African and Middle Eastern Australians (12%). According to chi-square tests conducted, these strong variations were statistically significant ($\chi^2(13, N = 2{,}086) = 47.29, p = 0.000$). As can be seen in Table 3.6, being a target of cyber racism was also unevenly distributed across religious affiliation. Muslim respondents were more likely than those associated with other religions to be targets of cyber racism (13.2%), though this was not statistically significant.

Table 3.5 Respondents targeted by cyber racism, by ethnicity, 2013 (*n* = 2,086)

	Total number of respondents within ethnicity groups	Percentage of respondents within ethnicity groups, who have been targeted online
	n	%
Aboriginal	58	20.7
North African and Middle Eastern	34	11.8
Southeast Asian	191	5.8
North and West European	201	4.5
South and East European	49	4.1
Australian	930	4
Other Oceania	28	3.6
Southern and Central Asian	142	2.8
Northeast Asian	265	2.6
North American	5	0
South American	7	0
African	4	0
Other	54	5.6
Multiple	118	9.3
Total (n)	2086	

Source: Cyber Racism and Community Resilience (CRaCR) Survey 2013

Table 3.6 Respondents targeted by cyber racism, by religious affiliation, 2013 (n = 2,141)

	Yes	No	Total
	%	%	%
Buddhism	3.6	96.4	100
Christianity	4.2	95.8	100
Hinduism	4.8	95.2	100
No religion	5.4	94.6	100
Judaism	9.1	90.9	100
Islam	13.2	86.8	100
Other	7	93	100

Source: Cyber Racism and Community Resilience (CRaCR) Survey 2013

Impacts of Cyber Racism

There is a set of substantial affective impacts on witnesses from cyber racism. Over two out of five witnesses said they felt angry, annoyed or frustrated (41.7%) in response to the racist content they witnessed online, and almost half said they felt disgusted by the perpetrator (48.9%) (Table 3.7). A small but significant number of witnesses had a strong physical response, with almost 4% reporting that they got a headache, got an upset stomach, felt tense, got a pounding heart or were sweaty as a result of this encounter.

For the targets of cyber racism, Table 3.7 shows that almost half indicated that the experience made them feel angry, annoyed or frustrated (48.5%). As may be expected, targets are more likely than witnesses to have a physical reaction to racism ($p < 0.003$). They are more likely to have a headache and an upset stomach, feel tense, get a pounding heart or get sweaty (9.7% compared to 3.8%). Disgust was higher for witnesses than targets ($p < 0.015$), while, at the same time, witnesses were less likely to be amused than targets ($p < 0.000$). Interestingly, targets are more likely to feel guilty than witnesses ($p < 0.000$). These findings indicate the potential cumulative effects of cyber racism, especially on targets, particularly for those who are vulnerable or at higher risk, such as Aboriginal Australians and those of North African and Middle Eastern background.

Table 3.7 Emotional response by both targets of, and witnesses to, cyber racism, 2013 ($n = 2{,}141$)

	Witnesses ($n = 744$)		Targets ($n = 103$)			
Emotional response	*n*	%	*n*	%	Z	P
Disgusted by the perpetrator	364	48.9	39	37.9	2.181	0.015
Angry, annoyed or frustrated	310	41.7	50	48.5	−1.256	0.105
Powerless, hopeless or depressed	107	14.4	18	17.5	−0.795	0.213
Sympathy or empathy for the perpetrator	90	12.1	12	11.7	0.159	0.437
Ashamed or humiliated	84	11.3	14	13	−0.655	0.256
Amused	73	9.8	26	25.2	−4.526	0.000
Anxious or fearful	35	4.7	9	8.7	−1.705	0.044
Headache, upset stomach, tense, pounding heart or sweaty	28	3.8	10	9.7	−2.706	0.003
Guilty	23	3.1	10	9.7	−3.226	0.000

Source: Cyber Racism and Community Resilience (CRaCR) Survey 2013

Table 3.8 Stress effects of cyber racism, targets and witnesses, 2013 ($n = 2{,}141$)

Stress	Witnesses		Targets			
	n	%	*n*	%	Z	P
Not at all	236	32	26	25.5	1.326	0.184
A little	284	38.5	36	35.3	0.622	0.535
Somewhat	153	20.7	26	25.5	−1.1	0.271
Very	46	6.2	7	6.9	−0.245	0.803
Extremely	19	2.6	7	6.9	−2.344	**0.019**
Total	738	100	102	100		

Source: Cyber Racism and Community Resilience (CRaCR) Survey 2013

When asked how stressed they felt after witnessing cyber racism, only 9% of respondents said that they felt extremely or very stressed (Table 3.8). However, a majority reported being a little or somewhat stressed (59.2%). Similarly, a majority of targets said they were a little or somewhat stressed by the incident (60.8%), while 14% were very or extremely stressed by being targeted. This is concerning given emerging evidence about the cumulative effects of cyber racism on targets. However, there is also evidence of resilience. Two-thirds of those who were targeted said that they were not stressed or only a little stressed after the incident.

Responding to Cyber Racism

In reflecting on their last encounter with cyber racism, witnesses were asked how they had responded. They were provided with a checklist of responses relevant to the platform where they had encountered the content. As Facebook was the platform in which the most racist content was encountered, results from that platform are examined in more detail here for responses and impacts. Almost half of the respondents said that they simply ignored the racism they witnessed (48.8%; see Table 3.9). This contrasts to the offline world, where 64–69% ignore experiences of race hate talk (Dunn et al. 2009). The international literature suggests that, for the offline world, witnesses or bystanders are more likely to ignore racist incidents, with two-thirds of respondents taking no action. Recent research on bystander action in Australia is more encouraging, suggesting that only half (48.6%) take no action (Nelson et al. 2011a). It is encouraging that half of the people who witness racism are likely to take action in the cyber world, which exceeds global averages, though it may reflect more active responses in Australia generally.

While almost half of witnesses ignored racism on Facebook, most of the remaining took some form of action, such as online disagreement through comments (25%) and reporting it to the platform (20%). The use of within-platform responses was strong, and these included reporting the racist content (20%), blocking (14%) and defriending (12%) the poster. The responses to racist content witnessed on platforms other than Facebook followed similar patterns. Complaints to other bodies (such as human rights commissions, police or other authorities) were less common as a form of response to racism witnessed on Facebook.

Close to 40% of targets also ignored the cyber racism they encountered. However, across the range of responses, targets were generally more likely than were witnesses to take action in response to racist content. For example, targets were more likely than were witnesses to make a comment disagreeing with the content (39.0%) and to block the poster (31.7%). Almost half of the targets reported the cyber racism they had experienced (49%).

Table 3.9 Response to racist content experienced by both targets and witnesses, Facebook, 2013 (n = 2,141)

Response on Facebook	Witness (n = 293)		Target (n = 41)	
	n	%	n	%
Liked it	3	1	2	4.9
Ignored it	143	48.8	16	39
Made a comment disagreeing with it	77	26.3	16	39
Reported it	59	20.1	20	48.8
Blocked the poster	42	14.3	13	31.7
Defriended the poster	35	11.9	9	2.4
Complained to another body	19	6.5	5	12.2
Shared it	10	3.4	5	12.2
Campaigned to remove the content	9	3.1	5	12.2
Made a comment agreeing with it	3	1	2	4.9
Made an active dissonant response[a]	165	56	28	68
Made an active dissonant response within platform[b]	159	54	27	66

Source: Cyber Racism and Community Resilience (CRaCR) Survey 2013
[a]Includes: made a comment disagreeing with it, reported it, complained to another body, campaigned to remove the content, blocked the poster, defriended the poster
[b]Includes: Made a comment disagreeing with it, reported it, blocked the poster, defriended the poster

These data are to be expected, but more witness action is needed so as to share the burden of anti-racism, which sits too heavily upon the shoulders of targets at present (Daniels 2013, p. 710; Rattan and Ambady 2014). Taking a more hopeful interpretation, we can see here untapped political resources; that is, scope for more witness (bystander) action.

Authors of Cyber Racism

Survey respondents were asked whether they had published any content online in relation to religion, race or culture. If they had, the follow-up questions asked whether this content was perceived by others as racist and whether the respondents themselves had considered this content racist. Just under 2% of the total respondents indicated that they had published content online that they considered to be racist or that others had perceived as racist (see Table 3.10).

Facebook was the most common platform where racist content was published (see Table 3.11). YouTube was the second most utilised platform for authors of racist content, followed by Twitter and online forums. Racist talk was the most common form of authored racist content, as was the case for racism experienced by witnesses and targets. However, there was a stronger proportional presence of racist exclusion and violent incitement than for witnesses and targets of cyber racism. This suggests that a small number of perpetrators are publishing especially, and intentionally, racist material.

Respondents who were identified as authors of cyber racism were most likely to publish this content quite regularly—daily or weekly in most cases (see Table 3.12). The number of authors is small, prolific. The reach and impact of these authors is substantial given the audience available through online platforms. The picture that emerges is of a small number of perpetrators publishing racist material at a high frequency that is intentional, exclusionary and hurtful.

We asked authors of racist content on Facebook what responses they had received to their racist content. Almost half said they had received "likes," and about one-third said they had received positive comments or that their posting had been shared (see Table 3.13). This suggests that there is a community of largely like-minded "racist" participants operating on Facebook (Daniels 2013, pp. 698–99; Bliuc et al. 2016). About one-fifth of the authors said that complaints had been made about them, and only about one in ten said that within-platform action had been taken against them (such as using a report button, dissonant comment or blocking).

The first two columns in Table 3.14 show that authorship of cyber racism is spread across nearly all ethnic groups, although sample sizes

Table 3.10 Participants who have authored cyber racism, 2013 (n = 2,141)

	n	%
Perceived by others as racist	36	1.7
Author considered content racist	24	1.1
Either of the above	38	1.8
Total	2141	100

Source: Cyber Racism and Community Resilience (CRaCR) Survey 2013

Table 3.11 Platform on which cyber racism was most recently published, and the form it took, 2013 (*n* = 2,141)

Most recent platform	*n*	%
Facebook	23	60.5
YouTube	6	15.8
Twitter	3	7.9
Forums	3	7.9
Comments—online news	1	2.6
Other	2	5.3
Total	38	100
Form	***n***	**%**
Talk	23	60.5
Exclusion	18	47.4
Violent incitement	14	36.8
Total	38	100

Source: Cyber Racism and Community Resilience (CRaCR) Survey 2013

Table 3.12 Frequency of publishing racist material, by platform, 2013 (*n* = 2,141)

Frequency	Facebook	Twitter	YouTube	Forums	Comments
Every day	8	3	6	3	2
Every week	6	5	3	4	1
Once a month	3	0	4	1	2
A few times a year	8	0	0	0	0
Once	5	0	0	0	0
Never	1	0	2	0	1
Total	31	8	0	8	6

Source: Cyber Racism and Community Resilience (CRaCR) Survey 2013

are small in some groups. There were more perpetrators of an Australian background than any other background, but this only constituted a small proportion of all the "Australian" respondents. Those from an African background were the most likely, as a proportion of their group, to be authors (25%). These variations had some statistical significance ($\chi^2(13, N = 2092) = 35.80, p = 0.001$). However, the number of authors was very small, and care should be taken in assuming this small sample of authors are representative of cyber racist authors more broadly.

Table 3.13 Responses to the author's racist content, Facebook (n = 23), 2013

	n	%
Liked it	12	52.2
Made a comment agreeing with it	8	34.8
Shared it	7	30.4
Reported it to the target	4	17.4
Complained to another body	4	17.4
Ignored it	4	17.4
Made a comment disagreeing with it	3	13
Reported it to Facebook	3	13
Blocked the poster	3	13
Defriended the poster	3	13
Campaigned to remove the content	2	8.7
None	3	13

Source: Cyber Racism and Community Resilience (CRaCR) Survey 2013

Cyber Racist Environments

Table 3.15 demonstrates that authors have a high likelihood of also being targets. Seventeen of the 38 authors had also reported being targets of racism (45%), compared to only 4% of non-authors of racist content who report being targets. This variation was statistically significant ($\chi2(1, N = 2{,}132) = 134.00, p = 0.000$). These data, and the above-mentioned information on a relatively small number of profligate authors, are suggestive of a minority of protagonists, who are involved in an environment of "to and fro" racist debate with each other, perhaps with those from other ethnic groups. Habtegiorgis et al. (2014, p. 178) found that "racist attitudes are related to racist behaviours among perpetrators that are, in turn, related to experiences of racial discrimination among targets." In the online world, there is evidence for such "micro-communities" of racist practice.

Earlier in this chapter we unpacked the attitude profiles of witnesses and targets of cyber racism. Witnesses of cyber racism were more likely to be pro-diversity, eschew stereotypes and be more tolerant of difference. There was no clear picture on the attitude profile of targets, although they are more likely to experience discomfort with cross-cultural encounters (19% for targets vs 11% for non-targets $\chi2(2, N = 2101) = 7.516; p = 0.023$).], which was also the case for authors (though more strongly so: 41%; $\chi2(2, N = 2101) = 30.312; p = 0.000$). There was a high level of

Table 3.14 Participants who authored cyber racist content, by ethnicity, 2013 (*n* = 38)

Ethnicity	Authored racist content		Sample	
	n	% of Authors	*n*[a]	%[b] of Group
Aboriginal	4	10.5	58	6.9
African	1	2.6	4	25
Australian	11	28.9	932	1.2
North American	0	0	5	0
North African and Middle Eastern	0	0	34	0
North and West European	1	2.6	203	0.5
Northeast Asian	7	18.4	266	2.6
Other Oceania	0	0	28	0
South American	0	0	7	0
South and East European	2	5.3	49	4.1
Southern and Central Asian	6	15.8	142	4.2
Southeast Asian	5	13.2	192	2.6
Other	1	2.6	54	1.9
Multiple	0	0	118	0
Total	38	100	2092	

Source: Cyber Racism and Community Resilience (CRaCR) Survey 2013
[a]Total participants in the sample who identify as belonging to the relevant ethnic group
[b]Percentage of participants within that ethnic group that had authored racist content

Table 3.15 Cross-tabulation of authors and targets of racism, 2013 (*n* = 2,141)

Authored cyber racism	Target of cyber racism		
	Yes	No	Total
Yes	17 (45%)	21 (55%)	38
No	86 (4%)	2,008 (96%)	2,094
Total	103	2,029	2,132

Source: Cyber Racism and Community Resilience (CRaCR) Survey 2013

self-acknowledgement of prejudice among targets (16% vs 8% for non-targets; $\chi 2(2, N = 2101) = 8.129$; $p = 0.017$), which is similarly shared, though more strongly, with authors (46% vs 7% $\chi 2(2, N = 2101) = 71.291$; $p = 0.000$). So, there is some similarity in attitudes between targets and authors. Cyber racism authors' attitudes are, however, distinct in their antipathy towards diversity. Authors were four times more likely to say they feel uncomfortable around people from other racial, ethnic or religious backgrounds than non-authors, five times more likely to admit being prejudiced, and four times more likely to be against inter-racial marriage. However, they were as likely to accept that Aboriginal Australians face greater challenges in society and that asylum seekers who are in need should be helped.

Attitudes on the Regulation of Cyber Racism

As show in Table 3.16, nearly half of those surveyed (47%) believe that freedom to speak your mind is more important than freedom from hate speech (neutral 32%; disagree 21%). And yet three-quarters of respondents agreed or strongly agreed with the need for protections against humiliation (74.1%), insult (71.5%) and intimidation (79.3%) based on race, culture or religion, and two-thirds (65.8%) with the need for protection against offence on that basis. Over two-thirds (69.2%) went so far as to say that incitement to racist violence should be punishable by gaol. An overwhelming majority (73%) put the onus on media platforms such as Facebook and YouTube to report the complaints they receive about racism to the relevant authorities. However, almost half of the respondents (51.8%) said that they were uncertain as to the current reporting guidelines, suggesting a lack of familiarity with the "within-platform" reporting mechanisms. Respondents indicated that they like the idea that people have a fair amount of freedom to say what they think, to express their opinions. But they also indicated concern for the more vulnerable and most often targeted groups in the community and prefer some regulation that would control racist speech, as well as supporting harsher penalties on racism that incites violence. Given the similar demographic and attitudinal profile of this survey with that of previous offline surveys, these findings suggest important trends in wider social attitudes towards online speech.

Key attitudes were analysed across demographic categories like age and gender, and also across those respondents who had witnessed cyber racism or been a target (see Table 3.17). The age variations show that older

Table 3.16 Attitudes towards regulation of racist speech, 2013 (*n* = 2,141)

	Agree %	Neither %	Disagree %	Total %
It should be unlawful to offend someone on the basis of their race, culture or religion	65.8	24.3	9.9	100
It should be unlawful to humiliate someone on the basis of their race, culture or religion	74.1	20	6	100
It should be unlawful to insult someone on the basis of their race, culture or religion	71.5	21.2	7.3	100
It should be unlawful to intimidate someone on the basis of their race, culture or religion	79.3	15.5	5.3	100
There should be laws against racial vilification	79.5	16.8	3.6	100
There should be laws against religious vilification	69.2	24.4	6.4	100
There should be more control over what is said on the Internet	64.7	22	13.3	100
Freedom to speak your mind is more important than freedom from hate speech	47.1	31.8	21	100
I don't have a clear understanding of the guidelines for reporting online racist content	51.8	30.5	17.6	100
Incitement to racist violence should be punishable by gaol	69.2	23.7	7.1	100
Internet services like Facebook and YouTube should report all complaints about racism to the relevant authority	72.7	20.6	6.7	100

Source: Cyber Racism and Community Resilience (CRaCR) Survey 2013

Australians (especially those aged over 55) are more supportive of laws against racist speech or content that offends (70%), the use of incarceration as a punishment for racist violence (77%) and the compulsory reporting of cyber racism to authorities by the media platforms (77%). Younger Australians were much less sure of these approaches, with many more in the neutral category. Younger Australians were also less likely, on average, to be able to express agreement or disagreement on the prioritisation of freedom of speech over freedom from racist hate speech. Female respondents were more positively disposed to laws against hate speech and for reporting of racist materials in cyber platforms. However, women were a little less supportive, than were men, of the use of gaol terms for

Table 3.17 Attitudes towards laws, 2013 (n = 2,141)

Age	Agree %	Neutral %	Disagree %	Total %	χ2; p	Gender	A	N	D	T	χ2; p	Cyber racism	A	N	D	T	χ2; p
It should be unlawful to offend someone on the basis of their race, culture or religion																	
18–34 (*n* = 489)	57.1	33.5	9.4	100	33.798; .000	Male (*n* = 1088)	64.5	23.8	11.7	100	7.865; .020	Witness (*n* = 734)	66.2	24.0	9.8	100	.094; .954
35–54 (*n* = 782)	66.0	23.1	10.9	100		Female (*n* = 1022)	67.1	24.9	8.0	100		Not witness (*n* = 1376)	65.6	24.5	10.0	100	
												Total	65.8	24.3	9.9	100	
55+ (*n* = 839)	70.7	20.0	9.3	100								Author (*n* = 38)	55.3	31.6	13.2	100	1.905; .386
												Not author (*n* = 2072)	66.0	24.2	9.8	100	
Totals	65.8	24.3	9.9	100		Totals	65.8	24.3	9.9	100		Total	65.8	24.3	9.9	100	
There should be laws against racial vilification																	
18–34 (*n* = 489)	78.3	18.2	3.5	100	2.687; .611	Male (*n* = 1088)	77.1	17.8	5.1	100	15.161; .001	Witness (*n* = 734)	83.5	13.2	3.3	100	11.412; .003
35–54 (*n* = 782)	631	127	24	100		Female (*n* = 1022)	82.1	15.8	2.2	100		Not witness (*n* = 1376)	77.4	18.8	3.9	100	
												Total	79.5	16.8	3.6	100	
55+ (*n* = 839)	79.1	16.6	4.3	100								Author (*n* = 38)	68.4	23.7	7.9	100	3.594; .166
												Not author (*n* = 2072)	79.7	16.7	3.6	100	
Totals	79.5	16.8	3.6	100		Totals	79.5	16.8	3.6	100		Total	79.5	16.8	3.6	100	

Table 3.17 (continued)

Age	Agree %	Neutral %	Disagree %	Total %	$\chi2$; *p*	Gender	A	N	D	T	$\chi2$; *p*	Cyber racism	A	N	D	T	$\chi2$; *p*
Freedom to speak your mind is more important than freedom from hate speech																	
18–34 (*n* = 489)	44.2	35.0	20.9	100	11.168; .025	Male (*n* = 1088)	50.7	30.4	18.8	100	12.874; .002	Witness (*n* = 734)	45.2	31.2	23.6	100	4.406; .110
35–54 (*n* = 782)	45.3	34.3	20.5	100		Female (*n* = 1022)	43.2	33.4	23.4	100		Not witness (*n* = 1376)	48.1	32.2	19.7	100	
												Total	47.1	31.8	21.0	100	
55+ (*n* = 839)	50.5	27.8	21.7	100								Author (*n* = 38)	68.4	26.3	5.3	100	8.682; .013
												Not author (*n* = 2072)	46.7	31.9	21.3	100	
Totals	47.1	31.8	21.0	100		Totals	47.1	31.8	21.0	100		Total	47.1	31.8	21.0	100	
Incitement to racist violence should be punishable by gaol																	
18–34 (*n* = 489)	58.9	32.1	9.0	100	46.429; .000	Male (*n* = 1088)	69.9	22.7	7.4	100	1.635; .442	Witness (*n* = 734)	67.4	23.2	9.4	100	9.382; .009
35–54 (*n* = 782)	67.8	24.8	7.4	100		Female (*n* = 1022)	68.5	24.9	6.7	100		Not witness (*n* = 1376)	70.1	24.1	5.8	100	
												Total	69.2	23.7	7.1	100	
55+ (*n* = 839)	76.5	17.9	5.6	100								Author (*n* = 38)	68.4	18.4	13.2	100	2.502; .286
												Not author (*n* = 2072)	69.2	23.8	6.9	100	
Totals	69.2	23.7	7.1	100		Totals	69.2	23.7	7.1	100		Total	69.2	23.7	7.1	100	

(*continued*)

Table 3.17 (continued)

Age	Agree %	Neutral %	Disagree %	Total %	χ2; p	Gender	A	N	D	T	χ2; p	Cyber racism	A	N	D	T	χ2; p
Internet services like Facebook and YouTube should report all complaints about racism to the relevant authority																	
18–34 (*n* = 489)	66.9	27.0	6.1	100	22.048; .000	Male (*n* = 1088)	68.6	22.1	9.4	100	31.921; .000	Witness (*n* = 734)	72.1	20.3	7.6	100	1.627; .443
35–54 (*n* = 782)	71.6	20.7	7.7	100		Female (*n* = 1022)	77.1	19.1	3.8	100		Not witness (*n* = 1376)	73.0	20.8	6.2	100	
												Total	72.7	20.6	6.7	100	
55+ (*n* = 839)	77.1	16.8	6.1	100								Author (*n* = 38)	60.5	23.7	15.8	100	5.768; .056
												Not author (*n* = 2072)	72.9	20.6	6.5	100	
Totals	72.7	20.6	6.7	100		Totals	72.7	20.6	6.7	100		Total	72.7	20.6	6.7	100	

Source: Cyber Racism and Community Resilience (CRaCR) Survey 2013

racist violence (with this variation being small and non-significant). Male respondents were more supportive of the prioritisation of free speech over freedom from hate speech (51%) than were women (43%). Witnesses to cyber racism were more likely to support laws against racial vilification (83% for witnesses, but only 77% for non-witnesses). Interestingly, witnesses were more likely to say that they were unclear on reporting guidelines for online racist content (54% vs 51% of non-witnesses), though this variation was not statistically significant. The other significant variations in attitudes to regulation, across exposure to cyber racism, were between authors and non-authors of racist content. Authors were less likely to agree there should be more control over what is said online (50% vs 65% of non-authors), more likely to agree that free speech is more important than freedom from hate speech (68% vs 47% for non-authors) and more likely to say they were unclear on reporting guidelines for online racist content (61% vs 52% for non-authors). Despite the latter uncertainty, they were, unsurprisingly, less likely to support the proposition that cyber platforms should report all complaints about racism to a relevant authority.

Discussion

The results of the CRaCR survey indicate that the prevalence of racism online is quite high, with one-third of those surveyed having witnessed cyber racism and 40% of these within a month prior to the taking the survey. Facebook, online news commentary and YouTube appear to have the highest prevalence of racist content, relative to exposure, Twitter and email being sites of lesser encounter.

Emotionally, the racism encountered online has significant effects on both witnesses and targets. It is interesting that witnesses very often feel disgust towards those who author the racist content they have encountered, more so than those who were targeted by the racism. Inversely, and perhaps to be expected, targets are more likely to feel anger as a response to the racist content directed towards them. Concerns lie in the way these encounters affect the well-being of targets, particularly those from vulnerable groups. In general, targets displayed a relatively high degree of resilience, with many of them able to brush off the incident. Research has shown that this kind of response can reduce the impact of racism on health (Paradies and Cunningham 2012). However, a considerable proportion of respondents were left at least somewhat or a little stressed. There is evidence to suggest a potential cumulative effect of racism experienced frequently over a period of time, which causes harm over and above individual instances (Nelson et al. 2011b). For those who are particularly at risk of encountering racism, there are serious concerns that the accumulation of such encounters is damaging to their sense of wellbeing as well as their sense of belonging. A stand out finding from these data, which aligns with data previously published on racial attitudes in the offline world (VicHealth 2014), is that Aboriginals are a focus of cyber racism in Australia, followed by Middle Eastern and Muslim Australians.

Racist material pervades social media. Just as in the offline world, there are a minority of people with exclusionary views who use such platforms to spread intolerance towards diversity and certain cultural groups. Authors of cyber racism as identified in the CRaCR data appear to be a deviant few. However, these few are frequent offenders, publishing regularly. The fact that these authors are themselves likely to have been targets of cyber racism hints at something like online "race wars," with to-ing and fro-ing between authors and members of other ethnic groups. Statistical analysis of the relations between authoring, witnessing and being a target of cyber racism confirms a close link between authorship and being a target. Although there are few authors, the frequent publishing of this racist material has potentially wide-ranging impacts.

The responses and reactions to cyber racism by both witnesses and targets are notable. Over half of the witnesses of cyber racism made active responses. This kind of response is higher online than in the offline world globally, although it aligns with Australian "off-line" realms (Nelson et al. 2011a). Additionally, targets of cyber racism are more likely to make active responses, compared to the offline world, where two-thirds of respondents do not respond to being a target of race hate talk (Dunn et al. 2009). The differences in responses of targets and witnesses are also worth noting. Targets of cyber racism are more (re) active than witnesses (with the exception of defriending tools), with over two-thirds of targets making an active response to the racist content targeted towards them.

The most common forms of response for both witnesses and targets are "within platform," that is, in the Facebook example, posting a dissonant reply (witness 26%; target 39%), reporting the content (20%; 49%), and blocking (14%; 32%) or defriending (12%; 2%) the author. A significant question that arises with regards to these in-platform responses is one that concerns data collection and reporting. What happens to all of this reporting? Is it recorded? How is it managed? What are the outcomes? There is a lack of transparency around reporting from these mechanisms and little if any feedback to the complainants. The lack of public reporting of within-platform actions is a serious impediment to regulatory racism online. It contravenes commitments under global instruments to record and monitor racism. Our results indicate a strong need for more sophisticated regulatory measures and, as we have shown earlier, Internet user opinion supports mechanisms for reporting online racism to the relevant authorities. Whilst significant numbers of users appear to be making use of the reporting functions within platforms or otherwise making active responses to the racism they encounter, the current practices around reporting and management of racist content are too idiosyncratic and opaque. Peer and community monitoring of Internet behaviour alone are unlikely to be sufficient for ensuring civil and healthy intercultural relations on the Internet.

References

Australian Human Rights Commission. (2012). *National Anti-Racism Strategy*. Sydney: Australian Human Rights Commission.

Australian Human Rights Commission. (n.d.). *2015–2016. Complaints Statistics*. https://www.humanrights.gov.au/sites/default/files/AHRC%202015%20-%202016%20Complaint%20Statistics.pdf. Accessed 12 Dec 2016.

Australian Multicultural Advisory Council (2010) *The People of Australia: The Australian Multicultural Advisory Council's Statement on Cultural Diversity and Recommendations to Government*, Canberra The Australian Multicultural Advisory Council.

Back, L. (2002). Aryans Reading Adorno: Cyber-Culture and Twenty-First Century Racism. *Ethnic and Racial Studies, 25*(4), 628–651.

Bliuc, A-M., Betts, J., Faulkner, N., Vergani, M., Chow, R.J., Iqbal, M., & Wada, T. (2016). *The Internal Dynamics of Racist Online Communities: How External Events Catalyse Group Cohesion Within Stormfront*. Paper Presented at the 2016 Annual Conference of the International Society for Political Psychology, Warsaw.

Daniels, J. (2009a). Cloaked Websites: Propaganda, Cyber-racism and Epistemology in the Digital Era. *New Media & Society, 11*(5), 659–683.

Daniels, J. (2009b). *Cyber Racism: White Supremacy Online and the New Attack on Civil Rights*. Maryland: Rowman and Littleman Publishers.

Daniels, J. (2013). Race and Racism in Internet Studies: A Review and Critique. *New Media & Society, 15*(5), 695–719.

Douglas, K. M., McGarty, D., Bluic, A.-M., & Lala, G. (2005). Understanding Cyberhate: Social Competition and Social Creativity in Online White Supremacist Groups. *Social Science Computer Review, 23*(1), 68–76.

Dunn, K. M., Forrest, J., Pe-Pua, R., Hynes, M., & Maeder-Han, K. (2009). Cities of Race Hatred? The Spheres of Racism and Anti-racism in Contemporary Australian Cities. *Cosmopolitan Civil Societies: An Interdisciplinary Journal, 1*(1), 1–14.

eco. (2016). *eco Survey: Every Third Person Has Encountered Racist Hate Speech Online*. eco: Association of the Internet Industry. https://international.eco.de/2016/news/eco-survey-every-third-person-has-encountered-racist-hate-speech-online.html. Accessed 14 Sep 2016.

eco. (2017). *eco—A Look Back at 2016: Internet Skepticism Dominates Digital Policy*. eco: Association of the Internet Industry. https://international.eco.de/2017/news/eco-a-look-back-at-2016-internet-skepticism-dominates-digital-policy.html. Accessed 2 Feb 2017.

Habtegiorgis, A. E., Paradies, Y., & Dunn, K. M. (2014). Are Racist Attitudes Related to Experiences of Racial Discrimination? Within Sample Testing Utilising Nationally Representative Survey Data. *Social Science Research, 47*, 178–191.

Hughey, M. W., & Daniels, J. (2013). Racist Comments at Online News Sites: A Methodological Dilemma for Discourse Analysis. *Media Culture Society, 35*(3), 332–347.

Lenhart, A., Ybarra, M., Zickuhr, K., & Price-Feeney, M. (2016). *Online Harassment, Digital Abuse, and Cyberstalking in America*. New York: Data and

Society Research Institute, and Center for Innovative Public Health Research. https://www.datasociety.net/pubs/oh/Online_Harassment_2016.pdf. Accessed 14 Jan 2017.

Lewis, T. T., Cogburn, C. D., & Williams, D. R. (2015). Self-Reported Experiences of Discrimination and Health: Scientific Advances, Ongoing Controversies, and Emerging Issues. *Annual Review of Clinical Psychology, 11*, 407–440.

Locke, J. (2012). Public Expressions of Private Sentiments: Unveiling the Pulse of Racial Tolerance Through Online News Readers' Comments. *Howard Journal of Communications, 23*(3), 235–252. doi:10.1080/10646175.2012.695643.

Nelson, J. K., Dunn, K. M., & Paradies, Y. (2011a). Australian Racism and Anti-racism: Links to Morbidity and Belonging. In F. Mansouri & M. Lobo (Eds.), *Migration, Citizenship and Intercultural Relations: Looking Through the Lens of Social Inclusion* (pp. 159–176). Surrey: Ashgate Publishing Ltd.

Nelson, J. K., Dunn, K. M., & Paradies, Y. (2011b). Bystander Anti-racism: A Review of the Literature. *Analyses of Social Issues and Public Policy, 11*(1), 263–284.

Paradies, Y., & Cunningham, J. (2012). The DRUID Study: Exploring Mediating Pathways Between Racism and Depressive Symptoms Among Indigenous Australians. *Social Psychiatry and Psychiatric Epidemiology, 47*, 165–173.

Paradies, Y., Chandrakumar, L., Klocker, N., Frere, M., Webster, K., Burrell, M., & McLean, P. (2009). *Building on Our Strengths: A Framework to Reduce Race-Based Discrimination and Support Diversity in Victoria. Full Report.* Melbourne: Victorian Health Promotion Foundation.

Rattan, A., & Ambady, N. (2014). How It Gets Better: Effectively Communicating Support to Targets of Prejudice. *Personality and Social Psychology Bulletin, 40*(5), 555–566.

Steinfeldt, J. A., Foltz, B. D., Kaladow, J. K., Carlson, T. N., Pagano, L. A., Jr., Benton, E., et al. (2010). Racism in the Electronic Age: Role of Online Forums in Expressing Racial Attitudes About American Indians. *Cultural Diversity and Ethnic Minority Psychology, 16*(3), 362–371.

Tynes, B., & Markoe, S. (2010). The Role of Color-Blind Racial Attitudes in Reactions to Racial Discrimination on Social Networking Sites. *Journal of Diversity in Higher Education, 3*(1), 1–13.

Tynes, B. M., Rose, C. A., Hiss, S., Limana-Taylor, A. J., Mitchell, K., & Williams, D. (2016). Virtual Environments, Online Racial Discrimination, and Adjustment Among a Diverse, School-Based Sample of Adolescents. *International Journal of Gaming and Computer-Mediated Simulations, 6*(3), 1–16.

VicHealth. (2014). *Findings from the 2013 Survey of Victorians' Attitudes to Race and Cultural Diversity*. Melbourne: Victorian Health Promotion Foundation.

4

Racism and the Affordances of the Internet

Why Racists Like the Internet

The Internet has become a dangerous place for encounters with racism, though, as Don Black of White supremacist online forum Stormfront has allowed many times, the Internet provides a welcoming hunting ground for racists (Richardson 1998; Jakubowicz 2012). As James Curran has noted in his summary of key myths about freedom on the Internet, "race hate groups were internet pioneers, with former Klansman Tom Metzger, then leader of White Aryan Resistance, setting up a community bulletin board in 1985 … From these cyber-frontier origins, racist websites have proliferated" (Curran 2012, p. 10).

In this chapter, we analyse how racists behave online, recognising, as we show in Chap. 6 on racist narratives, that not all racism has been constructed by people with overt racist agendas or a conscious sense of antipathy to other race or ethnic groups. As detailed in Chap. 1, commentators and scholars disagree about many of the phenomena that the term racism encompasses. When people talk about racism, they may be addressing unselfconscious micro-aggressions (Nadal 2011), overt hate speech (Elliott et al. 2016; Cohen-Almagor 2013) and/or

A. Jakubowicz et al., *Cyber Racism and Community Resilience*, Palgrave Hate Studies,
DOI 10.1007/978-3-319-64388-5_4

deliberate advocacy of violence against groups on the basis of their race, ethnicity or religion (Klein 2017). Such divergences are important to acknowledge, though most public comments tend to accept that racism promoting violence should not be allowed to circulate freely in modern democratic societies (except for overt racists who proudly boast of their stance). From that point on, there is significant disagreement about the scope, impact, appropriate response and regulation that should be imposed. In particular, there is little agreement over what personal factors constitute a "racist," as against what constitutes "racist behaviour."

In this chapter we identify a range of elements that constitute racist behaviour online. We explore what it means to be a racist online. We examine situations where racism is experienced by targets, though perpetrators often dispute that they are racist, rather claiming the privileges of freedom of speech. Three Australian cases are used:

1. Trolling Jews in social media;
2. Facebook "cyber swarming" and campaigning on anti-Muslim issues and
3. The use of online commentary through a news site and blog to vilify Indigenous Australians.

In each case the perpetrators have asserted they are not racist, while their victims/targets have experienced their harassment as racism.

The question of what counts as "racism" and to what extent it is or should be a problem of concern to government and the wider society has grown as a social issue in Australia ever since *News Limited* commentator Andrew Bolt was found to have breached section 18C of the Racial Discrimination Act (RDA) in 2011 (Gelber and McNamara 2013; Aggarwal 2012). Bolt had published a series of articles in the Melbourne *Herald Sun* and in his online blog attacking pale-skinned Indigenous people for claiming benefits that should, he asserted, have rightly been reserved for dark-skinned and more impoverished people who were the brunt of overt and sustained racism. The targeted named individuals had complained under the provisions of Section 18C that they had been insulted, offended, humiliated and intimidated by his words. He refused to apologise or withdraw his claims. Conciliation could not be achieved, and the complainants sought a Federal court ruling supporting their position. Bolt was unable to prove a defense under Section 18D of the Act, which provides for exemptions in relation to statements made in good faith. It is rare that a public figure could be so formally identified with race hate speech, though for Bolt the finding (it was a civil case, so not a criminal conviction) became a central coda of his future political performance.

Moreover, the case and its aftermath (it soon became a major focus for the many outlets of *News Limited* in Australia) (Albrechtsen 2017) have continued to stir passions and claim the attention of the Federal Parliament. Our research team has been regularly involved in the public debate over "18C," including drawing on our research in submissions to the Attorney General and the Parliament in their deliberations on whether the Act should be amended (Jakubowicz et al. 2017). At the time of writing, the government had failed to convince the Australian Senate to make substantive changes to the terms of the racial vilification legislation, and the changes had lapsed. The defeat of the government legislation was followed by a claim from its proponents, including Bolt, that racism against Whites had now emerged as a significant threat to the majoritarian culture (Bolt 2017).

Given that many of the complaints to the AHRC on issues of racial vilification have concerned online material, the legislation and surrounding debate reveal much about community attitudes to the apparent tensions between freedom of speech and freedom from hate online as well as in the offline world.

There is growing agreement that the Internet has played such a powerful role in building the edifices of contemporary racism due to three interacting factors (Klein 2017).

1. The political economy of the Internet favours freedom over control, facilitated by technologies that magnify the anonymity of racist protagonists;

2. The ideology of the Internet has long been pervaded by arguments that propose attachments to freedom without limits, allowing thereby boundless interactions, magnified by the overarching influence of the United States in support of unfettered freedom of speech.
3. The particular configuration of activity on the Internet that places billions of lone individuals before their screens, interacting effectively with anyone they choose, enhances the psychological dimensions of anonymity, disengagement and disinhibition, particularly where we find people whose personalities accord with the Dark Triad or Tetrad (Buckels et al. 2014).

The pursuit of targets online has come to be called "trolling" (Stein 2016). Trolling has become an omnipresent part of the Web experience for anyone of any celebrity status (or indeed anyone based on their appearance, clothing or opinions). It began its life when social media for the first time made it possible anonymously to identify and harass other people. The heartland of trolling has been the "Politically Incorrect" board of the website 4chan, essentially an old-style bulletin board which allowed people to post ideas totally anonymously, and then see where they ran (Hine et al. 2016). /Pol/, as ground zero for the rise of cyber hate in the period of social media, has been the development space for aggressive memes, for the capture and repurposing of Pepe the Frog—a meme asserting the dominance of White power online (Koebler 2016)—and the strategies for bypassing attempts by major Internet providers to develop bots to block race hate (Pearson 2016). Moreover, it has been the birthplace for the alt-right hashtag and cyber group, identified publicly as overtly racist when US Democratic presidential candidate Hillary Clinton outed the nascent movement in the US presidential elections in 2016 (Chan 2016).

While proponents of race hate have been avid users of the Internet since its inception, the explosion in racist demagoguery since at least 2010 has been fuelled by political, economic and social crises around the world. As identity politics has deepened along the fault lines of race, gender and sexuality, the Internet has become not merely a battleground but indeed a real weapon in the conflicts over resources, power and life choices (Klein 2017).

Trolling and Sadism: Why Racism Feels So Good to Some People Online

Racism online has exerted an (sometimes fatal) attraction on millions of people. They are drawn by the psychological reinforcement generated by the righteous anger it permits, and the disassociated and disinhibited outrage it magnifies. While personality theorists have long been aware of the profiles of people drawn to ideologies of extreme hate, it was the innovative work of Paulhus and Williams in 2002 in their articulation of the "Dark Triad" that really opened up the possibility of thinking about what makes a racist online (Furnham et al. 2013; Paulhus and Williams 2002). The Triad includes three personality traits—narcissism, Machiavellianism and sadism, to which has been added a fourth factor, sub-clinical psychopathy (Buckels et al. 2014) (producing thereby the Dark Tetrad). Each trait makes a significant contribution to the potential for racist behaviour online, while, if present together in the same person, they were unlikely to permit anything else.

A 2014 study of online behaviour asked people who admitted to trolling, about their motivations and satisfactions. A popular summary of this research (Golbeck 2014) claimed trolls were narcissists, psychopaths and sadists, while the more scholarly research on which the article was based pointed to sadism as the most powerfully aligned personality characteristic for trolls. Online users who commented frequently, enjoyed trolling and identified as trolls, were likely to be high on sadism, psychopathy and Machiavellianism. "Thus cyber-trolling appears to be an Internet manifestation of everyday sadism" (Buckels et al. 2014), though, as other analysts would note, enhanced through the added overlay of disinhibition due to the technology and potential for anonymity. Moreover, "all three of the Dark Triad admit prejudice against immigrants and, more generally, proclaim a social dominance orientation. All three are rated high in ruthless self-advancement" (Buckels et al. 2014, p. 207).

The sorts of people most likely to be trolls and to hold strong racist views (including antipathy to immigrants) may demonstrate significant though usually sub-clinical levels of psychopathy. In a study of Canadian students, Andrew Nevin concluded that:

> there are higher scores for cyber-psychopathy (primary, secondary, and total) than offline psychopathy counterparts, which suggests that the structural conditions of the internet facilitate increased expression of psychopathic personality traits. This finding is especially pronounced among males and individuals who are younger in age. Secondly, gender is a central social predictor of both cyber-psychopathy and online misconduct: males are most likely to have higher levels of cyber-psychopathy, be in the group that significantly increases in psychopathic expression online, and accordingly be more likely to condone and engage in misconduct behaviours on the internet (Nevin 2015).

Racists online, particularly those who indulge themselves in trolling behaviour, are likely to be young (often but not only) White men, excited by the thrill of the hunt, and callous in terms of the impact of their actions on others. The majority will be followers rather than leaders, lurkers and likers rather than posters. Whitney Phillips has argued that trolls are quintessentially of the mainstream culture, offering a "grotesque pantomime of dominant cultural tropes" (Phillips 2015). They seek, as one of her subjects reported, "lulz" (laugh out loud plural), which she described as "Amusement derived from another person's anger. Also the only reason to do anything." They seek to be disruptive, angling to find the most exploitable targets, whose distress provides them with satisfaction. As suggested earlier, trolls are sadists online, and "race" is one of the more satisfying and vulnerable spaces in which they can choose to act, following close behind gender (Anglin 2016; Back 2002).

The constellation of personality factors associated with online racism contributes to how the process of building and extending racism and racist discourse operates. It also points to why developing a momentum in opposing racism online has been so demonstrably difficult (Gagliardone et al. 2015). The Internet provides proponents of race hate with the three key elements that they need and desire in the communication process:

1. Opportunities to express their views;
2. Opportunities to directly address and have an impact on their targets and
3. Opportunities to recruit followers to their views among the initial bystanders to their messages.

Online racists typically adopt strategies of moral disengagement—the avoiding of distress, distancing from self-condemnation and minimisation of the impact of social sanctions. Groups that are opposed to racism tend to exhibit the opposite characteristics—namely, they experience and convey emotional distress, are aware of their own responsibilities and contradictions and are very sensitive to socially shared moral values.

Making a Racist and Doing Racism

In trying to decide what "makes" an online racist—that is, what factors contribute to the ideas, values and behaviours that find expression in racist discourses on the Internet—we need to be aware of societal, communal, familial and personal parameters. The historical pattern of race relations within a society and the hierarchies that those patterns produced will continue to have an impact over time. The values generated in that society coalesce with each other to form constellations that represent world views learned and shared at the community level. Community formation through shared values and orientations, linked to shared practices, illuminates how racism is performed every day. Individuals within communities with particular social and personality characteristics are more likely to be drawn into performing racism for the satisfaction it provides to them, both in group reinforcement and in personal emotional catharsis.

While the social status accorded to racism will affect the willingness of racists to acknowledge that their values are racist, or indeed that they are racists, the political debates over whether the values are indeed racist will also have an important impact. As our survey demonstrated in Chap. 3, racism is more likely to be called out by people who are less racist but recognise racism exists, while people who hold more racist values are less likely to recognise or acknowledge its presence. People who recognise racism are more likely to perceive it as a deviant and negative phenomenon, while people who do not recognise or acknowledge racism as being present are more likely to see the phenomenon as normal and unexceptional. Active racists seek to normalise their racist values through repetition in social media—the last person standing strategy. On the other hand, targets of racism are much more reluctant to expose themselves, and often try to withdraw from situations where they might experience recurrent racism (Dickter et al. 2012; Ellemers and Barreto 2015; Kahn et al. 2016; Nelson et al. 2011).

Racists are more likely to exist in societies that are racially demarcated, with histories of racist oppression, and in hierarchies where race is associated with privilege or liability. Metropolitan societies of former empires are very likely to generate racist hierarchies, though racially conflicted societies clearly also exist where the remnants of colonial regimes have left unresolved inequalities. Racially associated resentments can continue for many generations where it has proved as yet impossible to ensure equity between racial groups.

The USA, occasionally described after the election of President Barak Obama as "post-race," has more recently seen a re-ignition of racial conflict and racial antagonisms. Seeking to frame the origins and structural continuities in racism, especially for African Americans, Joe Feigin has noted:

> In the United States, racist thought, emotion and action are structured into the rhythms of everyday life. They are lived, concrete, advantageous for whites and painful for those who are not white. Each major part of the life of a white person or person of color is shaped directly or indirectly by this country's systematic racism (Feigin 2014, p. 47).

First published in 2000, Feigin's study tracks the structural impact of racism in the USA and explores how that structure affects the agency, opportunities and life outcomes for majority Whites, and a range of ethno-racial minorities. He denies though that the country can be truly analysed as post-racial, reflecting on the 97% of Americans who opted for a single racial label in the 2010 Census (p. 251). Feigin argues that the US society has been dominated by a White racial frame, such that the narrative of the society and the development of its institutions have been systemically bent towards the interests and perspectives of Whites, especially White men. While Australia has a much smaller historical involvement with slavery than the USA (mainly in relation to indentured workers from the Pacific), the idea of White framing that Feigin uses may well prove helpful in reflecting on Australian racism.

Australia has a particular constellation of histories of invasion, extermination, slavery, and exclusion—it was founded by the invasion from the UK by military forces and forced settlers (Jupp 2002). Thereafter, governments imposed various regimes of racial oppression and exclusion, mainly, but not only, on Indigenous people, Asians and Pacific Islanders, such that in the formal creation of the Commonwealth of Australia in 1901, racial hierarchy was central to the project and its legislative priorities (Williams 2012). Thus, for at least three generations after the establishment of the Commonwealth, White privilege drove the social, political and economic life of the country. Indigenous people were effectively without recognition until the mid-1960s, a time also when White Australia immigration restrictions began to erode. It was not until the late 1970s that the celebration and defence of White privilege was abandoned by government, though core values and world views of Whiteness remained circulating within the society, as an insistent counter-narrative (Jakubowicz 1997). Multiculturalism was adopted as formal policy in 1978 (Jakubowicz 2013), though there has been sustained resistance to it among many sectors of the society, which continue to bemoan the reduction in White privilege. Institutions geared towards countering racism have been in place since 1976, though not without resistance and sustained hostility (Soutphommasane 2014).

In Australia, a desire or nostalgia for White privilege will likely be a bedrock element for people who practice racism. In Chap. 6, we show how the narratives of belonging found on the Internet in sites celebrating Australian-ness are heavily inflected with White nostalgia. Other values associated with racism were explored throughout the survey in Chap. 3. These include:

- Denying that White privilege exists;
- Opposition to interracial marriage;

- Opposition to laws which limit freedom to insult, offend and humiliate people on the basis of race and religion and
- Opposition to multiculturalism.

In addition, these values are associated with a more general dislike of out-groups.

A Eureka Research study (N = 2000) carried out for the Australian government in 1998, during a period of significant social conflict over immigration from "non-White" countries (especially Indo-China), asked why respondents thought people were racist (Eureka Research 1998). Over 70% thought people were racist due to fear of loss, even though over half also agreed that "racist people are just ignorant and uneducated." The respondents were divided over how they defined racism. For the group (Anglo-Australians) who held the most negative and stereotyped views of other groups, racism, which was recognised as a socially stigmatised perspective, only encompassed extreme acts of violence against targeted groups (15%). For the group composed of "positive Anglo-Saxons" most Aboriginals and most migrants, racism included many everyday behaviours, such as name-calling, discriminatory treatment in provision of services and avoidance of contact (daily micro-aggressions) (85%). Eureka then developed a "racism index" of seven statements—non-racists would agree with the first five and disagree with the final two; racists would do the reverse. The study found that the more racist statements that one agreed to, the more likely it was that the extreme definition of racism would be supported. However, the report also argued that the hard liners adopted the:

> extreme definition partly as a defence mechanism against being labelled "racist". By assuming the definition that they do, they can "define themselves out of" any communications or messages about racism … because by their definition they are not racist (p. 26).

Indeed, the "softer" respondents were more likely to accept that they could be racist; they recognised that they could inadvertently engage in racist speech or behaviour, as a result of their socialisation in a society that exhibited structural racism.

The report argued that the three segments—subjects holding numerous racist views, some racists views (fence-sitters) or few racist views—were not clearly predictable on the basis of gender or ethnic background, though being older, less well educated, having lower income and having a working class occupation, and living outside the big cities made it more likely that people would hold more racist views. Yet holding to views that are racist does not necessarily make one a racist. For instance, people can recognise that their values or even their unintended actions may have racist content or impact, and seek to change their behaviour or limit its impact. Racists would be those people who behave in ways that implement their prejudiced values, intentionally or unintentionally, without caring about the impact. Also they would not voluntarily modify their behaviour when alerted to its impact. Moreover, their personalities might well display to some extent those characteristics of the Dark Triad referred to earlier. These traits include elements of the practices associated with Machiavellianism (a desire to manipulate and control others) and sadism (emotional satisfaction from causing pain and suffering to others).

Racism as a field of power generates many potential subject positions. As Orbe argued in his study of relations between members of co-communities, there are many relationships possible depending on the particular positions, strategies and reactions adopted by the participants in the encounters (Orbe 1998). In Table 4.1, we propose a three by three matrix, in which the spread used in the Eureka 1998 study of attitudes to multiculturalism forms the vertical axis, while the three forms of encounter with racism online, which we describe in Chap. 3, provide the horizontal axis. This produces nine potential power "cells," three each grouped by the form of the encounter—target, perpetrator and bystander. We note that racist, non-racist and mildly racist attitudinal clusters demonstrate different levels of salience—for both racists and non-racists, racism is recognisable and salient, while the middle cluster tends not to see racism as being a salient issue for them until it is activated. These cells are not still and unconnected. They all demonstrate internal energy and cross-border engagements and interactions. Once they are activated through the encounters, the system reveals the many ways in which racism online can be experienced, interpreted and responded to. In addition, the observed dynamics of each cell has implications for policy

Table 4.1 Attitudes to racism/experiences of racism: nine fields of power

Racism/Online	Target	Perpetrator	Bystander
Opponent/not prejudiced	Looks for alternative speech opportunities; opposes racism online through activity; defends self from attack	Asserts own group superiority, while decrying racism	Once alerted becomes increasingly aware and active; joins supportive spaces
Unconcerned/ not salient/ mildly prejudiced	Alerted to racism by being targeted; withdraws from exposure	Unaware amplifier; goes with the flow; often joke teller; normalisation of racist discourse	Doesn't recognise racism online or doesn't engage; avoids exposure
Proponent/ strongly prejudiced	Identifies enemies and builds racial fight back	Uses Internet to build support, advocate position, harass targets	Lurks to like; aware amplifier

responses, civil society reactions and development of resilience. We return to this matrix in our final chapter, when we delve into what strategies are required to engage with each of the cells in order to reduce the energy associated with racism and its impacts.

If we apply our matrix to some of the attitudinal questions used in the CRaCR survey discussed in Chap. 3, the distribution of the online user population across the nine cells can be identified (Table 4.2). Self-identified racists are those who accept that they have perpetrated racism online. They also were most likely to strongly agree with statements that expressed prejudice, though they would not necessarily see themselves as prejudiced. The question that attracted the strong agreement of most perpetrators concerned supposedly undeserved benefits accruing to Aboriginal Australians.

Using the question seeking agreement on the value of diverse communities modelling their behaviour on "mainstream Australians," the most aggressive perpetrators form less than 1% of our user population. Yet, in many ways they contribute to the energy that drives the agenda of prejudice.

Table 4.2 Fields of power in online racism encounters by seven attitudinal questions and four types of encounters

		Target	Bystander	Perpetrator	No encounter
I am prejudiced against other cultures	Strongly agree	2 2%	5 0.7%	2 5.4%	13 1%
	Agree, neutral, disagree	72 71.3%	527 71.9%	27 73%	1016 75.8%
	Strongly disagree	27 26.7%	201 27.4%	8 21.6%	311 23.2%
	Total	101 100%	733 100%	37 100%	1340 100%

		Target	Bystander	Perpetrator	No encounter
I feel uncomfortable around people from other racial, ethnic or religious backgrounds	Strongly agree	7 6.9%	27 3.7%	4 10.8%	26 1.9%
	Agree, neutral, disagree	65 64.4%	497 67.8%	25 67.6%	1032 77.0%
	Strongly disagree	29 28.7%	209 28.5%	8 21.6%	282 21.0%
	Total	101 100%	733 100%	37 100%	1340 100%

		Target	Bystander	Perpetrator	No encounter
I think it is better if people marry within their own cultural group	Strongly agree	6 5.9%	18 2.5%	3 8.1%	42 31.%
	Agree, neutral, disagree	78 77.2%	545 74.4%	28 75.7%	1108 82.7%
	Strongly disagree	17 16.8%	170 23.2%	6 16.2%	190 14.2%
	Total	101 100%	733 100%	37 100%	1340 100%

(*continued*)

Table 4.2 (continued)

		Target	Bystander	Perpetrator	No encounter
The Australian way of life is weakened by people from minority racial, ethnic or religious backgrounds maintaining their cultural beliefs and values	Strongly agree	16 15.8%	56 7.6%	6 16.2%	145 10.8%
	Agree, neutral, disagree	72 71.3%	550 75.0%	26 70.3%	1092 81.5%
	Strongly disagree	13 12.9%	127 17.3%	5 13.5%	103 7.7%
	Total	101 100%	733 100%	37 100%	1340 100%

		Target	Bystander	Perpetrator	No encounter
People from racial, ethnic, cultural and religious minority groups should behave more like mainstream Australians	Strongly agree	2 22.8%	89 12.1%	5 13.5%	255 19.0%
	Agree, neutral, disagree	69 68.3%	599 81.7%	29 78.4%	1063 79.3%
	Strongly disagree	9 8.9%	45 61.%	3 8.1%	22 1.6%
	Total	101 100%	733 100%	37 100%	1340 100%

		Target	Bystander	Perpetrator	No encounter
Being Aboriginal makes it harder to succeed in Australia today	Strongly agree	27 26.7%	103 14.1%	10 27.0%	89 6.6%
	Agree, neutral, disagree	63 62.4%	566 77.2%	24 64.9%	1130 84.3%
	Strongly disagree	11 10.9%	64 8.7%	3 8.1%	121 9.0%
	Total	101 100%	733 100%	37 100%	1340 100%

		Target	Bystander	Perpetrator	No encounter
Aboriginal people get more government money than they should	Strongly agree	24 23.8%	92 12.6%	10 27.0%	191 14.3%
	Agree, neutral, disagree	64 63.4%	566 77.2%	24 64.9%	1098 81.9%
	Strongly disagree	13 12.9%	75 10.2%	3 8.1%	51 3.8%
	Total	101 100%	733 100%	37 100%	1340 100%

Doing Racism in Australia: The Socio-legal Context of Racial Hostility

The Internet and the racial vilification provisions of the RDA serendipitously arrived in Australia at almost the same moment in 1996. Australia had signed up to the ICERD in 1965, accepting the definition that such discrimination encompassed:

> any distinction, exclusion, restriction or preference based on race, colour, descent, or national or ethnic origin which has the purpose or effect of nullifying or impairing the recognition, enjoyment, exercise, on an equal footing, of human rights and fundamental freedoms in the political, economic, cultural and any other field of public life (Langfield 2009, p. 139).

However, while Australia had signed the Convention in 1966 (thereby committing to the end of White Australia) (Tavan 2004), it did not ratify the document until 1975. Even on signing, the Australian government filed a reservation to Article 4, namely, that:

> the Government of Australia ... declares that Australia is not at present in a position specifically to treat as offences all the matters covered by Article 4(a) of the Convention. Acts of the kind there mentioned are punishable only to the extent provided by the existing criminal law dealing with such matters as the maintenance of public order, public mischief, assault, riot, criminal libel, conspiracy and attempts. It is the intention of the Australian Government, at the first suitable moment, to seek from Parliament legislation specifically implementing the terms of Article 4(a) (Moss and Castan 1991, p. 290)

This meant that it would not criminalise the elements identified in 4(a), namely:

- dissemination of ideas based upon racial superiority or hatred;
- incitement to hatred, contempt or discrimination against members of a group on grounds of their race, colour, descent or national or ethnic origin;

- threats or incitement to violence against persons or groups and
- expression of insults, ridicule or slander of persons or groups, or justifications of hatred, contempt or discrimination, when it clearly amounts to incitement to hatred or discrimination (UN Committee on the Elimination of Racial Discrimination 2013).

When the RDA as drafted by the Australian Labor Party government was finally adopted in 1975, a decade after the Convention had been signed, the legislation did not contain any provision about hate speech. Opposition from the conservative coalition parties and independent free-speech senators at the time ensured that senate support could not be secured for these elements. The draft legislation had included the following words, which were not enacted:

> A person shall not, with intent to promote hostility or ill will against, or to bring into contempt or ridicule, persons included in a group of persons in Australia by reason of the race, colour or national or ethnic origin of the persons included in that group —
>
> a) publish or distribute written matter;
> b) broadcast words by means of radio or television; or
> c) utter words in any public place, or within the hearing of persons in any public place, or at any meeting to which the public are invited or have access,
>
> being written matter that promotes, or words that promote, ideas based on
>
> d) the alleged superiority of persons of a particular race, colour or national or ethnic origin over persons of a different race, colour or national or ethnic origin; or
> e) hatred of persons of a particular race, colour or national or ethnic origin.
> Penalty: $5,000

As the 1991 Inquiry into Racial Violence Report notes, the reservation was in fact "a statement of future intention to enact legislation making specific provision to treat as offences all matters covered in Article 4(a)" (Moss and Castan 1991).

Once more with the return of the Labor Party to government in 1983, another attempt was made to move the matter forward. The Human Rights Commission proposed a series of amendments to cover incitement to racial hatred and racial defamation. If no criminal jurisdiction were to be introduced, the Commission offered a civil remedy. The law would then serve an educational purpose rather than a punitive one (Mitchell 1983). The Commission then proposed:

1. **Incitement to racial hatred.** A provision to make it unlawful for a person publicly to utter or to publish words or engage in conduct, which, having regard to all the circumstances, is likely to result in hatred, contempt or violence against a person or persons, or a group of persons, distinguished by race, colour, descent or national or ethnic origin: this provision should be drafted so as to ensure that certain valid activities are not brought within its scope, for example, the publication or performance of bona fide works of art, genuine academic discussion, news reporting of demonstrations against particular countries or the serious and non-inflammatory discussion of the issues of public policy.
2. **Racial defamation**. A provision to make it unlawful publicly to threaten, insult or abuse an individual or group, or hold that individual or group up to contempt or slander, by reason of race, colour, descent, national or ethnic origin (Moss and Castan 1991, p. 290).

The 1983 Report was not implemented. No Federal laws prohibiting hate speech were promulgated, though throughout the period after the first arrival of Indo-Chinese refugees in the wake of the end of White Australia after 1975, vigorous and racist public debate flourished, as the 1983 Report had well demonstrated. Moreover, complaints about rising racist violence also grew, covering Indigenous Australians, Asian Australians and Arab and Muslim Australians, in response to which the Human Rights and Equal Opportunity Commission announced its Inquiry into Racist Violence in 1988 (Moss and Castan 1991). One of the commissioned papers for the Inquiry addressed issues of racism, racist violence and the media, reflecting the situation just before the advent of the Internet (Jakubowicz 1990).

The Commission ultimately recommended all reservations on ICERD be removed and Australia accept all its obligations (p. 296). It further proposed that an amendment to the Federal Crimes Act be introduced to create the offence of racist violence and intimidation and incitement to racist violence and hatred (p. 297). In relation to the RDA, it was proposed that conduct that was "so abusive, threatening or intimidatory as to constitute harassment on the ground of race, colour, descent or national or ethnic origin" be made unlawful with civil remedies, as already available under the RDA for discrimination.

In terms of racist propaganda, the Inquiry proposed that the RDA should prohibit the incitement of racial hostility, though not at the expense of freedom of speech. It was not talking about "hurt feelings or injured sensibilities" but rather "the rights of people from different ethnic backgrounds to enjoy their lives free of harassment or violence" (p. 299). The report identified a range of freedoms of speech that should be protected, including private conversations and jokes, genuine political debate, fair reporting, literary and artistic expression, scientific and academic opinions and research and publications.

By 1993, the Australian Labor Government reported to the United Nations that it would seek to pass legislation that would remove the reservation on Article 4(a), and in 1995, drawing on the 1991 Human Rights and Equal Opportunities Commission (HREOC) report, it created Part IIA "Prohibition of Offensive Behaviour based on Racial Hatred" of the RDA of 1975 (Parliament of Australia 1995). Some 30 years after the signing of the Convention, the new Part offered a legal basis for implementing Article 4 even if in a reduced and non-criminal form.

However, resistance to the legislation never disappeared, especially when the Bolt case triggered heightened antagonism from the conservative side of Australian politics, leading to attempts to delete Part IIA, and then amend it so that the words "insult" and "offend" would be removed (Jakubowicz et al. 2017). In March 2017, the government introduced legislation that would remove the words "insult, offend and humiliate" from Section 18C of the RDA, replacing them with "harass." However, as noted above, the legislation failed in the Senate, and the *status quo*, apart from minor administrative changes, remained in place.

Where Does Racism Happen Online?

Australia has been tied to the Internet since its inception as a public system, well before the year 2000. This means there are realms on the Internet that are mainly about Australia and mainly tramped by Australians. On the other hand, there are many other realms where Australians mix with the world's Web denizens, bringing to it their values, cultures and orientations, and often taking the wider cultures of the Web back into Australian society. Before exploring our three case studies on online racism, it is useful to lay out in simple categories where one might find online racism, if one were to be so interested. For our purposes, in this overview, we are limiting the locations of online racist harassment or challenges to easily accessible sites.

Doing racially inflected "work" on the Internet may or may not be racism—that is, it may not reach the level of salience to tip it over into the space, for instance, where Facebook's community standards are excited, or suspect users have their pages frozen and their right to post withdrawn (banned). A study of Google's search delivery on names posited that "Black" first names were much more associated with a return query "Arrest record?" than "White" names. The author though was not sure whether this outcome was a consequence of unrecognised biases built into the search engine, or a revelation of how society treated people with Black names—suspiciously and reflecting the criminalisation of Black people by the American legal system (Sweeney 2013). In March 2017, there was widespread response by advertisers to the discovery by *The Times* of London's Investigative team that YouTube and Google were both serving advertising from mainstream companies and organisations on either video pages or search returns that contained material from race hate groups (and terrorist organisations). Neither YouTube nor Google at the time seemed to be able to easily prevent such occurrences, triggering a mass boycott of the sites by the advertisers (Watson 2017; Nicas 2017; Mostrous 2017).

Our research, reported in Chap. 3, indicates that Facebook has easily become the site of the most racist material, as identified by users. The other major sites include Twitter and YouTube, closely followed by the comments pages of the online media, and more recently the smaller online platforms such as Reddit, Instagram and so forth. There are also specific sites curated by racist organisations such as Stormfront, *The Daily Stormer* and the /b/ and /pol/ boards of 4chan. Anglin's 2016 Guide to the Alt-Right, by the publisher of *The Daily Stormer* (Anglin 2016), provides an insider's self-aggrandising view of this terrain. However, two more general examples are worth reviewing.

A single post on YouTube can go viral within minutes, and then accelerate into multitudes of strings across many platforms. When the Korean KPop group "Girls' Generation" won the Video of the Year award at the first YouTube Awards ceremony in November 2013 (they had accrued 74 million views on YouTube and 400,000 subscribers), a viral campaign against them (both racist and sexist) erupted around the tag "ching chong girls," supposedly fed by the enraged fans of competing White American singers Justin Bieber and Lady Gaga (Keeble 2013). "Ching chong" has a history as a term of racist abuse (Chansanchai 2011; Kohli 2013), and continues to be used on Twitter as a racist put-down, though in places has been reclaimed as an assertion of identity[1] (along the lines of African Americans referring to themselves as Niggas). For example, Instagram (December 2, 2016) listed 19,000+ tags for #chingchong, and over 960,000 for #niggas.

However, one of the most extensive locations for racial prejudice to be imposed and experienced online has emerged among dating sites. A US study in 2010 revealed that race was a significant issue of partner choice among men who were seeking male partners for sex online (MSM). Drawing on six focus groups with MSM from African American, Latino, Asian and Pacific Islander backgrounds in Los Angeles, the researchers described a range of experiences from "color preference" through to discriminatory and hostile interactions. The nature of the Internet, a private and solitary space, meant that "there was little to buffer the corrosive aspects of those negative experiences" (Paul et al. 2010). On the other side of the Pacific Ocean, an Australian study in 2016 of sexual racism among gay men, "a specific form of racial prejudice enacted in the context of sex and dating" found widespread experiences of racism, ranging from subtle forms to blatant and aggressive racial prejudice. The subjects reported responses ranging from disconnection to confrontation (Holt et al. 2016). Gay informants to our research report the widespread use of the tag "no rice or spice" on the gay dating site Grinder, proxy terms referring to distaste for (east and south, on the one hand, and north on the other) Asian sexual partners despite Grinder's rules forbidding racial categorisation.

In the heterosexual dating world, the reports were less dramatic. However, racial "pickiness" characterised a 2013 study by Firstmet.com (formerly the online dating site "Are You Interested?") of its clientele, which reported on the racial preferences of men and women segmented by US racial categories (Asian, Hispanic, Caucasian, and African American) (Firstmet.com 2013). Similarly, Lewis (2013) study of interracial online contacts and dating discovered that even though people seeking to "hook up" tended to search for their own ethnic or racial group, once they elicited a response from someone from a different group, they were more likely in the future to "cross over." Lewis argued this may have reflected preemptive discrimination, where, to avoid rejection, people went with their own, but if they gained a response they were not expecting, they were prepared to keep going (Lewis 2013).

These examples do not resolve any of the deeper questions, but they do point to the millions of racially encoded interactions taking place every day on the Internet, and how prejudice and fear mark the boundaries of race. We are also able to observe that these boundaries operate both within and between societies, suggesting that ideas of race and racial hierarchy circulate continually and widely on the Internet and feed "learning" about racial difference and the perceived, but perhaps not real, potential dangers of interracial intimacy and openness.

Doing Racism in Australia: Trolling Jews as the New Normal

Soon after the amendments to the RDA were promulgated in 1996, a small Holocaust denial site was established in South Australia, its birth advertised in the *News Limited* flagship newspaper *The Australian*. Created by Frederick Toben, the so-called Adelaide Institute was designed to "prove" the Holocaust as claimed by historians had never occurred, and the Holocaust industry had been constructed by Jews in order to gain leverage on the guilt of the West for personal and community financial gain. For his activities, Toben has been gaoled in Germany (where Holocaust denial is an offence); in Australia, he has been pursued under Section 18C of the RDA by the Executive Council of Australian Jewry (ECAJ) through their advocate Jeremy Jones (Executive Council of Australian Jewry (ECAJ) 2002). Jones was very persistent in his actions, given the circuitous processes available in civil law to avoid or bypass the intent of the legislature to minimise racist propaganda. Holocaust denial (Ames 2014) seeks to denigrate and humiliate people of Jewish faith and background, by claiming that there never was a systematic Nazi extermination plan, and that Jews are using the unfortunate events of the war to build their own post-war financial and political interests. It is widely accepted that Holocaust denial is a form of racism and antisemitism.

A Contemptible Timeline: ECAJ Versus the Adelaide Institute

1998: The case began when the ECAJ made a complaint to the Race Discrimination Commissioner that Toben's Adelaide Institute had published "malicious anti-Jewish propaganda" on its website (H97/120). The Commissioner found that the complaint was not amenable to conciliation, and it went to a public inquiry in 1998.

2000: The Commissioner ruled in 2000 that the complaint was well founded and that Toben should withdraw the material and apologise, commenting though that "determinations of this Commission are not of course enforceable." The request by Jones that Toben be ordered to undertake counselling by the Commission was rejected, as no such capacity existed in the Commission. Toben did not remove the content and did not apologise.

2002: Jones sought a Federal Court order, where he had to prove from the beginning the case that had already been proven to the Commission. Branston J found that the Adelaide Institute had behaved unlawfully under the RDA and ordered that the offending material be removed and Toben promise not to re-offend, although no order was made for an apology. Toben appealed, and lost. Toben, again, failed to remove the content.

2006: Jones began proceedings for contempt of court against Toben.

2007: At the end of 2007, the contempt was proven and Toben apologised for the contempt and agreed to take down the offensive material. Soon after, he indicated to the court he would not do so and the content remained on the website.

2008: In February 2008, Jones again took Toben to court for contempt.

2009: In April 2009, a further order was given stating that Toben had been in "wilful and contumacious contempt of court" on seven occasions, and that Branston's 2002 orders stood.

2009: In June 2009, Toben successfully sought an extension of time to appeal the contempt findings. In August 2009, Toben's appeal was heard and dismissed, and he was jailed for three months; he took down the offensive material as leaving it up would be a continuation of the contempt (and could have left him in gaol).

2010: Toben did not pay the order's costs and was taken to bankruptcy court by Jones.

2011: Five appearances later, in mid-2011, Toben finally committed to paying the $56,000 in legal fees incurred by Jones and the ECAJ, and did so, leaving only the remaining costs for the last cases. However, by then the Adelaide Institute site had a new owner, with new offensive material.

The details here are important as they point to the extraordinary difficulty and expense (Jones' time was unpaid and ran into hundreds of hours) that the average citizen would have in seeking redress for racism on the Internet, even when there is an identifiable perpetrator, an Australian site, and an acknowledgement of the offensive material.

The Toben case was not an isolated matter, but part of a wider phenomenon of antisemitism on the Internet that has continued to expand (Cohen-Almagor 2013), causing US Jewish community leaders to appeal to the Attorney General in February 2017 for some concerted action against real-world threats. Toben had been a part of an international movement in the 1990s that many saw as the last gasp of a failed antisemitic conspiracy, which included the re-emergence of the widely decried Protocols of the Elders of Zion (Hagemeister 2008), the confrontation by writer Deborah Lipstadt with David Irving and his bankruptcy over his failed theories of the Holocaust as a "Disneyland" and the agreed categorisation by the international community of Holocaust denial as a racist and antisemitic defamation (Ames 2014).

However, what had been hoped by its opponents to be a closure on a painful and outrageous calumny, a dying residue from the Nazi period in Europe, has come back to life (Stein 2016). Toben passed on his website to another protagonist of contemporary antisemitic and neo-Nazi propaganda, while Irving claimed, in the run up to the film of Lipstadt's legal bout with him *The Denial*, that a new generation had discovered him through social media. Interest in his work had grown exponentially, he delighted, with young people watching his 200 plus videos on YouTube, and sending him emails. Meanwhile, one of Lipstadt's lawyers reflected that the Internet, far from spreading the truth about the Holocaust and destroying Irving's reputation further, had in fact advanced his profile. The online trailer for the film had been trolled by over 4000 comments attacking the "Holohoax" and celebrating Irving's fight for "truth" (Cadwalladr 2017).

In February 2017, Melbourne, capital of the Australian state of Victoria, was "blanketed" with neo-Nazi posters featuring swastikas, anti-Jewish slogans and images of Adolf Hitler in "cool" aviator sunglasses. The organisation that claimed credit was one of many tiny groups engaged in competition with each other to seize the mantle of leading new fascist/ultra-nationalist groupuscule. The use of posters and graffiti resonates with the concerns expressed to the HREOC's Racist Violence Inquiry in the late 1980s, when racist posters were also spread across Melbourne. While the Internet presence of the group was limited, their use of overt events to raise a social media presence can be easily recognised.

Media reports of the poster campaign redacted its website address and blocked identifying material in published images.

Antisemitism online in Australia has a number of different origins and sources. They reflect the contemporary version of the array of antisemitic manifestations outlined in HREOC in the 1990 submission by the *Australian Jewish News* editor Sam Lipski (Moss and Castan 1991, p. 143), and previously noted in the HRC 1983 report (Mitchell 1983). These included threats or actual violent acts against Jewish communal property, verbal abuse of Jews in Jewish neighbourhoods, political agitation and propaganda by neo-Nazi groups, public expressions of hostility to Jews in the media, casual prejudice and slurs, workplace discrimination and targeting of Jews because of opposition to Israel.

The Online Hate Prevention Institute[2] (OHPI), a Melbourne-based Internet monitoring civil society organisation, began its life as a group responding to the rise of antisemitism online. Since then it has spread to encompass anti-Muslim, anti-Indigenous and other areas of anti-ethnic hate online, and beyond to gender, homophobic and trolling hate. Its 2016 report on antisemitism, prepared for the *Global Forum to Combat Antisemitism*, tracked some 2000 items over a ten-month period, and discovered that even when reported, only 20% were removed from social media. Each dimension of social media tended to concentrate particular aspects of antisemitism. For example:

- Facebook was least likely to promote violence against Jews or Holocaust denial and was most likely to take down material advocating violence;
- Twitter was most likely to promote violence and be more resistant to remove the advocacy of violence and
- YouTube was most likely to promote Holocaust denial, proving to be the most resistant to taking down traditional antisemitic material.

Social media is thus not an undifferentiated field, but rather one where antisemitism has carved out specific and effective presences.

In introducing the Executive Summary of the Report, OHPI's CEO (an author on this book) argued:

> Through the Internet, antisemitic content and messages spread across national borders, feeding not only anti-Jewish hate, but violent extremism more generally .... Responding to the rising social media incitement and very real consequences, German prosecutors opened an investigation into the possibility of criminal liability of senior Facebook executives in late 2015. Following this move an agreement was reached between the German Government, Facebook, Google and Twitter to see content that violated German law removed within 24 hours. Facebook has since gone further and announced a project to tackle online hate in Europe (Oboler 2016).

Concerns about the spread of antisemitism intensified after the emergence of the US alt-right (Pearson 2016) network in 2016, with its strong connection to traditional antisemitism, neo-Nazism and "lulz" strategies designed to intensify anxiety and fear among target communities. Despite the slight relaxation about social media that occurred after the German initiatives in early 2016 (Simon Wiesenthal Center 2016), the election of the Trump government in the USA, and the gravitation of alt-right strategists such as Steven Bannon of *Breitbart* to the presidential team, intensified fears in the US Jewish community, and, more widely, that antisemitism has been given new license, both online and in the real world (Amend and Morgan 2017).

Another example of the trolling/4chan approach, set up in Australia during the US presidential elections, was a group calling itself The Dingoes. Perched on a service provided by .xyz, The Dingoes exemplified all the various elements of state of the art antisemitic and racist online presences. The .xyz domain name was released to the general public in mid-2014, as part of a refreshing of the ICANN generic top-level domain name. Google adopted it for its corporate Alphabet site (abc.xyz), and by June 2016, it was the fourth most registered top-level global domain name after .com, .net and .org. The name is managed by Generation XYZ (http://gen.xyz), which describes itself as "a global community inspired by the Internet and its limitless potential … to connect with the world in a whole new way … you can focus on connecting with your audience anywhere in the world."

The Dingoes emerged online in mid-2016, joining Twitter in June. A number of the people associated with the group also joined about that time, including one tweeter whose display image contained the anti-immigration slogan "Fuck Off, We're Full." The Dingoes (once the name of a 1970s Australian music band that left for the USA) described itself as "#AltRight, but not in the way that violates #Rule1." Rule1 refers the 4chan /b/ rule 1, "Do not talk about /b/" (which is also rule 2.). As of February 2017, it had 1461 followers online, had posted 3640 posts, garnered 5507 likes and was following 442 other tweeters. It also followed a range of micro-nationalist groups, a raft of conservative online commentators and some "lulz" antisemitic posters, such as one identifying as "Goys just want to have fun," and another as "Dachau Blues," backed by an image of the Auschwitz "Arbeit Macht Frei" sign.

The 400+ twitter accounts that The Dingoes followed provided a helpful geography of the antisemitic old right and alt-right that show how fragmented, competitive and attention-seeking such groups can be. They consciously incorporate three other rules from 4chan/b/, numbers 13 to 15: "nothing is sacred; do not argue with a troll—it means they win; the more beautiful and pure a thing is, the more satisfying it is to corrupt it." These rules are compounded by two other insights (Rules 19 and 28) generated by the Dark Triad souls who drive the machine: "The more you hate it, the stronger it gets"; and "There will always be more fucked up shit than what you just saw." These views are nihilist rather than conservative, angry and pathological rather than intellectual or analytical.

The Dingoes were aligned with a number of ultra-Right sites in the USA, where they posted podcast interviews with former ALP leader and later conservative commentator Mark Latham[3] and National Party MP George Christensen[4] (hosted globally on The Right Stuff [TRS] (The Dingoes 2017). Querying the Christensen interview, one commentator (self-styled as "rabbijoeforeskinsucker") challenged the interviewers for focusing too much on Muslims (Christensen was rabidly anti-Muslim), declaiming: "Lot's of talk about Muslims, not much about the Jew. Are you guys kosher nationalists by any chance? Or are you just cowards?"

The slackbastard blog, an anarchist monitor of far-right politics in Australia, reported that TRS:

obtained its semi-popularity on the AltRight, in part, by its willingness to address The Jewish Question, ie to name the Jew as being responsible for All The (Bad) Things (@ndy 2017).

By February 2017, The Dingoes had become national news, when their TRS interview with Christensen became a point of attack by Jewish leaders on Prime Minister Malcolm Turnbull, on the day Benjamin Netanyahu, then Prime Minister of Israel, arrived in Australia for an official visit. BuzzFeedNews reported that:

> Dr Dvir Abramovich, whose Anti-Defamation Commission tracks antisemitism online … was angered to hear Christensen speaking to the podcast network. Abramovich said "The Right Stuff" network started one of the most prominent antisemitic memes on the internet, which involved racist trolls putting parentheses, or "echo" symbols, around the names of Jewish journalists. "While we do not know what is in Christensen's heart, for an elected official to be interviewed on a podcast that traffics in bigoted and demeaning stereotypes is deeply troubling," he said (Di Stefano 2017).

Which brings us to this research project, our book, and one of our researchers, Andrew Jakubowicz. Jakubowicz has a Jewish heritage, his immediate family being Holocaust survivors who escaped to Australia, via China, in 1946. Many members of his close family were murdered by the Nazis in Poland during World War II. Since 2011, Jakubowicz has been a regular contributor to *The Conversation*, a global website publishing articles by academics from universities across the world, written for an intelligent lay audience. Over this time, he has contributed over 40 pieces, mainly on multiculturalism and racism. On December 6, 2016, he published a piece on ethnocracy, applying to Australia the ideas of scholars who had analysed Israel and Northern Ireland in terms of their different populations' unequal access to democracy (Jakubowicz 2016a, b). The article attracted 5400 hits and 120 published comments. Among the comments (since removed by the community standards guardian at *The Conversation*) were a number of curious posts. The first, from a Clara Newsom, asked "To what extent are Jews like you overrepresented in positions of power?" She followed with: "I want to know how many university professors are Jews." Then she wrote: "Good take on this over at The Dingoes. 'Ethnocracy' is a pretty shabby concept tbh, not worthy of a real sociologist. Looks like more ethnically motivated, anti-Australian animus to me!" She then posted a link to The Dingoes Web page (posted on December 8), which referred to the "Skype-rat" jacubowicz [sic]. The comments continued—ending with "What if academics of a (((certain ethnicity))), e.g. are disproportionately guilty of sowing white ethnomasochist memes such as 'white privilege' .... Try this instead: the Dingoes link."

None of our team had, at that stage, heard about the moves made on 4chan or TRS to label people as Jewish and therefore a problem, or picked up on the bracket creep ((( ))). The (((echo))) device was first launched by the TRS blog in June 2016 (note its logo names it as an "Echo Free" site), as a way of capturing Jewish names "that echoed through history" (Fleishman and Smith 2016b). Soon after, a similar device was trialed by altrightmedia, which uploaded an extension for Google Chrome, called "Coincidence Detector." The extension would draw on a list of supposedly Jewish names, regularly updated, and wherever one of the names appeared during a search, the name would be echo-bracketed for that user. This process was designed as a "Jew detector for unknowing goyim," ensuring that those so inclined could see American media and politics draped in the brackets (Fleishman and Smith 2016a).

So, that explained for us the echo brackets that Newsom had used in her *Conversation* comment, and which re-appeared on The Dingoes attack on Jakubowicz. Indeed, we discovered that the story about the brackets had broken quickly in the USA and global media some weeks before, though in resistance many people adopted the brackets as a sign of solidarity with the targeted Jews.

But Skype-rat was something again. As it soon turned out, a new game was being tried with Google, where the trolls at 4chan had invented a strategy for identifying Jews (and Blacks and Mexicans) through attaching proxy labels, which were major commercial identifiers on the Internet. Hine and colleagues have studied 4chan, "the dark underbelly of the Internet," and its /pol/, the politically incorrect board. In particular they looked at "Operation Google," a response to Google's announcement of anti-hate machine learning-based technology and similar initiatives by

Twitter, designed to sabotage the then extant anti-hate strategies (Hine et al. 2016). The alt-right trolls proposed using "Google" to replace "nigger," and "Skype" to replace "Kike" (Jew). The call went out on September 22—that day "Google" increased its use of word count by five times on the Internet, while Skype doubled it. By September 26, the use had declined, though the words remained part of the /pol/ vernacular. Ten weeks later, the word "Skype" added to the old antisemitic label of "rat" was up and running in Australia.

Within two days of Jakubowicz's *The Conversation* article being published, the Skype-rat piece had been posted, over 2000 words focusing on the Jewish Marxist race traitor motif that has been a common trope of neo-Nazis (and indeed the original Nazis). The post opened with Jakubowicz's history laid out, the first being his (((Polish))) parents, under an image of a canopy of photographs of Jews killed in the Holocaust, taken from a Jewish memorial site, Yad Vashem.

By May 2017, The Dingoes had made it onto the front page of the *Sun Herald*, a leading Sydney newspaper. The article reported a forthcoming Dingoes conference, where the guest speaker was to be US neo-Nazi "MIke Enoch" (aka Mike Peinovich) founder of TRS and host of The Daily Shoah podcast. The article concluded by quoting Mammon, a Dingoes' spokesperson, as claiming Australia should become a White "ethnostate," a term curiously redolent of the Jakubowicz argument, which the site had attacked some months before (Begley and Maley 2017). While there is no more need to write on this here, the trajectory that produced this attack reflects the purpose of the alt-right.

Andrew Anglin, founder of the *Daily Stormer* (genuflecting in its name to the Nazi Party's *Die Sturmer*), described his group of alt-right thus:

> Trump-supporting White racial advocates who engage in trolling or other activism on the internet … The core concept of the movement, upon which all else is based, is that Whites are undergoing an extermination, via mass immigration into White countries which was enabled by a corrosive liberal ideology of White self-hatred, and that the Jews are at the center of this agenda (Anglin 2016).

The bursting forth of the alt-right into the sphere of antisemitism online has been paralleled in time with the rapid rise in threats directed against Jews in the real world, especially in the USA. It is important to recognise that the online trolling has close similarities with the growing public displays of antisemitism being tried out in cities and towns in the real world, be it the rapid spread of graffiti in trains and on walls, for instance "Jew" scrawled on images of rats stenciled on walls in Chinatowns in the USA to bring in the Year of the Rat (Chernikoff 2016) or the graffiti stickers that appeared in Melbourne.

4Channers have described themselves as:

> the people devoid of any type of soul or conscience. Products of cynicism and apathy, spreading those very sentiments daily. Anonymous is the hardened war veteran of the internet. He does not forgive or forget. We have seen things that defy explanations. heard stories that would make any god-fearing, law abiding citizen empty their stomach where they stand. We have experienced them multiple times and eagerly await their return.[5]

The bizarre worlds of trolls and cyber Nazis have become bound together with centuries-old antisemitism. The characteristics of the Internet contribute to hate, and the personalities of the trolls may well ensure that the hate will flourish.

Campaigning Against Islam and Slandering Muslims: Facebook at Play

The medium-sized rural city of Bendigo in Australia's state of Victoria has become the focus of a struggle over the place of Muslims in Australian multicultural society, and the acceptability of Islam as a faith able to be followed without harassment or fear. Muslims have been in contact with Australia since before the arrival of the British in the eighteenth century. Indigenous people in northern Australia would trade with Muslim sea farers from the Indonesian islands to the north; there were relationships established, families formed and lives intertwined. The arrival of the British and their seizure of the continent turned what had been a bridge between Asia and Australia into a moat that separated them (Bouma 1997).

Muslims later came to Australia from the British Empire, mainly from the frontier provinces of what is now Pakistan, and from Afghanistan. They made a major contribution to opening up the country, leading the camel trains that supplied the outback stations and the work-gangs that built transcontinental telegraph and railways. Small mosques were built where they rested. During World War I, a small group of Muslims took up arms against the Australians, fighting for the Turkish Sultan and the caliphate he headed, attacking a holiday train and shooting dead some picnickers.

While some Muslims continued to arrive in Australia after Federation, mainly from Europe including British Cyprus, Bosnia and Albania, and some from British Asia and Africa including Egypt, the numbers were limited until the post-war push for unskilled labour. Australia signed an immigration agreement with Turkey in the late 1960s, which included rights to religious practice and holidays (Humphrey 1987). With the end of White Australia, religion and race were no longer used as the basis for exclusion of immigrants. The second major Muslim intake included Lebanese Muslims during the 1976 Civil War, when refugees were first allowed to settle on the basis of claims lodged in Cyprus or elsewhere to international or Australian agencies (Jakubowicz 2007). By 2011, there were about 475,000 Muslims living in Australia, of whom 38.5% were Australian born (about 180,000). Muslims had increased by 69% since 2001, some 280,000. That decade of course was also the period from the Twin Towers attack in New York, the invasion of Iraq and Afghanistan and the deadly attacks on Australians in Bali and on Westerners elsewhere.

In the 1998 Eureka study of attitudes to racial groups, 30% of respondents identified Muslims as having "strange ways," though only 3% thought Muslims were failing to live in harmony with the wider society (Eureka Research 1998). That is, prior to the rise of global terrorism when the distant enemy became a target for attacks on the infidel, Muslims were seen generally as acceptable, if slightly strange, neighbours who seemed to be settling well (Humphrey 1987). The momentary unease with Muslims that had arisen with the first Gulf War in 1993 and the question of their loyalty had dissipated. However, the recalibration of Muslims as the first enemy accelerated rapidly, especially in Sydney, after a series of violent crimes in 2000, and the New York attacks of September 2001 (Jakubowicz 2007; Poynting and Noble 2004; Costello 2006).

While there was an increasing public debate in Australia about the Muslim presence after the sequence of events associated with 9/11, the Bali bombings and the 2005 Cronulla riots (Sheehan 2006), the state of the Internet precluded widespread campaigning. For instance, most of the organising in relation to the December 2005 Cronulla beach attacks on Middle Eastern-looking youth was undertaken either through message boards like Stormfront Downunder (Bliuc et al. 2011) or by text messaging. However, the public availability of Facebook after 2006 rapidly changed the possibilities for anti-Muslim "swarming," or the amassing of followers on anti-Islamic websites.

The Bendigo Mosque: A Swarming Case Study

In 2014, considerable public interest in Australia was focused on the decision of the local council in Bendigo, Victoria, to approve the application by a Muslim organisation to build a mosque in the area. A number of anti-Islamic groups joined the counter-campaign, which blossomed during 2015 to seek the sacking of the Council. One anti-mosque group created a Facebook community site called "Bendigo Mosque—reasons to oppose it." Since that original impetus, the site has spawned other pages, and has focused on building links across Australia and globally to other similar Facebook pages. New associations have also been promoted with the Reclaim Australia Facebook page and its spokesperson. The page has also featured posts by the founder of the anti-mosque site criticising multiculturalism and sharing anti-Islamic material collected from various sources. The case marks a significant growth in anti-Islamic protest, the first so strongly amplified by its use of online strategies.

While Islam as a religion is not protected under Commonwealth law from vilification, it does have protection in Victoria under religious vilification provisions. The practice of anti-Islamic activism has many of the same features as racism, and for our purposes of understanding how hate speech can grow, the Bendigo mosque events and the Facebook sites "Bendigo Mosque—reasons to oppose it[6]" (and "Stop the Mosque in Bendigo")[7] provide useful cases. This once-local issue went national, and indeed now has international interest. It also has real on-ground political activities associated with it, and it shows how a digital hate swarm builds and retains its base.

The Facebook pages have different but complementary goals—the "reasons to oppose" page claims "to educate people on the dangers of Islam and the adverse effect it will have on the community if the proposed Bendigo Mosque goes ahead." The page posts irregularly itself, though it collects material from around the world that supports its views. By December 2016, it had collected some 13,000 likes, though it posted only once a fortnight or so after a more intense period during the October 2016 local elections. "Stop the Mosque" has a more activist edge, and posts far more regularly, sometimes many times a day. Over 50,000 likes (February 2017) and followers indicate it had collected a significant part of the anti-Islamic movement in Australia, even though it claimed not to be a Hate Site; "there is a BIG difference between being anti-Islam (macro level) and against individual Muslims (micro-level)."[8]

The Stop the Mosque page has hosted one of the posts to be charged under Victoria's religious vilification provisions: a woman had responded to the Sydney Lindt siege, a 2014 hostage situation in a Sydney café where three people died, and stories of mosques burning in Sweden with a Facebook post calling for "all mosques to be burnt down with the doors locked at prayer time" (Mannix 2015). The Stop the Mosque page posted an article (written by The Age) about this woman's post, which received over 80 likes, though the original post itself had been taken down. In responding to its users' comments, the page owners went on to say that they would take down any post found to be in breach of the law, though if the action against the unnamed woman succeeded, then perhaps it could serve as a precedent against the Koran. However, the case did not proceed to trial.

A comparison of the two Bendigo anti-mosque pages shows some commonalities, though it also reveals that the more regular and sensational the posts, the stronger is the following. In order to build the swarms, and retain them to sustain the movement, sensationalist posts act as "sticky spots." They attract interest by arousing emotions, which are then fed by the other comments. On these pages we have traced cycles of:

- Righteous anger;
- Systematic stigmatisation of Muslims;
- Attacks that are designed to undermine Islam as a religion and
- Arguments that Islam is essentially a political ideology bent on world domination that has to be resisted by Australians for reasons of survival.

While it is possible to derive lists of names of posters and track down their online activities, it is more important to build a social ecology of online hate speech. When our team analysed some 500 Stop the Mosque posts in 2013, we were able to show that there were about six active posters, with most followers limiting themselves to liking the posts of others—some reused posts on their own pages—which forms one of the ways ideas move through the Internet.

The Stop the Mosque site provides a case study of the way in which a race hate speech cyber-swarm can develop and grow. Prior to the mosque proposal becoming public in January 2014, the site did not exist. It started with a concern about the impact of Muslims on the harmony and beauty of Bendigo, and then began to grow as the founders built relationships with more overtly Islamophobic sites. It started to introduce Australian nationalist images (i.e., a fist inside an Australian flag glove) and continued to build its following through the introduction of global anti-Islamic materials. By April 2014, it had drawn in posts from the Patriots Defence League Australia, and had become a major channel for anti-Islamic information. A year later, the site had over 3000 likes, and was strongly involved with support for Reclaim Australia rallies.

The site therefore provides a good example of the growth of an online community built around the use of hatred of Muslims in cyberspace as a means of mobilising offline activism. The Charlie Hebdo events in Paris in January 2015 provided the opportunity for the host to post what was to become one of the most "liked" comments, on 6 January. Using that post we have explored the pattern of likes and the links between them using the Facebook application NetVizz and the data visualisation software Gephi. Essentially, NetVizz creates a spreadsheet of likes and how they link to other likers of the comment (two circles of contact), while Gephi processes that data into patterns that allow visual representation.

Figure 4.1 lists sequentially an excerpt of users, who had made comments on the Facebook page of "Stop the Mosque" early in its life in 2014. Using an app that can take the user ID and track it to their Facebook page, it is possible to build a picture of the "community" that is forming around the particular post or "sticky spot." Most of the commenters were locals. The post by Stop the Mosque asserts the importance of defending Bendigo from the influx of Muslims, which would follow the erection of the mosque. Twelve discreet and discoverable lurkers "liked" the post and reacted to it in some way, including reposting. Ten of the 12 were males, from a range of occupations. Most were trades-

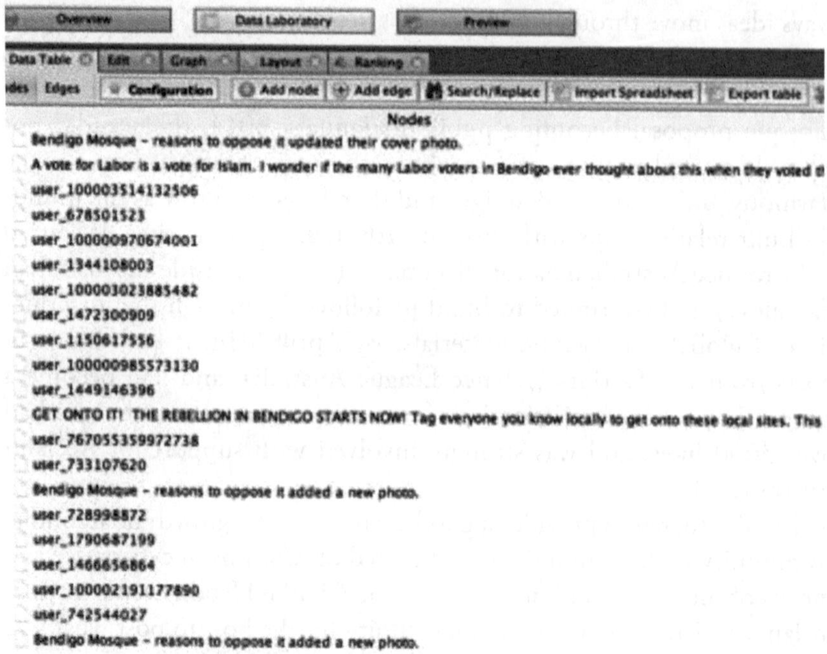

Fig. 4.1 NetVizz capture of then recent 20 posts Bendigo mosque opposition site

people, ranging from a wedding photographer to a number of mechanics. A few had already retired. About half had an interest in car racing or motorcycles. About a third held strong anti-Muslim views, including being public opponents of Halal certification, opponents of Muslim slaughtering methods and people who described and reposted comments about Islam as a religion of "filth." A few were in contact with the right wing political party One Nation, while others had flagged their relationship with the United Patriots Front. One was an ex-serviceman with past history of fighting "Muslim enemies" in Afghanistan or the Middle East. Two made most of their impact through reposting requests for signatures from the petition-site Change.org, where they typically supported anti-Islamic perspectives, opposition to the adoption of Sharia law, while calling for the closure of mosques and the expulsion of Muslims.

One point of critical conflict involved a tweet by an anti-mosque councillor to a pro-mosque member of the public that showed a picture of mutilated female genitalia, the alleged future consequence of any Muslim presence in Bendigo. Two Change.org petitions were uploaded, one supporting the councillor and opposing Islamic teachings and Sharia law (De Witte 2013), and the other condemning the councillor (Jordan 2013), with support running at 2:1 for the councillor. The Change.org pages provided lists of the supporters, revealing the range of attitudes and approaches to the relationship between freedom of speech, intolerance of Islam and support for multiculturalism. The support for the Islamophobic councillor was global, including anti-Islamic posts from Germany, while the opposition to her perceived Islamophobia was concentrated in the local Bendigo community. The "sides" were those that re-emerge in the Australian environment where Islamophobia has become widespread. A major confrontation between a group "No Room for Racism" (23,000 likes, February 2017) and the radical anti-Islam group "United Patriots Front" (112,000 likes), both spurred on by Facebook supporters, took place in Bendigo in June 2016, after the High Court of Australia effectively granted final planning permission for the Islamic Centre (Holmes 2016).

Facebook has become one important locus for these onward struggles over Islam in Australia (Oboler 2015). As Imran Awan has argued for the UK (Awan 2016), the boundary between online threats, incitement and abuse, and the confrontations that have happened in the real world, can no longer be clearly defined. For the victims, they are experienced as a linked and continuing harassment, with environments both dangerous and threatening. While Facebook has proven to be a very effective avenue for the promulgation of vicious antipathy to Muslims and Islam, it has been less responsive to arguments that it should take greater responsibility to police its own service standards. An OHPI report from late 2015 "Spotlight on Anti-Muslim Internet Hate" examined over 1000 items of anti-Muslim hate. According to the study, anti-Muslim hate accelerated sharply during 2015, mainly on Facebook (Oboler 2015). Despite claims by Mark Zuckerberg, Facebook head, that: "If you're a Muslim in this community, as the leader of Facebook I want you to know that you are always welcome here and that we will fight to protect your rights and create a peaceful and safe environment for you," the reality appeared rather different; even so, his Facebook post making this claim secured over 1.5 million likes, 200,000 shares, and 75,000 comments by April 2017 (Zuckerberg 2015).

In fact, the notification and removal of offensive material has proved very difficult, and Facebook has been very slow to act. Of the 1100 items assessed by the OHPI network as anti-Muslim in December 2015, over 90% were on Facebook. Key categories included:

- Muslims as a cultural threat;
- Demonisation of Muslims;
- Muslims as a security risk;
- The incitement of anti-Muslim violence and
- Anti-refugee xenophobia based on being Muslim.

Many of the items were clearly designed to be extremely offensive to Muslim users, often through the use of pig imagery, while others were tied to anti-halal claims. One particularly effective trope linked halal killing of animals to gross animal cruelty, thus touching on the concerns of animal lovers who might have no other interest in the Islamic question.

Facebook provides swarming opportunities, it builds communities and it allows like-minded people to find each other and reinforce each other's prejudices. Because posting can be both a cooperative and a competitive process, in an environment where everything points to disinhibition, there are no effective processes for holding anyone to account, other than through ultimately taking down posts or banning users and closing their pages. Yet, as we saw in Chap. 1, the underlying driver for Facebook, the value produced when users interact, necessarily moderates any social or cultural press against freedom to post. While Facebook has increasingly responded to local concerns about racial slurring and antipathy to refugees, especially after the German ultimatum in late 2015 (and Zuckerberg's 2016 meeting with young Germans in Berlin [Zuckerberg 2016]), Australia's rather indeterminate governance regime complicated by the majority Party push for lessening constraints on this freedom and lack of clear local priorities leave Facebook uneasily seeking guidance from government agencies for decisions that perhaps it might best make itself.

From Old Media to New Media: In Pursuit of Soft Targets Among Indigenous Australia

In this chapter, we have so far explored trolling of Jews and Facebook haunting of Muslims; now we turn to the online attacks in the online news media on the Indigenous Australian struggle for survival and self-respect. *News Limited* commentator Andrew Bolt in 2011 wrote in the old media and posted in the new about a scam he claimed was being promulgated where light-skinned Indigenous people secured special support supposedly provided for "disadvantaged" Aborigines. He named a number of people for whom he reserved his critiques. Later that year, a number of his targets decided that rather than suing for defamation, they would take Bolt to the Human Rights Commission to lodge a complaint under the racial vilification Section 18C of the RDA. Conciliation, which sought an apology and an agreement to desist from these sorts of attacks, failed, and the matter went to the Federal Court. At stake were two articles published in the *Herald Sun* newspaper, on its online site, and two blog articles by Bolt on the *Herald Sun* website. The Court (Bromberg 2011) found that the newspaper material was likely to offend, insult, humiliate or intimidate and that because of the "erroneous facts, distortions of truth and inflammatory and provocative language," the justifications offered by Section 18D did not apply. Two key elements relate to our analysis of online racism—that Bolt believed anyone in the public arena was "fair game" for his critique, and that the targets of his attacks were using their "racist and trivial" claims to indigeneity as excuses to divide Australian society on unfounded racial grounds. "At its worst, it's them against us," he stated. Neither Bolt nor the *Herald Sun* was required to apologise to the targets, nor required to commit not to continue to publish such stories in the future. The online archive copy of the stories was to remain on the Internet, with the court finding prominently attached.

This combination of factors suggests that the next series of attacks by Bolt on an Indigenous person were driven by similar concerns, namely, that it was his belief that a privileged Indigenous person who sought to "racialise" Australian society should be called out. Australia should operate as a post-racial country where individuals should not use race to divide and threaten others (Bolt 2016).

The wider attack on Adam Goodes, a leading Indigenous footballer (also discussed in Chap. 6), who himself called out racism against him, and who pantomimed spear-throwing during an Indigenous round of the Australian Football League (AFL), has received widespread coverage (Matamoros-Fernandez 2016). Here we want to drill down into the way in which Bolt addressed Goodes' behaviours online, to explore what the deepest level of racialised debate looks like. In particular, we are interested in how Bolt's blogs and the comments they elicited contribute to the constitution of narratives of unacceptable Aboriginality. Bolt provides a useful case study for three reasons:

1. He has been one of the only mainstream journalists to fall foul of the RDA and not have had an acceptable justification;
2. He has continued to write and speak with a similar argument about the legitimacy of Indigenous claims to a separate identity and its negative consequences and
3. His career has grown through the period, with him now a major commentator on cable television and radio, as well as online and in print.

> Bolt continued to be the main protagonist in the campaign to have the RDA, or at least Part IIA, which contains Section 18C, withdrawn.

To put Goodes in context, he holds an elite place in AFL history as a dual Brownlow Medallist, dual premiership player, four-time All-Australian, member of the Indigenous Team of the Century and in representing Australia in the International Rules Series. He was also Australian of the Year in 2014. For Bolt, Goodes represents the sort of success a non-racial Australia accorded excellence—he was particularly incensed by Goodes asserting his Aboriginal identity and suggesting thereby that Australia was a racist society.

The Bolt–Goodes relationship, played out on TV and across the Internet, turns closely on the issues of freedom of speech, the meaning of racism and who has the authority to specify that meaning. Our analysis of the interchange online around Goodes' identity as an Indigenous person, the way in which he performed that identity and his pillorying by many online commentators, most significantly by Bolt, demonstrates the dynamic of making racial identity into online racism.

Bolt provides a useful example of a mass media celebrity, who, additionally, already having been found to have breached the RDA by the Federal Court in another case involving Indigenous Australians, continued to develop his arguments about the authenticity of Indigenous public figures. In a series of posts, Bolt criticised Goodes for calling out a teenage girl at a game as racist, after she called him an "ape." Bolt extended his critique of Goodes' actions during an AFL Indigenous round game, in which he did a mock war dance, shaking a make-believe spear after a goal was scored. Bolt posted 43 times on Goodes in the seven months from February to August 2015.

A conceptual analysis of the posts relating to the spear incident was undertaken using Leximancer, a thematic visualisation program. The conceptual map of Bolt's posts displayed the following clustering of key themes:

- "Goodes" was closely associated with "Aboriginal," "booing" and "racist" and

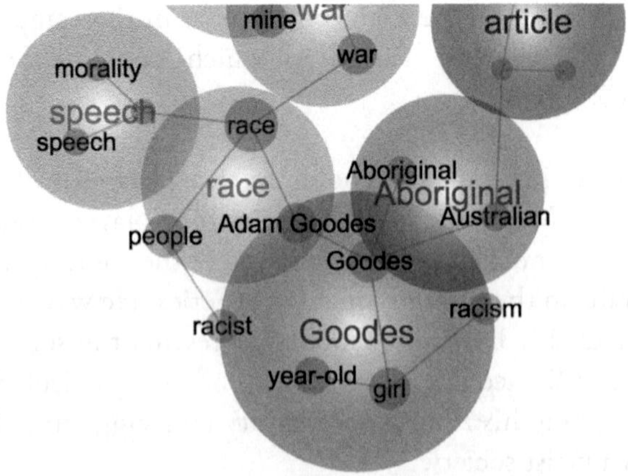

Fig. 4.2 Leximancer concept map of Andrew Bolt's 2015 posts on Adam Goodes

- "War" was closely associated with "race" and less closely, though directly, with "people" (Fig. 4.2).

The rhetoric of the posts was driven by Bolt's assertion to the effect that any signifier of race, such as Aboriginality, would necessarily create intergroup conflict by drawing people to identify with a racial group, rather than as a citizen. Once this occurred, the individuality of people would be subsumed by the racial descriptor. This descriptor would call up aggressive, if not violent, tendencies, and reinvigorate within the Indigenous community a perception that they were still at war. For Bolt, the war was long over, and the Europeans had long ago won, so no purpose could be served by continuing to propose an Indigenous identity, especially not one associated with warriorhood. Moreover, successful athletes such as Goodes were already fully accepted and rewarded. Both meanings are accommodated by Bolt's comment that "he [Goodes] is fighting a war that has already been won."

The recurrent use of "war" and the turning of "racist" to be used as a description for Goodes point to the way in which new racisms recognise the protected discourses associated with targeted minorities. They then use these discourses to dispute the logic of anti-racism, turning the language against the targets so that the victims become the cause of the problem (of racism). Indigenous people who claim Aboriginality are thus the reason that racism exists—for if they eschewed their Aboriginal identity and never performed being Indigenous, then the discourses about race would dissipate and, by implication, racism with it. Such discursive constructions spread very rapidly online, where they are liked and reposted.

Racism aims to systematically erode the self-belief of minorities in narratives of their group and its value. The attack on Goodes, in which Bolt played a key part, demonstrates exactly how dramatically such an engagement can emerge, and how potentially destructive it can be.

We turn now from these cases of "doing race", and the contentious questions about whether they are also always about "doing racism," to the next chapter where we meet some of the targets of these racisms and how they experience, interpret and respond to the racism they encounter.

Notes

1. https://twitter.com/hashtag/chingchong
2. http://ohpi.org.au
3. https://radio.therightstuff.biz/2017/02/07/the-convict-report-episode-62-mark-latham-2-dingo-boogaloo/
4. https://radio.therightstuff.biz/2017/02/16/the-convict-report-episode-63-interview-with-george-christensen/
5. http://www.urbandictionary.com/define.php?term=4chan
6. https://www.Facebook.com/Bendigo-Mosque-reasons-to-oppose-it-767055359972738/
7. https://www.Facebook.com/StopMosq/
8. https://www.facebook.com/pg/StopMosq/about/

References

@ndy. (2017, February 11). Bonus Latham! *slackbastard*. http://slackbastard.anarchobase.com/?p=40895. Accessed 23 Feb 2017.

Aggarwal, B. (2012). The Bolt Case: Silencing Speech or Promoting Tolerance? In *More or Less: Democracy and New Media* (pp. 238–257) [Online]. Version.

http://www.futureleaders.com.au/book_chapters/pdf/More-or-Less/Bibhu_Aggarwal.pdf. Accessed 8 Jan 2016.

Albrechtsen, J. (2017). There Is No Room for Section 18C in a Truly Liberal Australia. *The Australian*, News Ltd. http://www.theaustralian.com.au/opinion/columnists/janet-albrechtsen/there-is-no-room-for-section-18c-in-a-truly-liberal-n-australia/news-story/4a3edf4f69e7e3eb42928caad537950e. Accessed 22 Feb 2017.

Amend, A., & Morgan, J. (2017). Breitbart Under Bannon: Breitbart's Comment Section Reflects Alt-Right, Anti-semitic Language. *Hatewatch*. https://www.splcenter.org/hatewatch/2017/02/21/breitbart-under-bannon-breitbarts-comment-section-reflects-alt-right-anti-semitic-language. Accessed 23 Feb 2017.

Ames, M. (2014). As Reason's Editor Defends Its Racist History, Here's a Copy of Its Holocaust Denial "Special Issue". *Pando*. https://pando.com/2014/07/24/as-reasons-editor-defends-its-racist-history-heres-a-copy-of-its-holocaust-denial-special-issue/. Accessed 14 Mar 2017.

Anglin, A. (2016). A Normie's Guide to the Alt-Right. *The Daily Stormer*. http://www.dailystormer.com/a-normies-guide-to-the-alt-right/. Accessed 15 Jan 2017.

Awan, I. (2016). Islamophobia on Social Media: A Qualitative Analysis of the Facebook's Walls of Hate. *International Journal of Cyber Criminology, 10*(1), 1-20.

Back, L. (2002). Aryans Reading Adorno: Cyber-Culture and Twenty-First-Century Racism. *Ethnic and Racial Studies, 25*(4), 628–651.

Begley, P., & Maley, J. (2017). *White Supremacist Leader Mike Enoch to Visit Australia*. http://www.smh.com.au/federal-politics/political-news/white-supremacist-leader-mike-enoch-to-visit-australia-20170513-gw46fn.html. Accessed 13 May 2017.

Bliuc, A.-M., McGarty, C., Hartley, L., & Muntele Hendres, D. (2011). Manipulating National Identity: The Strategic Use of Rhetoric by Supporters and Opponents of the 'Cronulla Riots' in Australia. *Ethnic and Racial Studies*. doi:10.1080/01419870.2011.600768. Accessed 14 Mar 2017.

Bolt, A. (2016). Politicians Must Acknowledge That Culture Counts—It Must Be a Factor When Deciding Who We Let into Australia. *Bolt Report/Sky News*. https://www.facebook.com/theboltreport/videos/343440509347418/. Accessed 28 Nov 2016.

Bolt, A. (2017, April 5). Why This Anti-White Racism? *The Bolt Report*, Herald Sun. http://www.heraldsun.com.au/blogs/andrew-bolt/why-this-antiwhite-racism/news-story/bb36667e0294a427db9954b57af9bb81. Accessed 11 Apr 2017.

Bouma, G. D. (1997). The Settlement of Islam in Australia. *Social Compass, 44*(March), 71–82.

Buckels, E., Trapnell, P., & Paulhus, D. (2014). Trolls Just Want to Have Fun. *Personality and Individual Difference, 67*(September), 97–102.

Cadwalladr, C. (2017). Antisemite, Holocaust Denier … Yet David Irving Claims Fresh Support. *The Guardian*. https://www.theguardian.com/uk-news/2017/jan/15/david-irving-youtube-inspiring-holocaust-deniers. Accessed 22 Feb 2017.

Chan, E. (2016). Donald Trump, Pepe the Frog, and White Supremacists: An Explainer That Cartoon Frog Is More Sinister Than You Might Realize. www.hillaryclinton.com, 2017(20 February). https://www.hillaryclinton.com/feed/donald-trump-pepe-the-frog-and-white-supremacists-an-explainer/. Accessed 21 Feb 2017.

Chansanchai, A. (2011). UCLA Student's 'Asians' Rant Ignites YouTube Responses. *Today*. Accessed 25 Feb 2017.

Chernikoff, H. (2016, August 5). In D.C., Chinese Zodiac Symbol Becomes Jew Rat Graffiti. *Forward*. http://forward.com/news/346985/in-dc-chinese-zodiac-symbol-becomes-jew-rat-graffiti/. Accessed 28 Mar 2017.

Cohen-Almagor, R. (2013). Freedom of Expression v. Social Responsibility: Holocaust Denial in Canada. *Journal of Mass Media, 28*(1), 42–56.

Costello, P. (2006). *Worth Promoting, Worth Defending: Australian Citizenship, What It Means and How to Nurture It*. Speech to the Sydney Institute. http://www.treasurer.gov.au/DisplayDocs.aspx?doc=speeches/2006/004.htm&pageID=005&min=phc&Year=2006&DocType=1. Accessed 10 Oct 2012.

Curran, J. (2012). Reinterpreting the Internet. In J. Curran, N. Fenton, & D. Freedman (Eds.), *Misunderstanding the Internet* (pp. 3–33). London: Routledge.

De Witte, R. (2013). *We the Undersigned Residents, Ratepayers and Interested Parties of Bendigo Formally Support Elise Chapman's Position as Councillor on the Bendigo City Council and Insist That She Does Not Resign or Be Removed from Her Position*. https://www.change.org/p/to-bendigo-city-council-we-the-undersigned-residents-ratepayers-and-interested-parties-of-bendigo-formally-support-elise-chapman-s-position-as-councillor-on-the-bendigo-city-council-and-insist-that-she-does-not-resign-or-be-removed-from-her-position?just_created_and_shared=true. Accessed 11 Apr 2017.

Di Stefano, M. (2017). Jewish Leaders Are Disgusted That a Government MP Appeared on an "Anti-Semitic" Podcast. *BuzzFeedNews*, BuzzFeed. https://www.buzzfeed.com/markdistefano/jewish-leaders-condemn-christensen?utm_term=.sxWLvKlANq-.xjpXj8VD45. Accessed 23 Feb 2017.

Dickter, C. L., Kittel, J. A., & Gyurovski, I. I. (2012). Perceptions of Non-Target Confronters in Response to Racist and Heterosexist Remarks. *European Journal of Social Psychology, 42*, 112–119.

Ellemers, N., & Barreto, M. (2015). Modern Discrimination: How Perpetrators and Targets Interactively Perpetuate Social Disadvantage. *Current Opinion in Behavioral Sciences, 3*, 142–146.

Elliott, C., Chuma, W., El Gendi, Y., Marko, D., & Patel, A. (2016). *Hate Speech: Key Concept Paper* (Working Papers in the MeCoDEM). Leeds: University of Leeds. http://www.mecodem.eu/wp-content/uploads/2015/05/Elliot-Chuma-ElGendi-Marko-Patel-2016_Hate-Speech.pdfw. Accessed 21 Feb 2017.

Eureka Research. (1998). *The Anti-racism Campaign: Quantitative Market Research to Guide Campaign Development.* Canberra: Multicultural Affairs Branch, Anti-racism Campaign Unit, Department of Immigration and Multicultural Affairs. https://andrewjakubowicz.files.wordpress.com/2011/11/dimiaantiracismreportquant1998.pdf. Accessed 28 Nov 2016.

Executive Council of Australian Jewry (ECAJ). (2002). *Human Rights and Equal Opportunity Commission of Australia.* Racial Hatred Cases: ECAJ. http://www.ecaj.org.au/case_study/human-rights-and-equal-opportunity-commission-of-australia/. Accessed 12 Mar 2017.

Feigin, J. (2014). *Racist America: Roots, Current Realities, and Future Reparations* (3rd ed.). New York: Routledge.

Firstmet.com. (2013). Online Dating: Does Ethnicity Matter? *Firstmet.com*, 2017(24 February). http://www.firstmet.com/dating-blog/online-dating-ethnicity-racial-preference-interracial-dating-ayi/. Accessed 25 Feb 2017.

Fleishman, C., & Smith, A. (2016a). "Coincidence Detector": The Google Chrome Extension White Supremacists Use to Track Jews. *Tech.Mic*. https://mic.com/articles/145105/coincidence-detector-the-google-extension-white-supremacists-use-to-track-jews-.piTy94fCs. Accessed 23 Feb 2017.

Fleishman, C., & Smith, A. (2016b). (((Echoes))), Exposed: The Secret Symbol Neo-Nazis Use to Target Jews Online. *Tech.Mic*. https://mic.com/articles/144228/echoes-exposed-the-secret-symbol-neo-nazis-use-to-target-jews-online-.pQLaSlPIC. Accessed 23 Feb 2017.

Furnham, A., Richards, S., & Paulhus, D. (2013). The Dark Triad of Personality: A 10 Year Review. *Social and Personality Psychology Compass, 7*(3), 199–216.

Gagliardone, I., Gal, D., Alves, T., & Martinez, G. (2015). *Countering Online Hate Speech.* Paris: UNESCO. http://unesdoc.unesco.org/images/0023/002332/233231e.pdf. Accessed 21 Feb 2017.

Gelber, K., & McNamara, L. (2013). Freedom of Speech and Racial Vilification in Australia: 'The Bolt case' in Public Discourse. *Australian Journal of Political Science, 48*(4), 470–484.

Golbeck, J. (2014). Internet Trolls are Narcissists, Psychopaths, and Sadists: Trolls Will Lie, Exaggerate, and Offend to Get a Response. *Psychology Today*. https://www.psychologytoday.com/blog/your-online-secrets/201409/internet-trolls-are-narcissists-psychopaths-and-sadists. Accessed 21 Feb 2017.

Hagemeister, M. (2008). The Protocols of the Elders of Zion: Between History and Fiction. *New German Critique, 35*(1), 83–95.

Hine, G. E., Onaolapo, J., De Cristofaro, E., Kourtellis, N., Leontiadis, I., Samaras, R., Stringhini, G., & Blackburn, J. (2016). *A Longitudinal Measurement Study of 4chan's Politically Incorrect Forum and Its Effect on the Web (version 3)*. https://arxiv.org/pdf/1610.03452v3.pdf. Accessed 20 Feb 2017.

Holmes, A. (2016). High Court Rejects Mosque Appeal. *Bendigo Advertiser*. http://www.bendigoadvertiser.com.au/story/3969395/high-court-rejects-mosque-appeal/. Accessed 14 Mar 2017.

Holt, D., Callander, M., & Newman, C. E. (2016). 'Not Everyone's Gonna Like Me': Accounting for Race and Racism in Sex and Dating Web Services for Gay and Bisexual Men. *Ethnicities, 16*(1), 3-21.

Humphrey, M. (1987). Community, Mosque and Ethnic Politics. *Journal of Sociology, 23*, 233–245.

Jakubowicz, A. (1990). Racism, Racist Violence, and the Media. *Quoted in HREOC Report Racist Violence 1991*. Sydney: Unpublished Paper Commissioned by the National Inquiry into Racist Violence.

Jakubowicz, A. (1997). In Pursuit of the Anabranches: Immigration, Multiculturalism and a Culturally Diverse Australia. In G. Gray & C. Winter (Eds.), *The Resurgence of Racism; Howard, Hanson and the Race Debate, Monash Publications in History*. Clayton: Department of History, Monash University.

Jakubowicz, A. (2007). Political Islam and the Future of Australian Multiculturalism. *National Identities, 9*(3), 265–280.

Jakubowicz, A. (2012). Cyber Racism. In *More or Less: Democracy and New Media* [Online]. Version. http://www.futureleaders.com.au/book_chapters/pdf/More-or-Less/Andrew_Jakubowicz.pdf. Accessed 16 July 2016.

Jakubowicz, A. (2013). Comparing Australian Multiculturalism: The International Dimension. In A. Jakubowicz & C. Ho (Eds.), *'For Those Who've Come Across the Seas…': Australian Multicultural Theory Policy and Practice* (pp. 15–30). Melbourne: Australian Scholarly Press.

Jakubowicz, A. (2016a). First the Word, then the Deed: How an 'Ethnocracy' Like Australia Works. *The Conversation*. https://theconversation.com/first-the-word-then-the-deed-how-an-ethnocracy-like-australia-works-69972. Accessed 12 Mar 2017.

Jakubowicz, A. (2016b). Once Upon a Time in … Ethnocratic Australia: Migration, Refugees, Diversity and Contested Discourses of Inclusion and Exclusion. *Cosmopolitan Civil Societies: An Interdisciplinary Journal, 8*(3), 5239.

Jakubowicz, A., Dunn, K., & Sharples, R. (2017). Australians Believe 18C Protections Should Stay. *The Conversation*. https://theconversation.com/australians-believe-18c-protections-should-stay-73049. Accessed 17 Feb.

Jordan, P. (2013). *Remove Elise Chapman from Council*. Change.org. https://www.change.org/p/bendigo-council-remove-elise-chapman-from-council. Accessed 16 Mar 2017.

Jupp, J. (2002). *From White Australia to Woomera: The Story of Australian Immigration*. New York Port Melbourne, Vic.: Cambridge University Press.

Kahn, K. B., Barreto, M., Kaiser, C. R., & Rego, M. S. (2016). When Do High and Low Status Group Members Support Confrontation? The Role of Perceived Pervasiveness of Prejudice. *British Journal of Social Psychology, 55*(1), 27–43.

Keeble, E. (2013, November 5). Justin Bieber and Lady Gaga Fans Unleash Racist Hate on Girls Generation. *Gigwise*. http://www.gigwise.com/news/85739/justin-bieber-and-lady-gaga-fans-unleash-racist-hate-on-girls-generation. Accessed 15 Feb 2017.

Klein, A. (2017). *Fanaticism, Racism and Rage Online: Corrupting the Digital Sphere*. Cham: Palgrave Macmillan.

Koebler, J. (2016, September 15). Hillary Clinton Is Right: Pepe Is a White Supremacist. *Motherboard*. https://motherboard.vice.com/en_us/article/hillary-clinton-is-right-pepe-is-a-white-supremacist. Accessed 21 Feb 2017.

Kohli, R. (2013). Unpacking Internalized Racism: Teachers of Color Striving for Racially Just Classrooms. *Race, Ethnicity and Education, 17*(3), 367–387. http://www.tandfonline.com/doi/abs/10.1080/13613324.2013.832935?src=recsys&journalCode=cree20. Accessed 25 Feb 2017.

Langfield, M. (2009). Indigenous People are Not Multicultrual Minorities': Cultural Diversity, Heritage and Human Rights in Australia. In M. Langfield, W. Logan, & M. Craith (Eds.), *Cultural Diversity, Heritage and Human Rights: Intersections in Theory and Practice*. New York: Routledge.

Lewis, K. (2013). The Limits of Racial Prejudice. *Proceedings of the National Academy of Sciences of the United States of America, 110*(47), 18814–18819.

Mannix, L. (2015). Woman Charged After Facebook Call to Burn Down Mosques Goes Viral. *The Age*, Fairfax. http://www.theage.com.au/victoria/woman-charged-after-facebook-call-to-burn-down-mosques-goes-viral-20151222-gltqz9.html. Accessed 13 Apr 2017.

Matamoros-Fernandez, A. (2016). Platformed Racism: The Mediation and Circulation of an Australian Racebased Controversy on Twitter, Facebook and YouTube. *Association of Internet Researchers Annual Conference*. http://eprints.qut.edu.au/101370/1/AoIR2016_platformed_racism.pdf. Accessed 14 Mar 2017.

Mitchell, R. (1983). *Proposal for Amendments to the Racial Discrimination Act to Cover Incitement to Racial Hatred Act Racial Defamation*. Canberra: Human Rights Commission. Available at: https://www.humanrights.gov.au/sites/default/files/HRC_report7.doc.

Moss, I., & Castan, R. (1991). *Racist Violence: Report of the National Inquiry into Racist Violence*. Canberra: Human Rights and Equal Opportunity Commission. https://view.officeapps.live.com/op/view.aspx?src=https://www.humanrights.gov.au/sites/default/files/document/publication/NIRV.doc. Accessed 6 Sep 2016.

Mostrous, A. (2017, March 17). YouTube Hate Preachers Share Screens with Household Names. *The Times*. http://www.thetimes.co.uk/article/youtube-hate-preachers-share-screens-with-household-names-kdmpmkkjk. Accessed 25 Mar 2017.

Nadal, K. (2011). The Racial and Ethnic Microaggressions Scale (REMS): Construction, Reliability, and Validity. *Journal of Counselling Psychology, 58*(4), 470–480.

Nelson, J., Dunn, K., & Paradies, Y. (2011). Bystander Anti-Racism: A Review of the Literature. *Analyses of Social Issues and Public Policy, 11*(1), 263–284.

Nevin, A. (2015). *Cyber-Psychopathy: Examining the Relationship Between Dark E-Personality and Online Misconduct* (Electronic Thesis and Dissertation Repository). University of Western Ontario, MA.

Nicas, J. (2017). Google's YouTube Has Continued Showing Brands' Ads With Racist and Other Objectionable Videos: PepsiCo, Wal-Mart, Dish Network Say They are Suspending Google Non-Search Ads. *The Wall Street Journal*. Accessed 25 Mar 2017.

Oboler, A. (2015). Anti-Muslim Hate Interim Report. *The Online Hate Prevention Institute*, OHPI. http://ohpi.org.au/anti-muslim-hate-interim-report/-sthash.AWmDJ6ii.dpuf. Accessed 11 Apr 2017.

Oboler, A. (2016). Measuring the Hate: The State of Antisemitism in Social Media. *The Online Hate Prevention Institute*, OHPI. http://ohpi.org.au/measuring-antisemitism/

Orbe, M. (1998). From The Standpoint(s) of Traditionally Muted Groups: Explicating a Co-cultural Communication Theoretical Model. *Communication Theory, 8*(1), 1–26. https://wmich.pure.elsevier.com/en/publications/from-the-standpoints-of-traditionally-muted-groups-explicating-a--3.

Pat Eatock v. Andrew Bolt and The Herald and Weekly Times Pty Ltd (ACN 004 113 937) (Eatock v. Bolt [2011] FCA 1103).

Paul, J., Ayala, G., & Choi, K.-H. (2010). Internet Sex Ads for MSM and Partner Selection Criteria: The Potency of Race/Ethnicity Online. *The Journal of Sex Research, 47*(6), 528-38.

Paulhus, D. L., & Williams, K. M. (2002). The Dark Triad of Personality: Narcissism, Machiavellianism, and Psychopathy. *Journal of Research in Personality, 36*, 556–563. Accessed 21 Feb 2017.

Pearson, J. (2016). Scientists Invented a Tool to Expose 4chan's Racist Trolling Campaigns. *Motherboard*. https://motherboard.vice.com/en_us/article/block-4chan-to-stop-the-alt-right-from-spreading-racist-memes-scientists-say. Accessed 21 Feb 2017.

Phillips, W. (2015). *This Is Why We Can't Have Nice Things: Mapping the Relationship Between Online Trolling and Mainstream Culture*. Cambridge, MA: MIT Press.

Poynting, S., & Noble, G. (2004). *Living With Racism: The Experience and Reporting by Arab and Muslim Australians of Discrimination, Abuse and Violence Since 11 September 2001*. http://www.hreoc.gov.au/racial_discrimination/isma/research/UWSReport.pdf

Racial Hatred Act. 1995. No.101 (An Act to Prohibit Certain Conduct Involving the Hatred of Other People on the Ground of Race, Colour or National or Ethnic Origin, and for Related Purposes).

Richardson, S. (1998). Net Spreads Message, Draws in Recruits. *Montgomery Advertiser*. https://www.stormfront.org/dblack/press062898.htm. Accessed 28 June 1998.

Sheehan, P. (2006, January 16). How the Politics of Sheer Populism Led to Racial Riots. *Sydney Morning Herald Online*. http://www.smh.com.au/news/opinion/how-the-politics-of-sheer-populism-led-to-racial-riots/2006/01/15/1137259943539.html/. Accessed 6 Jan 2010.

Simon Wiesenthal Center. (2016). Simon Wiesenthal Center Applauds Social Media Giants' Commitment to Combat Online Hate in Europe. *Simon Wiesenthal Center*. http://www.wiesenthal.com/site/apps/nlnet/content.aspx?c=lsKWLbPJLnF&b=8776547&ct=14851471. Accessed 15 Dec 2016.

Soutphommasane, T. (2014). Unconscious Bias and the Bamboo Ceiling. https://www.humanrights.gov.au/news/speeches/unconscious-bias-and-bamboo-ceiling. Accessed 24 Oct 2016.

Stein, J. (2016, August 18/September 12). How Trolls Are Ruining the Internet. *Time*. Accessed 21 Feb 2017.

Sweeney, L. (2013). Discrimination in Online Ad Delivery. *Data Privacy Lab*. http://dataprivacylab.org/projects/onlineads/1071-1.pdf. Accessed 25 Mar 2017.

Tavan, G. (2004). The Dismantling of the White Australia Policy: Elite Conspiracy or Will of the Australian People? *Australian Journal of Political Science, 39*(1), 109–125.

The Dingoes. (2017, February 7). The Convict Report Episode 62: Mark Latham 2—Dingo Boogaloo. *The Right Stuff Radio: Echo Free Entertainment*. https://radio.therightstuff.biz/tag/alt-right-australia/. Accessed 11 Apr 2017.

UN Committee on the Elimination of Racial Discrimination. (2013, September 26). General Comment 35—Combating Racist Hate Speech. *UN Doc: CERD/C/GC/35*, (Para. 13).

Watson, L. (2017). A Bunch of New Companies Just Pulled Their Google Ads Over Racist Videos. *Gizmodo*. https://www.gizmodo.com.au/2017/03/a-bunch-of-new-companies-just-pulled-their-google-ads-over-racist-videos/-Ld2tOgwTFKhL5mek.99. Accessed 25 Mar 2017.

Williams, G. (2012). Removing Racism from Australia's Constitutional DNA. *Alternative Law Journal, 37*, 151–155. https://ssrn.com/abstract=2144763. Accessed 25 Mar 2017.

Zuckerberg, M. (2015, December 10). I Want to Add My Voice in Support of Muslims in Our Community and Around the World. *Facebook*. Accessed 17 Apr 2017.

Zuckerberg, M. (2016). Mark Zuckerberg to Hold Facebook Q&A Session in Berlin. *Zuckerberg Transcripts 189*. http://dc.uwm.edu/zuckerberg_files_transcripts/189. https://www.youtube.com/watch?v=5ZAMx-1mDmI. Accessed 23 Feb 2017.

5

Targets

Introduction

What happens to communities or members of ethnic or religious groups when they are targeted by online racist abuse? In this chapter, we explore two dimensions of this question in the Australian context. Firstly, we explore the dynamics of response among six highly targeted Australian communities—Indigenous, African, Middle Eastern, Chinese, Jewish and Muslim—as individuals from these communities share their experiences of managing racism and hate online. Secondly, we report on two empirical case studies from Sydney, NSW, where members of faith communities, Muslim and Jewish, decided that the level of the threat posed by the online harassment to which they were exposed, required them to act. This chapter is about the everyday Internet users who suddenly discover they have wandered into the Dark Side.

A. Jakubowicz et al., *Cyber Racism and Community Resilience*, Palgrave Hate Studies,
DOI 10.1007/978-3-319-64388-5_5

Targets of Hate: Six Groups of Australian Online Users

As part of the research strategy for the CRaCR Project, six focus groups were convened in 2016 with Australian nationals of Indigenous, African, Middle Eastern, Chinese, Jewish and Muslim backgrounds. These groups were identified as the most highly targeted communities in the CRaCR online survey undertaken in 2013. We report here the outcomes of these focus groups with members of target communities, and then reflect on their implications.

The question structure for the groups followed the key issues raised in the 2013 survey, though concentrating on their experience of, and reaction to, online race hate speech. Through these discussions, each group found its own voice and narrative (Table 5.1).

The Indigenous Australian Community

Indigenous Australians today are the surviving descendants of the many nations that populated the Australian continent prior to the seizure of the country by the British Crown in the period after 1770. It is estimated that the Indigenous population at the time of the declaration of British Crown settlement in 1788 stood at about 750,000, speaking over 700 languages.

Table 5.1 Participants in focus groups from "target" communities

Cultural background	Males	Females	Total
Indigenous	2	3	5
Muslim—mix of ethnicity including Pakistan, Eritrea, Egypt, Bangladesh and Somalia	2	4	6
Jewish—mix of ethnicity including South African, German, Russian and Israeli	3	3	6
Arab/Middle East—Syrian and Lebanese	1	5	6
African: Congolese/West African/Sierra Leonean	2	3	5
Chinese (mix of Mandarin and Cantonese speakers)	3	2	5
Total	13	20	33

The 2011 Census identified about 550,000 people as being of Indigenous descent (with estimates of up to 770,000 to include those who had not answered the Census question); the community had a median age of 21 years (Australian Bureau of Statistics 2012). The 2016 Census found that the "typical Aboriginal and/or Torres Strait Islander" was 23 years old, compared with the "typical Aussie" in the general population who was 38 years old.

In general, the Indigenous population has poorer health outcomes, poorer education outcomes and far greater incarceration than the general population. While many have adopted European lifestyles and become more middle class, significant numbers remain impoverished. Discrimination against Indigenous people remains widespread, with very high rates of reported racism.

Indigenous Australians are the most harassed group online, and yet social media is very popular among them. All our respondents had a Facebook page, while two also had Instagram. Facebook is their main source of news. It provides them with access to their social networks and links them to local buying and selling information on sites like eBay. Most have links to Aboriginal cultural pages, and they use Facebook to alert their networks to issues of domestic violence and assault. They also use the Internet to research issues affecting Aboriginal people and have high levels of self-confidence about their Internet skills. The Internet serves them as a means to reinforce Black community, share information and, in some cases, discuss Aboriginal art. One respondent, a practicing Christian, would report anti-Aboriginal memes and also Muslim slandering of Christians.

Although the Internet brings myriad benefits, it does expose Indigenous people to a flow of racist memes and abuse based on stereotypical hatreds. This includes false claims about Indigenous mendacity, alcoholism and their access to Commonwealth benefits. Race, poverty and life on the edge of society regularly confront Indigenous people via social media and mainstream media comment threads. One participant referred to refugees getting financial support while Indigenous people lived in overcrowded and ramshackle accommodation on controlled debit cards (BasicCard). Another used Facebook to highlight the impact of methamphetamine (ice) on young Aboriginal people, sending it to his networks as a warning. Yet, many also used Facebook to simply "hang out with family" and friends. Many said they will not mix with people who refuse to put their faces on Facebook. "If you've got nothing to hide, you'd have your face there," commented one participant.

One of the most controversial issues discussed by the group concerned the online game Survival Island 3, created by a Russian programmer Krisitna Fedenkova in 2015, and available on Apple's Appstore, Google Play and YouTube gaming. This game, it was claimed, requires gamers to hunt down and kill Aboriginal people (PV Reviews 2015).

When alerted to the existence of this game, a number of group members were so upset they contributed to a Change.org petition, "Killing Indigenous Australians is not a game," to have it removed. It was only after 90,000 people had signed that Apple and Google, then Amazon, removed the game (Mantle 2015).

So, what happens when hate speech arrives in their space? The group immediately picked up on the psychological gains of hate speakers: "They've got that little knife going in and they're twisting that knife," said one participant. Members of this focus group believe that Facebook should be the entity responsible for removing the hate.

> Facebook should be accountable for removing that person so they can't have an account. They should be named and shamed, although when you report, they [the haters] make up a fake [identity] … and come back.

When they are harassed, tricked or intimidated, members of this group did use Facebook devices to "report, delete, block" other people and their posts. They also mentioned being able to brush it off:

> Overall there's not that much [abuse and racism] … somebody might put a smartarse comment on Aboriginals of Australia or something, and you sort of think, well you're an idiot, and then you move on. The first five people [to respond] might really give it to them … [then] they back right off.

Reflecting on how racism has worsened, one member said:

> Since computers have started the racists have become bolder. When I was young they would still have a go at me, but at least I could stand up for myself. Now they're invisible attackers.

Another man commented:

> For me, it seems to hurt more because you can't have your say back. I've got very upset and very depressed and cried. It's made me feel terrible.

Yet, despite the dangers, the Internet provides some safety. A number of groups agreed that sites like Cultural Survival Australia provide an affirmative environment to explore their culture and participate in conversations about their traditional world and its contemporary meanings. (Cultural Survival is a global website where First Nation people can find support for their struggles to survive. Its cultural affirmation marks it as a crucial player, and one that is well protected from cyber racism.) Group members also mentioned that they belong to a variety of local, private, online Aboriginal groups that discuss familial, political and cultural issues. Importantly, posting a comment on these sites often results in supportive feedback that acknowledges their contribution and helps refine their perspectives.

Even though the participants were quite active in their communities, they were not able to identify resources that might defend them on the Internet, nor Indigenous organisations that could act for them online. Their informal networks are strong, but their involvement in formal networks far weaker, suggesting that supportive intervention might need to come from elsewhere.

Australia's African Community

African immigration to Australia has grown quite substantially in the past 20 years, and is one of the more significant changes to occur since the end of the White Australia Policy in the 1970s. Before 1976, the intake was primarily from South Africa and White (42% of all African-born residents in 2006) or from Mauritius (7.3% in 2006) and Egypt (13.5% in 2006) and Christian or Jewish (Australian Bureau of Statistics 2008). By 2006, there were almost 250,000 African-born people living in Australia, accounting for 5.4% of the total overseas-born population. Communities whose members mainly (90% or more) arrived after 1996 include those from Liberia, Sierra Leone and Sudan. Participants in our focus group were from Sierra Leone and the Democratic Republic of the Congo.

The group was highly technology-savvy, quoting their use of Facebook (nearly all), Instagram, WhatsApp, Viber, Imo, Snapchat and LinkedIn. They accessed news through proprietary sites (like the *Sydney Morning Herald* and CNN), news compilation services on Apple or news on Facebook from the main commercial channels, the Guardian, SBS and the BBC. Using Facebook to engage with issues of racism varied from scarcely ever, to regularly. One of the most attractive personalities to gather support has been Waleed Aly, an Egyptian Australian academic who also now features in a prime-time commercial talk show. He was controversially selected as the Golden Logie winner in 2016, a prize voted on by audiences for the most popular TV personality. He is African, dark-skinned and Muslim, and therefore acts as something of a role model for Africans who see no other Muslims of colour as leading prime-time TV personalities.

During 2016, the issue of Africans in Australia continued to grow in the public mind, especially where conflict or crime appeared to be rising. The issues became highly contentious, with traditional mainstream Australian commentators increasing their rhetorical clashes, and young Africans looking for ways to push back against the racism and xenophobia.

One example given by the group was of public "fake news" in which a non-African woman claimed that her baby had been seized by an African, when it turned out she had actually slain the infant. The group swapped stories about the alleged murder, and the slanders of Africans that had erupted on Facebook in its wake. One response to the defamations was quoted in the group, describing the reaction of an African woman discussing the man who had been accused She said that the woman should be charged with defamation, as her comments were so different to the reality of African men, and were a slander on all African men.

Another example of successful pushback referred to a 2005 case of a university professor who campaigned against African refugees; he eventually was forced to resign (Dick 2005). Every member of the group was able to share experiences of public online racism against Africans, including the 2007 false criticism of Somalis and Sudanese by the then conservative Australian Immigration Minister Kevin Andrews (Jakubowicz 2010).

Pushback, unfortunately, has another side to it—the issue of the power relationships between racists and targets. The angrier the targets become, the more successful perpetrators feel they have been (Anglin 2016). Discussing whether he would respond to racism online, one group member said:

> It depends on how angry I get. Sometimes I just don't want to give them that power over me, over my emotions. I'm stubborn that way. I don't want them to be able to see the impact they've had on me. But sometimes I have, I admit, made a statement to say "this is what I think", but on my terms rather than in responding to something.

The most angering aspects are the anonymity possible on the Internet, the rapidity of comments and their short lifespan on screen as they are soon overwhelmed by later posts. The participant suggested it is better to use social media to bring about change "to create something real and approach people with respect."

Exposure to online racism is not a new phenomenon solely associated with contemporary social media. One group member recalled a time when he was a schoolboy participating in an ICQ forum (an open-source instant messaging computer program). Suddenly, discovering he was African, the group turned on him, calling him a monkey and hurling derogatory remarks. Some tried to defend him, but he became so upset he logged out of the group. "To be attacked in that way was horrible. That was probably the first time." Yet, advances in technology have only intensified the dangers—with Facebook Live allowing real-time abuse and hostilities between people to be aired widely.

One group member, a community worker, said that she took action when hostilities and fights between Tongan and African girls were videoed and aired on Facebook. She reported it to Facebook who took it down as the gratuitous violence clearly breached the ISP rules, and bullying is banned.

Alternatively, one Congolese member spoke of the value to be found in a closed Facebook group designed for young Congolese only. Importantly, these forums keep people in touch transnationally, wherever the diasporic flow of Congolese or other Africans has taken them. Moreover, they also serve to build resilience by affirming African qualities that are undermined in the wider social media.

> Young people enjoy that space to express themselves because they are very active on social media. They can tweet when they play soccer, when they play basketball, they can share their videos and make comments.

Another participant commented on a particular social media channel they followed for its positive messaging.

> There's an Instagram account that I deliberately follow. It's a magazine that deliberately includes people with darker skin. All their models are dark skinned and it talks about celebrating melanin and loving all the different skin shades. It is also against the skin bleaching and shares positive messages about the beauty of dark skin.

And yet, the awareness of racism being "everywhere on the Internet" persists. "And sometimes it's the subtle ones that get me as well.... Because white people are unconscious of their racism," one woman concluded.

Australia's Middle Eastern Community

Under the categories used by the Australian Bureau of Statistics, the Middle East is included in a region that also encompasses the Arabic-speaking countries of North Africa. The Middle East stands as a proxy for overlapping but importantly different ways of identifying populations—it includes Muslims from the region who are not Arabs (e.g., Turks and Iranians), Arabs from the region who are not Muslims (e.g., Eastern Catholic, Orthodox and Coptic Christians) and also Jews who were born in Israel.

At the 2011 Census there were some 220,000 people born in the countries of the Middle East, up from 184,000 in 2006 (Australian

Government 2014). The major sources were Lebanon (76,000), Iraq (48,000), Iran (34,000) and Turkey (33,000). A majority of the Lebanese are Christian, while minorities from the other countries include Ba'hai, Assyrian Christians, Yazidis and Alawites.

One of the important distinctions made about race hate speech refers to whether religion *per se* can be included in an analysis of racism, given that faith is a choice and race is not. In the Australian context at the national level, religion is not a protected characteristic in racial vilification legislation. In order to explore what the relationship might be between religion and "race" or ethnic appearance and culture, we created a focus group drawn from people of Middle Eastern background. In a number of Australian police jurisdictions, "Middle Eastern appearance" is used as an identifier for alleged criminal perpetrators. This label has caused widespread distress, given its highly stereotypical implications and its impact on anyone who might fit such an imagined phenotype. In Sydney, this has been a particular issue where the NSW Police Middle Eastern Organised Crime Squad (MEOCS) has been blamed for stigmatising whole communities who might fit the label (Australian Arabic Council 2003).

One particular and difficult issue has been the additional tension created for the Christian and Muslim Lebanese communities, where some Christians have argued the label really should be dropped as the criminals are more likely to be Muslim, yet the label also gets attached to their supposedly more law-abiding community. Our focus group contained three Syrians (all Muslim) and three Lebanese (all Christian); the majority of the group was female, as the issue of gender has been noted as a key dimension of online racism (Ellemers and Barreto 2015)

All members of the group had Facebook accounts, with one having Twitter and another YouTube. All used Facebook to access news and used satellite TV for international news. Most used their Facebook pages to follow the news feeds of friends and maybe connect with friends. A central focus for all participants was the politics of the Middle East including the Palestine Israel conflict, an issue of significant controversy and conflict online. They all noted that their engagement with politics online has become more muted as interactions have become angrier. One participant also noted that any undue interest in these issues, especially accompanied by a desire to go to the area, could be identified by security officials.

In this group, the emphasis, however, was on communicating ideas about peace and unity, rather than difference and conflict. One Lebanese woman noted:

> Yes, I share things online, like a wise saying or something about religion and how it should, is meant, to unite people, not separate or incite them against each other.

Another Lebanese woman commented:

> I avoid or ignore things that are racist or about discriminating against someone else. But if something is talking about harmony and peace, I will share it.

One of the Syrian women reflected on how she tried to avoid entering into online arguments about her points of view, but would "like" things in order to signal her position:

> People might "like" something but not necessarily want to engage. I think I tend to do that now. If I agree with something, I'll hit "like", but I don't comment because I expect someone to respond, and I don't really want to have that.

One example discussed was the emergence of Syrian Christian sites, which the Syrian Muslim women saw as anti-Muslim. "I found it very degrading very humiliating," one commented. She also referred to her horror at another anti-Muslim site she had seen:

> They put up a lot of pictures, like an Arab having sex with a camel, and they say, "This is really what happens in Islam."

A Lebanese Christian woman said this was "Disgusting!" and the Lebanese male responded:

> At the beginning of the war in Syria there was a site: "Fuck Assad"… and it was so disgusting. It encouraged people from other Muslim religions to

> abuse a particular sect [not identified—could have been Alawi, or Syrian orthodox or Catholic] … I was one of the many people involved in a petition to close it down but it actually stayed up for a long time.

The Lebanese man, who works for a Muslim Organisation, continued:

> To be honest, I've seen some terrible stuff online, and I've complained a lot. But I notice on Facebook they just don't take them off, even when you complain. I don't understand the censorship [rules] that they have, and yet they took down a woman feeding her baby because she showed her boobs! So I don't understand how that works. … Because I work for a Muslim organisation and I actually do workshops on Islam, people post me very anti-Islamic, very shocking stuff because they know I'm not Muslim. I unfriend such people but sometimes, I don't know whether the people repost on their pages or share things, because I get people making comments and I don't know who they are.

The Syrian Muslim women had all encountered anti-Muslim hate speech, and their strategy was mainly to ignore it because of bad past experiences of responding.

> I've just commented, "Are you serious?" or something like that, but then it keeps going on and on. That's why I don't do it anymore because there's no point.

When asked about their attitudes to the balance between freedom from hate and freedom of speech, the Syrian women were critical of Australian laws that allowed supporters of ISIS to rally freely, and appear on television apparently, to them, without constraints. The group agreed that censorship of Facebook was needed, though they disagreed on how this should happen. The youngest Syrian woman, a student, noted:

> You can't permanently remove something, it's always going to come back. So, something like this with Facebook even when you do remove it, it's not stopping people from posting the same thing next time, and next time, and next time…

However, in response:

> I know Facebook blocked anyone sending photos of Osama, but by the time it's taken, thousands of people have already seen it and can share it themselves. Facebook's not going to block 100,000 people for sharing a post.

One woman, who had previously been active on Facebook posting information about the Syrian conflict, reflected:

> I think I used to get up and talk about a lot of this stuff, and then I just backed away because anytime you mentioned something you're straight away put into a group about it.

Their friendship groups online reflected the different life histories of the groups—the Syrian women had very mixed circles of friends, some drawn from school days, others through work. The Lebanese tended to have more of their close circle from their own community. Even so, they recognise the Internet as a dangerous place:

> I get scared of saying the wrong thing to someone on Facebook, because I'm scared for myself and my family. I'm very, very cautious of what to say and try and play along with everybody so as not to upset anyone.

While community hostility is one factor, government surveillance can also lead to fear.

> I have a relative who posts about Palestine, she is against the Israelis and against Islamic things. I'm always careful not to respond to her and she has messaged me so many times saying, "You never messaged me back or shared my posts." I say, "Either I didn't see them or I don't like them," to her because I'm really concerned and worried living in Australia. You don't know whether the ASIO [Australian Security Intelligence Organisation] will come.

Anonymity has its "rewards":

> I don't wear the hijab, but I am Muslim, so people don't immediately know I'm Muslim. The amount of anti-Muslim things that have been said and

shared with me, because people don't know I'm Muslim, is incredible… It's kind of like this silent racism.

But the opposite also occurs:

People think I'm Muslim because I work in a Muslim agency and run sessions on Islam. I get Muslims coming up to me saying, "Come on sister, let's go. These people are no good, we'll take you out for coffee", so I get exactly the opposite. I think, actually, that people hide behind social media.

How then does the religion versus race perspective appear to the group members? One Lebanese woman contributed:

I actually think that yes, there is racism or discrimination against Islam, but I think it's more a discrimination against Arabs, not all Muslims. I don't hear comments about Afghani or Dari or even Turks in the same way as I do about Arabs. I think it's discrimination against Arabs. It's anti-Arab campaign.

A Syrian woman continued:

So, once you say Muslim, it'll become an Arab automatically…. They're not educated enough. As soon you say, Muslim, "Oh, he's a black Arab." Or say an Arab, "Oh, he's a Muslim." But is it ignorance, or something else?

A Lebanese woman noted:

I don't think that's ignorance; I don't think it's uneducated. It's more that they know what they are doing.

People do choose spaces in which they feel safe. On issues such as Israel/Palestine, group members chose pages where they expected people to have views similar to their own. When on pages they encountered "friends" who did not share their views, they would simply avoid engaging. However, the sense of the surveillance society still intrudes. Says one: "But where is the safe place if the government monitors everything?"

Another voice:

> Personally, I get very scared of voicing my opinion. I'm very, very careful. I'm scared that ASIO will knock on my door. There are many, many people like myself. I would ignore things rather than voice my real opinion.

This sense of wariness seems particularly acute among communities associated with the Middle East conflict:

> I've told people in my family that they probably shouldn't put certain things on Facebook. I haven't said anything online, but I've told them [face to face]. I've done it to my sister. I've said, "I really don't think you should be posting that sort of stuff on your Facebook. You work in the public sector. If someone sees that screenshot it will affect your job."

Even with as experienced a set of social media users as this group, knowledge about the availability of external advocacy and support on cyber racism was limited. There was some awareness of government anti-discrimination bodies. In one interchange, comments were made about the Australian Arabic Council, which one member believed "the Jewish shut down." Following on this theme, the group discussed the difference between the Jewish Israelis versus the Arabs in terms of "our ability to organise and stay together." One Syrian woman noted: "they tend to be very collective, very nationalistic; whereas the Arabs, for some reason, are in this divide and conquer [mode] all the time, right?"

During the Lindt café siege in Sydney in December 2014 (Semmler 2014), the Internet was alive with posts about Muslims, especially when it was revealed that the perpetrator had declared himself a follower of Daesh (ISIS). One post a year later proposed that in retribution Muslims should be locked into mosques and burnt to death (Mannix 2015). A number of the focus group members had complained to Facebook about this, demanding the video post be taken down. Facebook assessed the complaints but did not take down the post. It was only after an online petition was launched and attracted 20,000 signatures that Facebook acted.

Yet, such responses have a downside according to one participant: "Don't feed the fire! With something like that it's great that Facebook pulled it down, but if 20,000 people signed it 20,000 people watched it." The woman who posted the video from Sweden and "liked" it was in fact charged under the Victorian section 25 of the Racial and Religious Tolerance Act 2001. Ultimately, the case did not proceed.

Talking of anti-Arab sentiment, one of the Syrian women said:

> I feel like the only time Arabs are going to get a break, is when the next group of migrants come in from somewhere completely different and everyone gets over the Arabs like they got over the Wogs. For me it's almost like it's just sit and wait. Someone else will come and it'll be their turn.

Australia's Chinese Community

Australia's ethnic Chinese communities are drawn from across the diaspora in Asia—the mainland People's Republic, the Taiwanese Republic, Vietnam, Malaysia, Indonesia, Singapore, Hong Kong and elsewhere (Australian Bureau of Statistics 2013). While some 320,000 were born in mainland China, there may well be 100,000 more when all possible permutations of what it means to be Chinese are taken into account (Jakubowicz 2011).

Unlike people from Africa and the Middle East, most Chinese in Australia have not been refugees, apart from the 40,000 accepted in the wake of the Tiananmen Square violence and its aftermath in 1989 and 1990. The past generation of Chinese immigrants has been far more fashioned by nationalism and racial identity politics in China than by either Communist or anti-Communist ideologies and struggles. The focus group of five (three men, two women) spoke either Mandarin or Cantonese as their first language and were aged from 18 to 38 years. All had their own Facebook account. Most also had Instagram accounts and Twitter, which they rarely used, while two of the men had Reddit accounts. Facebook is used primarily for news, to keep up with friends or to find news that the Australian media do not broadcast. However, for at least one member, Facebook is used for staying in contact with Australian friends and clients, while Weibo and WeChat are primarily used in their interactions with Chinese friends, family and for culturally relevant information. Another uses Twitter to commend writers whose ideas he likes, while another uses MSN and also the Chinese international students' site "6park."

There was considerable reticence among the group to post anything controversial; for example, during the Australian Federal election of mid-2016, an evangelical Christian Chinese group began a campaign on WeChat to bolster support for the conservative Liberal Party, using communal fears of gay marriage and sex education in schools to raise the salience of the election choices (Deare 2016).

> I have a friend, not that close, who did share something that was sensitive about two weeks before the election. She posted an article saying the Liberals were closer to the Chinese, vote for the Liberals and that Labor was not good.... After a few days, she posted another article saying that the Labor was going to have a gay school or something like that. It was really weird and I was very uncomfortable.

Much of the racist material this group encountered online, they identified as being about American Blacks and Muslims, far less about Chinese. However, bursts of anti-Chinese chatter do occur.

> You get things like: "You people like eating dogs" which was a big thing a few months back. Then you see a lot of memes of Chinese people in America or Australia letting their two-year-olds go to the bathroom in the public part of a store. Or you see memes where Chinese people go to buffets and take a whole plate of lobsters. The racism towards Chinese people isn't like "oh my god, they are getting killed or they are terrorists". It's "oh my god, what are they doing! Look how uncultured they are". That's the kind of racism I have seen and I don't think it's ever taken that seriously because it all gets turned into jokes... I think what Muslims or Black people go through is way more serious than what Chinese people go through.

Real-world experiences of racism may be fed back into the community through WeChat. One member told of her partner who had gone for fishing, and how his friends were harassed by the inspectors for their licenses, while the Australians nearby were not. He posted this story onto WeChat, where it entered a long line of posts about unfair treatment of Chinese.

The group also discussed whether they felt the racism they encountered was intentional.

> If I see someone make a racist comment, … It's not something I take personally, it's just something I chalk up to ignorance rather than direct hate. I think it comes down to ignorance and the misunderstanding of culture rather than someone actively trying to anger me by insulting my race or culture. I think there is a very small minority that is actively trying to get a reaction by being racist.

Overall, the group tended to disregard offensive material—they distinguish between what is "objectively racist" and their reaction, "Whether you take offense to it is a different issue."

Is there some truth, they asked, in stereotypes? One group member identified a number of stereotypes prevalent on the Internet about Chinese in Australia. Their poor driving records, their purchase of apartments driving up prices, their purchase of baby formula to send back to China emptying stores of product and their queuing up to buy the first iPhones for dealers to send back to China were all commonplace. One member noted that he knew that aspects of these stereotypes had validity, and while they did not apply to the majority of Chinese people, they seemed valid for some. So he argued that they should just be ignored. On the other hand, he was adamant about racism:

> Although I am not African American, I feel their pain more than anything else… [talking of Black Americans being shot by police]. For me, that racism is causing more problems than the Chinese stereotypes.

Discussing their common experiences of racism online, a number of group members noted some real-life experiences of being harassed, but little direct online harassment personally. They would encounter occasional general anti-Chinese comments or memes, but they had not been personally targeted. Perpetrators, they thought, were people who "had too much pressure in their lives. Maybe they want to release this anger by posting racist things online, thinking they can say whatever they want."

Online they tend to stick to their own friendship or family groups, so they can be protected from exposure to material they find offensive. Sometimes though, on Facebook, there may be pages where strangers can enter and post offensive material, but even then it is rarely targeted personally, and they agreed it was usually the consequence of ignorance. They expressed a preference for WeChat where participants were all family, or friends or friends of friends, a much safer place than Facebook. On Weibo, because of Chinese government supervision, "no one posts negative things … anything bad will be deleted within one or two minutes."

Among the Asian communities, different national groups use different apps—WeChat and Weibo for the Chinese, the Koreans use KaoKao Talk, Japanese and Thai use Line. One group member says: "You can ping people nearby on WeChat, so you can connect with them. It's more inside the community… whereas for Facebook and Instagram, they are so public."

The group members agreed that they felt racism online was not "as bad" as it had been in the past. It appears that their experience of online racism has declined because their behaviour online has adjusted—they set access to "private," they avoid public pages or offering their opinion where it might trigger an undesired reaction from anonymous posters and they ignore casual slurs that do not touch them personally. Moreover, they are far less the targets of racism than in the past, as the focus of online racist predators has shifted to other groups, especially Muslims.

Australia's Jewish Community

The Jewish community has come under increasing pressure on the Internet as conflicts in the Middle East have intensified and opponents of Israel have spread their targets to Jews in general, while, at the same time, neo-Nazi and White power groups have intensified their attacks on Jews as race traitors and destroyers of the White race (Anglin 2016). The Australian Jewish community numbers about 110,000, with a pre-war segment, a post-Holocaust refugee segment and more recent waves of immigrants from Russia, South Africa and Israel (Graham 2014).

The Jewish focus group comprised three men and three women, half aged under 27, half over 45 years, from many different national backgrounds. All used Facebook, and most had some other accounts—Instagram, LinkedIn and Hot Copper among those listed. They ranged from individuals who visited Facebook many times a day, to those who preferred to mainly communicate using email. Apart from friend and family links, they all follow a variety of news pages on Facebook. Some also follow Reddit and Buzzfeed for the news "without getting all too heavy."

Unlike most of the other groups, some members were avid readers of the online version of news and the comments attached. The comments were described as "quite heated," where people were often reluctant to comment because of fear of exposure. "In terms of open debate out there, I don't think it properly exists… it just puts me off wanting to get involved with strangers." They describe the comments lists as increasingly places for people to attack each other and "call each other idiots."

However, the group was aware of the political nature of social media. One recalled at school being told in class that everyone should go to a particular anti-Zionist page and report it to Facebook, so it would be taken down. "I've seen racist pages that have a very small group of vocal people following," suggesting that such conflicts occur among very committed activists from different political positions, rather than mass engagement from the wider society. By implication, people who engage there do so knowing they will be entering a space of torrid interactions. On the other hand, one member reported her daughter being in a University Facebook group where, during a Palestine–Israel conflict, the language became very heated and aggressive among Muslim and Jewish students until one of the students called "time out" to allow tempers to cool.

Another woman referred to the rising tensions in Europe being caused by Syrian refugee immigration, intensified by the Daesh conflict. She linked that into how the Internet had become increasingly intense on issues about race and religion, specifically about Muslims in Australia. She identified the controversy around local television personality Sonia Kruger who had publicised her views condemning Muslim immigration to Australia, and the associated rise of the One Nation party and the election of its anti-Muslim Senators led by Pauline Hanson (Jakubowicz 2016). She reflected that the extent of social media attention on the refugee crisis and its Australian implications had increased dramatically, with the language and the arguments getting more intense. "It's getting a bit scary and has definitely increased over the last few years."

People who did get involved in political debates about world events reported that once the comments began to flow, attention wavered from the issue at hand to become personal. Sometimes they were attacked because of being Jewish and their opinions criticised from that perspective; in one case concerning political events in Europe, one group member said "it went out of control." She described the follow-up: "People who had been involved in the discussion wrote to the person who had made the [specific] comment and said it was really, really inappropriate. So, he actually took it off a couple of days later."

One of the men responded that the world of online interaction can be very brutal:

> If you comment on something, you frequently expect to be attacked. I certainly found, at times, you get comments where they're digging. I once made a very neutral comment on a news site about the Paris fight and started getting questions like, "oh, you're obviously a Jewish B" and stuff like that. Then of course other people start joining in. I have a really thick skin, I went to a tough school and I'm used to environments where you have to stand up for yourself … [but] probably not a month goes by without seeing something where you actually think, "oh, hang on, that's not good! If something is not done this could become the norm. Should I be upset about it? Should I be thinking this is unacceptable?". It's quite profound.

His comment points to the way in which norms of online civility are generated through what is accepted and what is rejected. Given the millions of people involved and the reluctance of many people to be hammered by opposing opinions, especially if the responses are racist or antisemitic, the norms are likely to drift towards being increasingly discourteous and personal.

Discussion moved to the online attack on Aboriginal Senator Nova Peris in May 2016 when she had been racially abused by a White man on Facebook. Although his name was on the post, he clearly expected to remain unidentified and not called to account (Killalea 2016). Initially, he claimed he had been the victim of a vicious cyber hack, but this defense soon foundered and he was convicted of a criminal offence. Commenting on the case, one woman noted that the anonymity and disinhibition offered by the Internet were readily apparent, with racism a deeper issue.

> There's a lot of stuff [in comment threads] that I don't believe most people would actually say in person. When they are hiding behind a keyboard and screen they think they can say whatever they want. Peoples' racist undercurrent surfaces frequently though. I think if you look at anything that has a few hundred comments (and I don't mean something completely innocuous like a photo of food), if it's anything political, by comment 50 it has become racist and goes in every direction. The more comments there are, the more racist it becomes.

This case led directly into discussions around freedom of speech. The group members supported freedom of speech as a value, one saying:

> There's a point at which making sure people aren't hurt by the things that are said, becomes more important. I know there's the argument that you don't want to silence people just because someone is going to get offended, but there are, I feel, quite clear lines about where people are going to get offended.

Once more the issue of anonymity came into play, with one member arguing:

> There should be a difference between having that cloak of invisibility as an anonymous blogger and actually using your name. If you are prepared to have your name and identity out there then you should have the freedom to say exactly what you want.

But what happens to those opinions? One member expressed a concern with the impact of hate speech on social cohesion; however, another thought that the real issue was how anything said online could be taken by "millions of publishers… and cast differently," so now people are much less willing to say anything "because there isn't really a freedom of speech." Another came back supporting the right to say whatever people wish if they are identified, urging civility as a norm: "I think you should own what you say, and when you speak try to think about other people and not hurt their feelings." In order to create a safe place, one group member has a private page where only friends can engage, and where freedom to speak one's mind in a civil way is much more the norm.

We have already noted the complexity of the relationship between anti-Zionism, criticisms of Israel's policies and antisemitism. This is not the place to rehearse that debate or make a judgement call, but only to note that perspectives on these issues are heated, and political struggles over where the line is between legitimate critique of Israeli policies and antisemitism remain a focus of concern within the community and are an element of global political contestation. Asked about their personal experiences of online antisemitism, the group identified that distinction as critical. One member said:

> Most of the antisemitism I'm seeing is actually people who are anti-Zionist and they are calling out the Jews through that, which is not actually the right thing to do…. The comment section [on major international stories], are always racially driven and that's invariably where you see antisemitic activity.

With the advent of the alt-right in the USA (Amend and Morgan 2017), attention has been focused on its attempt to formulate undetectable online methods to label Jews, Blacks, Muslims and other minorities (also discussed in Chap. 4 in more detail). Jews, for instance, are either identified by triple brackets (e.g., (((Jakubowicz)))) or with the label "Skype," as these cannot be easily distinguished by Google's anti-racist strategies and blocked. When the group began to discuss this emerging issue, one member became very emotional. He referred to how Jews try to successfully merge into the societies in which they live, and keep their religious beliefs personal, familial and communal. Being called out as

Jewish and having his religious beliefs made public in a negative way brought back to him the horrors of antisemitic societies, such as Nazi Germany. A number of members agreed there was a communal norm that Jewish people would try not to draw attention to themselves in the wider society, challenged, however, by those who would call out antisemitism by wearing religious symbols and identifying very publicly with their culture and faith.

Much antisemitism occurs in unexpected quarters online, for instance, among Premier League football clubs in the UK. One group member, who emigrated from England a decade before, noted that there was intense racism and antisemitism in the discussion threads of the Premier League clubs, where Tottenham Hotspur was known as the Jewish club. "Jewish" clubs and their players were labelled as Jews as a form of hate. Similar insights were offered when discussion turned to cricket in Australia, where a Jewish player was often condemned as unsuitable to represent Australia even though he captained Western Australia. There was general agreement this was similar to the negative assessments Aboriginal footballers had to survive.

In discussing what individuals can do if they find themselves a target of antisemitic attacks, the group was not optimistic. One, who had been abused in broadcast emails at her work, had been very disheartened by the lack of avenues for recovery. Another, a lawyer, said that her advice was always the default—"Don't bother. It's a massive waste of time." Responding, another woman said:

> Facebook is the ultimate arbiter… it sits above the law, because it can dictate what you see and how it gets distributed to you. It's essentially running the law in so many countries… Facebook should be able to name and shame … embarrassment, I think, is extremely powerful.

As with the other groups, the Jewish group was modifying its use of social media to provide safer spaces. Its members recognised that antisemitism was rife, probably unbreakable, and thus best dealt with by avoidance. Interestingly, no one mentioned Jewish community organisations or online anti-racism groups like the Online Hate Prevention Institute as sources of support. One mentioned using the Australian

Human Rights Commission, with her case taking five years to reach resolution. The remarkable aspect was the pervasiveness of antisemitism as a discourse, not dissimilar to the sense of danger expressed especially by the Indigenous group.

Australia's Muslim Community

Islam was present in Australia before European settlement, through Indonesian fishermen who traded with indigenous people along the northern Australian coast. Thereafter, there has been regular contact with Muslims during the period of the exploration of Australia (Humphrey 1987). In 2011, about 470,000 people self-identified as Muslim.

Our Muslim group, specifically chosen from Islamic communities outside the Middle East, comprised African and Asian Muslims. Islamophobia has become a major area of concern on the Internet, with a number of Australian civil society organisations having been established to monitor and counter it . The group comprises three men and three women: three from Africa (Egypt, Somalia and Eritrea) and three from Asia (Bangladesh and Pakistan). They were aged between 24 and 37 years and are very active on social media. All have Facebook accounts, though one closed hers recently. Amongst them, they also have Snapchat, Tumblr and LinkedIn, and two have their own YouTube channels and blogs for their professional activities. Their primary source of news is online, with one member following key sources on Twitter and others using a range of sources including Al Jazeera, Russia Today and other country of origin online and satellite news services.

The woman who had given up Facebook did so following a period of racial abuse of her and her friends. She felt too exposed:

> Facebook is about your profile, whereas for Twitter, I felt like I could disguise myself as someone else, or didn't have to connect to other people.... I use Twitter as a reprieve. Facebook is very intense and serious, and brings a lot of anxiety in comparison. On Twitter I follow musicals, and Broadway stars.

However, one of the other Twitter users says the opposite:

> I am there for political issues, and feel comfortable being able to communicate in that way. There is a lot of racism, because people have no shame.... and I would get slandered on there.... but that's where I got the truth, specifically from people on the ground who would share what's happening on the minute, every minute.

Another member used Tumblr to connect with a group of artists from her home region, enabling her to write. This writing grew into a creative blog and resulted in a supportive online community whose members interacted with each other.

Another participant cited the use of YouTube to publish an interview with an elderly Muslim revert, and then closed it. "The comments were brutal.... People have no respect.... We ended up deleting and disconnecting the comments."

The group has very different attitudes to posting controversial material. Some will post stories about gay Muslims or Shia and Sunni partying together. Others feel these issues are too hard. "I am very sensitive. I either get really into my little box [i.e. withdraw] or I explode." The same woman described a major internal fight she had in her community on Facebook, where she had been attacked for entering a Sikh temple, as she was a Muslim. Since then, after "my whole community pushed me aside, even if I want to say something, I don't."

Controversial issues about religion included an online debate over whether Muslims should participate in a government Iftar dinner, given the continuing incarceration of hundreds of Muslim asylum seekers by the government. One member said she was accused on Facebook of trying to divide Muslims when they were trying to fit in.

Another woman mentioned an occasion where she had been attacked on Facebook for wearing hijab, on a page devoted to herbs and nutrition, when she contested claims about nutrition she thought were incorrect. A more political group member told of how she had gradually closed off her Facebook page, so that the only people on it now were people she could trust and who had values similar to her own. Yet another demonstrates great persistence on Facebook. "So as a friend, and between friends of friends, I will comment until I'm actually deleted. That's happened on only two occasions."

One woman said she would not reply directly to offensive comments, "I don't want to get into it because I don't have enough experience to deal with it," but did describe how she dealt with some material she found very offensive (an ISIS beheading and unpleasant comments) on Facebook by reporting it. Facebook deleted the material and blocked the commentator from posting to the page.

In discussing the strategies of responses to racist material, one member stressed the importance of philosophical calmness. Another described the escalation in an interaction that made her very angry when a friend's partner who had appeared pro-Palestinian then proved himself to be very anti-Black.

> I was having a conversation with him on something the Israelis were doing in Africa, then he shared that this is what happens when you enter these countries [Black African]. I tried to have a conversation with him about that, and he obviously got very angry.... When I get angry, I feel like I get nowhere. It just caused more problems.

As with the African focus group, the awareness of surveillance affects how members of the Muslim group use social media.

> ASIO contacted me. I was doing community work with certain groups and I was very vocal on Facebook about how I felt....They got all my information from Facebook... so I just felt trapped and thought it was better to get off.

One of the other women offered a different perspective.

> I now curate a Facebook page where there are people who are supportive of me and my cause. Previously, I would share things about women's issues or queer Muslims, but people started calling me in the middle of the night to threaten me. It can really get to the point where it is very dangerous.

Thinking about how to handle these situations she continued:

> Everyone I know has had a phase in their life where they're having online debates until 3am, and having lots of anger. Then you move towards a

> more balanced way of responding to people… in a way that will actually get you what you want. I think anger is really important to begin with, I don't think it's a negative thing, but there needs to be a time when you move on from that and start discussions that are beneficial for the community.

The group recognised the anger reaction as a major challenge.

> I go through those articles [in online conservative newspapers] to read what comments come up, and it is amazing what you see. The level of racism, you can't even put a level to it! You just sit there and think, "do such people really exist in this world?" Which is scary, absolutely scary. It's in my nature to neutralise or try to bring some understanding to some of these people, but what I've realised, is that they'll go onto the next person who's attacked or insulted them in some way, and you'll be completely ignored.

What then is the value of freedom of speech? Several in the group felt that freedom of speech was not equally available—that Muslim communities were far more restricted in their freedom, while majority and White communities were far freer, especially in relation to racism against Muslims. Commenting on former Australian Prime Minister Tony Abbott's criticism of Muslims, the Somali participant argued:

> Freedom of speech is a tool of white supremacy, to keep the system the way it is…. We're all operating under a white supremacist system. It wouldn't make a difference whether you're in England or France or Australia, you still turn the TV on and hear people talking about your religion, talking about your race. I'm a black Australian. I identify with the issues that Indigenous Australians face. Freedom of speech is definitely reserved for one area of the community, and it's a way to silence a whole other area of the community.

The woman who had left Facebook continued on her theme that she had been exposed to harassment on Facebook, and in the minds of the authorities she was associated with more dangerous people, which made her community work more difficult. "What I was doing for the community ended up being questioned, but all the people saying all this hatred stuff were never questioned."

She discussed a friend, an Indigenous Muslim woman, who was blogging about her faith. Her friend also wanted to quit Facebook because of the hatred and abuse she experienced, but found it hard to leave because she also had many supporters:

> Her whole intention was to get people to think, and to get people to understand what's going on, and be a voice for people who weren't going out and speaking. But yet it was tearing her apart.

Another woman interjected: "There are some, especially men, online that are incredibly violent and incredibly hateful in the way that they write things."

While Facebook was the main arena for this discussion, one member pointed to the comments lists on online news sites (in this case MSN).

> The kind of hate I've seen there is scary. You don't want to step in, because once you do, you know for a fact it's not going to end. … I might have someone inbox message me and say eff off, you Muslim, it's happened before. … When I come off social media for a while, I find inner peace. I've got more time to myself.… When I go back on it's overwhelming.

The discussion moved to the issue religion or race, and its role in building community online. The Somali woman has a global friendship network, majority of whom are not Muslim, but are Black.

> I feel like I have more in common with the black community around the world than with Muslims. I don't really like Muslims online, a lot of times.

However, for people who were less willing to get into conflict, Facebook serves as a way of creating communities of like-minded people and excluding people who might not agree or would be offended.

> I have Muslim and Christian Facebook friends, but when two Jewish people approached me to be friends I said no. I am very anti-Israel and pro-Palestine … I didn't want them because I post stuff that is anti-Israel, anti-Zionist. So I don't want to offend friends in public.

The more activist member said:

> Because I use my Facebook page primarily for political views, I don't care if I offend you. If this is something that I feel is right … I don't have a problem having a discussion.

She had been "boxed" (messaged to her inbox) by a Muslim Somali woman who said she had found a recent post incredibly offensive. This conversation happened via email, as they argued back and forth, rather than in the open on Facebook.

One example demonstrates the interaction between the online and offline worlds. One member, a Pakistani community worker, was leading a group project comprising Indian men. She was harassed online by activists from the Pakistani Muslim community who believed she should not interact with Indians or men. Her husband was also harassed. She completed the project but then withdrew from community work. "I was really worried. I have got young kids, and at this age they are vulnerable…How can I fight? What am I fighting for? Is it worth it?"

Another woman was abused because her Facebook profile image showed her with hijab, but she was seen without one in the real world. She was castigated, because she was not "wrapped" like other Muslims.

> They were his exact words. You're not wrapped like the other Muslims. He said incredibly demeaning and borderline violent things about who I am as a person, and my character.

Yet, she also said the real world is where the real danger lurks:

> Real life Islamophobia and racism is probably the scariest thing I've ever experienced in my life…. I was going to my sister's house with a bag, and three guys on the train said to me, "Oh my god, she has a bomb in the bag." No one on the train said anything…. It's so much scarier. Especially when it comes from men that look like they would act out the violence that they're threatening.

Individuals in this group joined various online groups that provided safe spaces to discuss issues such as race in Australia, women of colour, activism and self-care. Close groups are formed when more open discussions generate feelings of threat or experiences of harassment and abuse. The Islamophobia Watch list was named as a community actor, but not as a site that could protect people. "Nothing protects us," added one woman. "Once you step into the online world, there is nothing to guard you. You are there to guard yourself, simple."

There was a desire, though, to see the Human Rights Commission be given more active powers, especially in relation to online bullying:

> To set up guidelines for Facebook, so that Facebook is not setting the laws for itself. A moderator on Facebook telling me what I have reported is not racist enough for it to be taken down isn't good enough.

What Do the Target Communities Share?

As we can see from these discussions, the Internet has become a dangerous and treacherous place, where proactive engagement in public spaces will almost always trigger an attack from some predator looking for the opportunity. It varies, group to group, as to what they perceive as the direct threats and how they develop defense strategies; nevertheless, there are commonalities. All groups think of the dangers of racism as having at least two sources. The first, and less common, comprise organised, sustained attacks using Facebook, Twitter, Instagram and myriad other platforms, by groups for whom landing racist punches on their targets is the primary aim. The second and more common experience is casual racism. This emanates from individuals who are angry about cultural, racial or religious difference and express this rage by finding random targets they can irritate, enrage and intimidate.

While for people of colour, much of the source lies in the White communities, there is also evidence of intergroup racism and harassment. Whichever source generates specific experiences, responses appear be in line with each other. Most people do not want to be harassed or intimidated, so they self-censor online in order to avoid provoking an attack, even if this undermines their right to freedom of speech. On issues they belief require strong advocacy, some will put themselves "out there," hardening their digital skins to the blows that may soon be landed. Sometimes it has proven valuable, where online conversations that initially proved difficult are gradually resolved into more useful explorations of issues from different perspectives.

Discussion

The transcripts of the group discussions were cleaned and combined into files for processing through Leximancer (Biroscak et al. 2017). The key concept that was selected was "racism," with "race" and "racism" as related concepts. A concept map was then generated that identified the most closely linked concepts to these in the conversations by the six focus groups (some 34 cases) (Fig. 5.1).

Commencing with "race" as the primary concept, the Leximancer concept map demonstrated the links between the core concept and other concepts that were linked to them through their shared presence in the

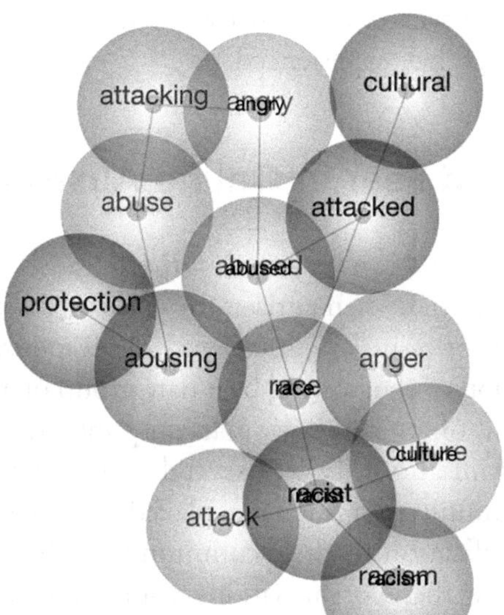

Fig. 5.1 Leximancer Concept map of targets' common experiences of online racism

field. The group members felt angry and abused by their exposure to racism. Moreover, they reiterated that these emotions—anger generating anger—consistently permeated their awareness of hate speech. Often, they would seek to withdraw from engagement, yet were engulfed by the associated emotions and labels.

The experience of targets demonstrates once more the complexity of the universe that has been created by the Internet, and the level of energy and activity that now fills it. It seems impossible to enter that universe and not encounter racism, so there are three implications, which we explore in later chapters. Firstly, individuals join or find communities as a protection, with each community evolving its own (somewhat similar) practices to enable resilience. Secondly, the platforms are coming under increasing pressure to ensure their spaces are safer to navigate and they take more proactive initiatives to help create these spaces, lifting, where possible, the responsibility from the hands of the targets themselves. Thirdly, the regulators are being forced to engage with the platforms to try to ensure civility, even though more libertarian pressures are trying to release constraints on online behaviour.

In the next section we explore how two such "communities of targets" addressed the threats they experienced.

Faith, Hate and Fear in Online Australia

There are similarities that can be found in the actions taken by the NSW Jewish Board of Deputies when antisemitic content was posted on a popular high school website in late 2009, and the response of the Muslim Women's Association to the posting by a hate group of a photo, and subsequent vilification, of a Muslim woman on public transport in Sydney in 2014. Both organisations, two of several representing the Jewish and Muslim faith communities in Australia, played proactive roles in combatting the online vilification of their communities. By tracing the extent and nature of how these organisations countered vilification on the Internet, we can identify the processes through which a resilient practice can be formed by groups targeted by hate.

As a complex, global, electronically powered communication environment that supports access through a range of mechanisms, the Internet embodies the notion of a "networked information environment" (Savolainen 2002, p. 11). The vast majority of organisations and institutions now have an Internet presence, including many operated by faith communities. With the Internet a transborder phenomenon, information and communication moves at lightning speed across the globe. As global faiths, both Judaism and Islam are embedded in such transnational networks and open to whatever marauding harassments discover their local presence.

As already discussed, racism covers the use of terms reflecting claims about the superiority or inferiority of groups defined by their race (in the vernacular sense), ethnicity, skin colour, descent, national origin and immigrant (i.e., non-nativist) status. Racism, for some purposes, covers claims about groups defined by religion where it has similar connotations, such as the NSW discussion of ethno-religious discrimination or the Victorian specification of religion as a protected category within racial vilification laws. Racial discrimination, the practice of excluding, distinguishing, restricting or preferring individuals or groups based on race, colour, national or ethnic origin, can, to all extents and purposes in these two jurisdictions, be regulated by civil law. The International Convention on the Elimination of All Forms of Racial Discrimination refers to such practices as intended to impair recognition of, or access to, human rights and freedom on an equal footing. The case studies we will soon describe point to the complexity and contradictory nature of Internet regulation. Under the Commonwealth civil law (the RDA), for example, religion is not protected. However, Islam is protected in Victoria, but not in NSW. Jewish people are protected as a "race" at the Commonwealth level, and as an ethno-religious group in NSW, as are Sikhs, but not Muslims. In neither case is the protection for religious believers a proxy for blasphemy, which is a protection for belief, and no longer sits on the Australian statute books.

Racial vilification covers material that is likely to affect a member of a targeted community or category of people through (in increasing order of seriousness) offending, insulting, humiliating, intimidating and advocating the expulsion from the country of, or advocating violence against, that person. In different Australian jurisdictions, these may be civil or criminal offences. Conversely, we use the phrase "counter-racism" to refer to the development and application of strategies designed to resist, divert, diminish or reverse the processes of racist harassment and vilification.

Theoretical Assumptions

The use of online media has made it possible for religious communities to create an active e-public sphere of discussion about issues of faith, practice and interfaith relationships. This discourse and the people who engage in it extend beyond conservative or traditional approaches and interpretations of religion. As it is based on electronic instead of face-to-face communication, contributors are able to overcome theological objections to their participation in public life because of their gender or cultural/political considerations. Jewish communities have, for example, found opportunities to address the Middle East conflict with Arab youth (Yablon 2007), though often these topics trigger "flame wars"—a lengthy, to and fro of angry and abusive messages between users of an online forum. Communities of Orthodox and Haredi Jews have been found to use the Internet in ways that spanned religious, communal, personal and educational purposes including the maintaining of websites for theological and social information, while the online presence of these communities was found to reflect the breadth of Jewish religious diversity (Lerner 2009). The negotiated use of the Internet has also been observed among female members of ultra-Orthodox Jews for whom the medium is officially frowned upon as a carrier of secular values (Livio and Weinblatt 2007).

The Australian media's coverage of Islam and Muslims has been described as inherently biased and reliant on stereotypes and hysteria (Kabir 2006). The responses of Muslims to this coverage have been varied. Some Australian Muslims have called for long-term engagement with the mainstream media, while some have rejected the idea of any cooperation with an industry they view as anti-Islamic. Others support the notion of alternative or independent Muslim-run or Muslim-focused media in which alternative discourses about Islam and its role in modern

Australia can be engaged in (Bahfen and Wake 2011). Aly (2007) has looked at the ways in which Australian Muslims both engage in and refute the dominant narratives related to Islam found in mainstream media. For British Muslims, the Internet quickly became an important communication tool for the expression of Islamic identity (Bunt 1999). Members of young Muslim minorities in Western, non-Muslim countries such as the USA use the Internet to engage in the formation of Islamic community and identity on the basis of visibility, individual choice, transnationalism and social ethics (Schmidt 2004). Young people in both Muslim majority countries and members of Muslim diasporas are adept at Internet communication and use and mediate their religion online in the process of obtaining both spiritual and material aims (Echchaibi 2009), such as using the Internet in overcoming geographical boundaries and cultural barriers when seeking a spouse (Lo and Aziz 2009). It is not yet clear what the intersection of Islamic faith communities and the Internet will produce in terms of new Muslim identities, relationships or qualitative experiences—or even if these things will be occurring at all, given that the migration of Muslims online might serve to merely reflect existing offline Islamic cultures and practices (Bunt 2004).

However, online media serve to illustrate the wide range of religious expression that can be found in Muslim and Jewish faith communities (Campbell 2004). Because the majority of racist activity by far-right organisations can be found online (Loewenstein 2014), cyber racism represents a key new front in a socio-cultural ideology war. Efforts to strengthen the communities who are victims of such activity need to be concentrated in the area of engagement online as well as offline. Faith communities can engage in discourse about cyber racism and online hate, particularly where these pertain to attacks on their communities, and be supported in the process of obtaining the required levels of cultural literacy that will permit such engagement (Islamaphobia online 2016). Pushback by communities against hate speech and vilification has an important role to play in resisting being pressed into a position of victimhood, and will avoid more dramatic and dangerous actions in the real world. Our research suggests that affirming social identity and strengthening cultural capital in the face of online hate is difficult or impossible without a community involving itself in the process of developing resilience to cyber racism.

Case Studies: Response and Resilience

The approach taken to analyse the two examples here is an inductive approach, using a socio-culturalist perspective. In this approach, the functions of the Internet in social life are emphasised, and social factors are acknowledged as an influence on the production and reception of content found online. It, therefore, is suited to the nature of the research, which explores how two religious community organisations reacted to, and attempted to, combat racism as experienced by members of their faith groups online. The theory this research is attempting to generate is how resilience by religious communities in the context of online racism might be defined and implemented.

The qualitative methodology selected for this project was textual analysis, which is used to evaluate how the faith groups reacted or responded to the racism directed at their members. Textual analysis was chosen because of its potential to explore the ways in which these responses construct and represent their communities. It is a methodology that provides depth of analysis crucial for inductive qualitative research and allows for an understanding of the role played by online content in cultural construction.

Case Study 1: The NSW Jewish Board of Deputies' Response to Online Antisemitism

In December 2009, postings on the popular high school students' online forum, Bored of Studies, contained antisemitic remarks such as "Kill all Jews" and "I hope Iran nukes them big time," in addition to posts that discussed the locations of Sydney synagogues and provided instructions on how to make Molotov cocktails (Edelson 2009). Following complaints about the violent and threatening nature of the posts to the publishers or owners of the site, the users who were responsible were banned from participating in the forum, and the offending content was removed.

The organisation that was instrumental in bringing the vilification to the attention of the forum's administrators was the NSW Jewish Board of Deputies, the peak Jewish body in NSW. In an interview with the Jewish news and community site J-wire in 2009, CEO Vic Alhadeff explained that the motivation for attempting to get the content removed was an apprehension that such attitudes could be expressed on Web forums (which represented, at the time, a popular medium for discussing news and opinion), especially one that was primarily used by high school students. Such comments, he said, constituted racial hatred, which the Board of Deputies was concerned about whether it was directed at Jews or any other group (Ross 2009). Although he praised the actions of the website operators in quickly banning the offending users and removing the content, Alhadeff noted that preventative measures were preferable, such as tighter monitoring or controlling of the forum's discussion space. "Taking action to moderate after the comments are posted is too little, too late, because the damage has been done" (Ross 2009: np).

The Muslim Women's Association and an Anti-Islam Hate Group

In April 2014, several Australian media outlets reported on the intimidation and online vilification campaign run by the Australian Defence League (ADL), involving photographing and filming Muslim schools and Muslim women for the purpose of posting the content online in order to invite derogatory and abusive comments. The effect of the campaign extended beyond the women or schools targeted to encompass many members of the local Muslim community.

One Sydney-based Muslim woman described to a *Guardian* writer and prominent Jewish author the fears held by many within the community following the mainstream media reporting of the incidents. The ADL simply reacted by reposting the photo onto Facebook along with a status asking its members to take photos of random Muslim women to humiliate them online… "Hearing that definitely made me anxious. My commute home that day was very uncomfortable. I kept glancing around me, keeping an eye out for anyone who might be trying to snap a photo of me due to my hijab" (in Loewenstein 2014).

Students at a Muslim school that had been filmed said they felt threatened, while the principal alerted the Australian Security Intelligence Organisation (ASIO) and the police (Rubinsztein-Dunlop 2014). Following the Facebook posting of a photo, taken without her knowledge, of a mother of three children and the abuse and vilification of the victim online, the NSW police launched an investigation into a number of ADL members (Olding 2014), and the photo was taken down, although the group set up another of its myriad social media groups (Fleming 2014).

In interviews with Australian media outlets, one victim said she had to attend counselling following the cyberattacks, and expressed her fear of taking public transport as a consequence. She also indicated that she had reached out to the Muslim Women's Association (MWA) for support.

According to Maha Abdo, the executive officer of the MWA, the organisation reacted on two fronts by "supporting the young woman and taking up this issue of cyber bullying and intimidation with the appropriate authorities." Abdo explained that the issue of harm minimisation arising from the online vilification of the woman (and threats to blow up her child's school) extended beyond the individual victims, adding that the welfare of the woman was not her organisation's only focus. The MWA had to look at the impact of the attack on the women of the Muslim community in Australia as a whole.

> The public humiliation and vilification on the basis of someone's faith is not acceptable…It is defamatory and hurtful. We believe there is a real concern for the welfare of not just this young mother, and the fact that she is now too afraid to catch the train to get to work because it [the online stalking or harassment] might happen again. But we are worried about the other adverse effects this may have on other Muslim women in Australia. (Maha Abdo, September 2016, personal communication)

She described the actions of the hate group as an attempt to vilify "not only the woman herself, but also Australian Muslim women generally." Abdo said that she believed the online harassment of the woman was not an issue of freedom of speech but of "security and safety…that is owed to us in a multicultural nation like Australia."

Discussion

The two case studies examined here illustrate that, given the rapid pace at which online communication evolves (from centralised spaces such as online forums to extremely decentralised individual content on social media), legislative and other approaches to combating hate online tend to play a game of catch-up, and that older frameworks for dealing with hate speech (such as the criminalisation and restriction of such speech or limits on its distribution) need adaptation. Here, we note the prime difference in the Australian contextual interpretation of free speech as a nation-state where no bill of rights exists and where free speech is only alluded to in the constitution (and even then, limited to political matters). Although Australia—like the USA—supports international treaties designed to broadly protect free speech, the extent to which free speech is privileged and accorded legal protection in Australia differs markedly from that in the USA (Gelber 2002).

Comparative studies of the US and European approaches to hate speech in the online sphere have concluded that some nation-states (such as France or Germany) (Akdeniz 2001) have more vigorously attempted to enforce laws and constraints, with varying degrees of success. These studies also found that legal discussion of such cases often rests on whether it is possible (given the globalised nature of online communication means) to determine where the hate speech took place or was disseminated, and therefore whether such actions were subject to the relevant laws of the nation-state (Nemes 2002; Timofeeva 2002; Douglas-Scott 1998).

The antisemitic hate in the first case study constituted an example of hate content on a website that was locally run, within a geographically and demographically defined target user base (students in their final year of secondary schooling in the Australian state of NSW). The group that brought the hate speech to the attention of the website operators could point to the specific state and national laws that such comments could have been demonstrated to have broken, had their approaches initially to the website operators been rebuffed.

By comparison, in the second case study, by the time the MWA attempted to support one of the victims of the ADL online hate campaign in 2014, the most popular methods of online communication had shifted from website forums to social media. Here, it is arguably much harder—though not impossible—to hold site operators to account over hate speech. Groups such as ADL are constantly playing cat and mouse with social media networks in their attempt to establish a presence despite being banned, having pages deleted but constantly developing new pages. This indicates working with social media website operators to report users who engage in spreading hate works to an extent, despite sites such as Facebook operating on seemingly arbitrary definitions of what constitutes "hate speech" or incitements to violence. As Oboler (2008) points out, social network sites in the age of WFeb 2.0 are not prepared to deal with the deluge of cyber hate their platforms inadvertently paved the way to.

The lessons that can be drawn from these case studies suggest that there are a few key approaches to take when responding to racism or vilification online:

a) Approach the platform responsible for publishing the hate online such as website operators or social networks, irrespective of the chances of success. Most major social networks now have built in reporting mechanisms as a result of public pressure and criticism;
b) Document and report instances of cyber hate to the relevant authorities and organisations who specialise in researching online hate groups, such as the Online Hate Prevention Institute or specific community-based groups and
c) Promote the accessibility of community organisations (such as the two involved in the case studies discussed here) that can lobby on behalf of victims and provide support to counter the emotional or psychological negative effects of hate and vilification online.

To properly constitute "resilience" by the victims of cyber hate, we argue that these approaches need to be combined and pursued in conjunction with community education about the threat of online hate. The antisemitic hate on the high school Web forum in the first case study took place against a backdrop of rising violence targeting Jews (Ross 2009), while the second case involving the online and offline harassment of Muslim women and schools occurred in the context of the well-documented spike in Islamophobia and attacks against Muslims in recent years (e.g., Bouma 2011; Ho 2007). One of the key skills required to use the Internet has now become how to identify, report and deflect hate speech in order to minimise its impact and build individual and group resilience.

The legislative framework controlling hate speech erects, as it were, a fence within which people rightfully feel they would be protected. In the case of cyber racism, however, this fence is neither well embedded nor well defended. In a situation where social media providers are reluctant to proactively police the posts on their platforms, and the state seeks to limit its intervention in the name of free speech, the onus then shifts towards civil society. Civil society organisations, therefore, are very important groups in combating this problem. They can collaborate with well-meaning providers to have the offending content removed, they can impel the state to use the law where it exists and they can engage with industry to act against hate speech. Also, critically, they can support victims to resist the negative impact that hate speech might have on their lives. They can help the community to identify race hate, understand its impact, deflect its trajectory, report its presence and recover from its pain.

The next chapters take up these themes as we explore questions of resilience, narrative and regulation.

References

Akdeniz, Y. (2001). Case Analysis of League Against Racism and Antisemitism (LICRA), French Union of Jewish Students, v Yahoo! Inc.(USA), Yahoo France, Tribunale de Grande Instance de Paris, Interim Court Order, 20 November 2000. *Electronic Business Law Reports, 1*(3), 110–120.

Aly, A. (2007). Australian Muslim Responses to the Discourse on Terrorism in the Australian Popular Media. *Australian Journal of Social Issues (Australian Council of Social Service), 42*(1), 27–40.

Amend, A., & Morgan, J. (2017). Breitbart Under Bannon: Breitbart's Comment Section Reflects Alt-Right, Antisemitic Language. *Hatewatch*. https://www.splcenter.org/hatewatch/2017/02/21/breitbart-under-bannon-breitbarts-comment-section-reflects-alt-right-antisemitic-language. Accessed 23 Feb 2017.

Anglin, A. (2016). A Normie's Guide to the Alt-Right. *The Daily Stormer*. http://www.dailystormer.com/a-normies-guide-to-the-alt-right/. Accessed 15 Jan 2017.

Australian Arabic Council. (2003). *Ethnicity & Crime in NSW: Politics, Rhetoric & Ethnic Descriptors*. http://www.aac.org.au/report1/. Accessed 30 Apr 2015.

Australian Bureau of Statistics. (2008). *'Census 2006 – People Born in Africa', Commonwealth of Australia*. http://www.abs.gov.au/AUSSTATS/abs@.nsf/Lookup/3416.0Main+Features32008. Accessed 26 Apr 2009.

Australian Bureau of Statistics. (2012). 2011 Census Counts—Aboriginal and Torres Strait Islander Peoples. *2075.0—Census of Population and Housing—Counts of Aboriginal and Torres Strait Islander Australians, 2011*. http://www.abs.gov.au/ausstats/abs@.nsf/Lookup/2075.0main+features32011

Australian Bureau of Statistics. (2013). People in Australia Who Were Born in China. *2011 Quickstats Country of Birth*, ABS. http://www.censusdata.abs.gov.au/census_services/getproduct/census/2011/quickstat/6101_0. Accessed 23 Sep 2016.

Australian Government. (2014). *The People of Australia: Statistics from the 2011 Census*. Canberra: Department of Immigration and Border Protection. https://www.youtube.com/watch?v=OctRQoCodVw

Bahfen, N., & Wake, A. (2011). Media Diversity Rules: Analysing the Talent Chosen by Student Radio Journalists Covering Islam. *Pacific Journalism Review, 17*(2), 92–108.

Biroscak, B., Scott, J., Lindenberger, J., & Bryant, C. (2017). Leximancer Software as a Research Tool for Social Marketers: Application to a Content Analysis. *Social Marketing Quarterly, 23*(3), 223–231.

Bouma, G. D. (2011). Islamophobia as a Constraint to World Peace: The Case of Australia. *Islam and Christian–Muslim Relations, 22*(4), 433–441.

Bunt, G. R. (1999). Islam@Britain.net: 'British Muslim' Identities in Cyberspace. *Islam and Christian- Muslim Relations, 10*(3), 353–362.

Bunt, G. R. (2004). *Religion Online: Finding Faith on the Internet*. London/New York: Routledge.

Campbell, H. (2004). Challenges Created by Online Religious Networks. *Journal of Media and Religion, 3*(2), 81–99.

Deare, S. (2016, July 13). Federal Election: Same Sex Marriage and Gay Rights Targeted in How-to-Vote Pamphlets Written in Chinese. *Northern District Times*. Available at: http://www.dailytelegraph.com.au/newslocal/the-hills/federal-election-same-sex-marriage-and-gay-rights-targeted/news-story/4b930ea4141f2e310d95a333e52be62c

Dick, T. (2005). Academic Stirs Fight Over Race. *Sydney Morning Herald*. Available at: http://www.smh.com.au/news/national/academic-stirs-fight-over-race/2005/07/15/1121429359329.html

Douglas-Scott, S. (1998). Hatefulness of Protected Speech: A Comparison of the American and European Approaches. *William and Mary Bill of Rights Journal, 7*(2), 305–346.

Echchaibi, N. (2009). Hyper-Islamism? Mediating Islam From the Halal Website to the Islamic Talk Show. *Journal of Arab & Muslim Media Research, 1*(3), 199–214.

Edelson, D. (2009, December 31). Australian Youth Site: Kill All Jews. *Y-net*. http://www.ynetnews.com/articles/0,7340,L-3827855,00.html. Accessed 4 Jan 2017.

Ellemers, N., & Barreto, M. (2015). Modern Discrimination: How Perpetrators and Targets Interactively Perpetuate Social Disadvantage. *Current Opinion in Behavioral Sciences, 3*, 142–146.

Fleming, A. (2014, January 29). Who Are the Australian Defence League? *New Matilda*. https://newmatilda.com/2014/01/29/who-are-australian-defence-league. Accessed 4 Jan 2017.

Gelber, K. (2002). *Speaking Back: The Free Speech Versus Hate Speech Debate*. Amsterdam/Philadelphia: John Benjamins Publishing.

Graham, D. (2014). *The Jewish Popualtion of Australia: Key Findings from the 2011 Census*. Sydney: JCA.

Ho, C. (2007, August). Muslim Women's New Defenders: Women's Rights, Nationalism and Islamophobia in Contemporary Australia. *Women's Studies International Forum, 30*(4), 290–298.

Humphrey, M. (1987). Community, Mosque and Ethnic Politics. *Journal of Sociology, 23*, 233–245.

Jakubowicz, A. (2010). *Australia's Migration Policies: African Dimensions Background Paper for African Australians: A Review of Human Rights and Social Inclusion Issues*. https://www.humanrights.gov.au/publications/african-australians-project-australia-s-migration-policies-africandimensionsaustralia: Australian Human Rights Commission.

Jakubowicz, A. (2011). Empires of the Sun: Towards a Post-Multicultural Australian Politics. *Cosmopolitan Civil Societies: An Interdisciplinary Journal, 3*(1), 65–86.

Jakubowicz, A. (2016). First the Word, Then the Deed: How an 'Ethnocracy' Like Australia Works. *The Conversation*. https://theconversation.com/first-the-word-then-the-deed-how-an-ethnocracy-like-australia-works-69972. Accessed 12 Mar 2017.

Kabir, N. (2006). Representation of Islam and Muslims in the Australian Media, 2001–2005. *Journal of Muslim Minority Affairs, 26*(3), 313–328.

Killalea, D. (2016). Nova Peris Owns Racist Troll Who Posted Shocking Facebook Tirade. *news.com.au*. http://www.news.com.au/technology/online/social/nova-paris-owns-racist-troll-who-posted-shocking-facebook-tirade/news-story/b234d9c204a77cb50c1ffbb8f495f33e

Lerner, H. G. (2009). Internet Resources for Researching Orthodox Judaism. *Journal of Religious & Theological Information, 7*(3–4), 161–169.

Livio, O., & Tenenboim Weinblatt, K. (2007). Discursive Legitimation of a Controversial Technology: Ultra-Orthodox Jewish Women in Israel and the Internet. *The Communication Review, 10*(1), 29–56.

Lo, M., & Aziz, T. (2009). Muslim Marriage Goes Online: The Use of Internet Matchmaking by American Muslims. *Journal of Religion and Popular Culture, 21*(3), 5–5.

Loewenstein, A. (2014, May 6). Australia's Far-Right Fringe Groups Must Not Be Allowed to Become Mainstream. *The Guardian*. https://www.theguardian.com/commentisfree/2014/may/06/australias-far-right-fringe-groups-must-notbe-allowed-to-become-mainstream. Accessed 4 Jan 2017.

Mannix, L. (2015). Woman Charged After Facebook Call to Burn Down Mosques Goes Viral. *The Age*, Fairfax. http://www.theage.com.au/victoria/woman-charged-after-facebook-call-to-burn-down-mosques-goes-viral-20151222-gltqz9.html. Accessed 13 Apr 2017.

Mantle, G. (2015). *Killing Indigenous Australians Is Not a Game!* https://www.change.org/p/amazon-killing-indigenous-australians-is-not-a-game

Nemes, I. (2002). Regulating Hate Speech in Cyberspace: Issues of Desirability and Efficacy. *Information & Communications Technology Law, 11*(3), 193–220.

Oboler, A. (2008). The Rise and Fall of a Facebook Hate Group. *First Monday, 13*(11). https://doi.org/10.5210/fm.v13i11.2254.

Olding, R. (2014). *Anti-Islamist Nathan Abela Charged Over Shooting at His Home.* http://www.smh.com.au/nsw/antiislamist-nathan-abela-charged-over-shooting-at-his-home-20140418-36viw.html. Accessed 4 Jan 2017.

PV Reviews. (2015). *Survival Island 3: Australia*. YouTube Android Gameplay. https://gaming.youtube.com/watch?v=pH9GJt0J0qs. Accessed 12 Nov 2016.

Ross, M. (2009). *Student Race Hate Posts Anger Jewish Group*. http://www.abc.net.au/news/2009-12-29/studentrace-hate-posts-anger-jewish-group/1192842. Accessed 4 Jan 2017.

Rubinsztein-Dunlop, S. (2014). *Inside the Far-Right Group Spreading Fear Through Muslim Communities*. http://www.abc.net.au/news/2014-04-21/inside-the-far-right-group-spreading-fear-through/5402494. Accessed 4 Jan 2017.

Savolainen, R. (2002). Network Competence and Information Seeking on the Internet: From Definitions Towards a Social Cognitive Model. *Journal of Documentation, 58*(2), 211–226.

Schmidt, G. (2004). Islamic Identity Formation Among Young Muslims: The Case of Denmark, Sweden and the United States. *Journal of Muslim Minority Affairs, 24*(1), 31–45.

Semmler, C. (2014, December 12). Could the Sydney Siege Have Been Predicted and Prevented? *The Conversation*. https://theconversation.com/could-the-sydney-siege-have-been-predicted-and-prevented-35608. Accessed 17 Jan 2017.

Timofeeva, Y. A. (2002). Hate Speech Online: Restricted or Protected? Comparison of Regulations in the United States and Germany. *Journal of Transnational Law & Policy, 12*(2), 253–285.

Yablon, Y. B. (2007). Feeling Close From a Distance: Peace Encounters Via Internet Technology. *New Directions for Youth Development, 2007*(116), 99–107.

6

Racist Narratives Online

Introduction

By the time Australia Day was commemorated across most of the country in January 2017, the debate over Australia's identity and the interpretation of its founding moments had reached a crescendo. While most of the country had a public holiday that celebrated the arrival of the British armed forces and their possession of the lands of the Eora people in 1770, many Indigenous people and their supporters were commemorating what they saw variously as Invasion Day and Survival Day.

This fundamental tension over the narrative of the nation marks much of the conflict around questions of race and the role of racism in Australian public discourse. These two archetypes of Australian national identity constantly confront each other (Bliuc et al. 2012; Fozdar et al. 2015). One narrative celebrates the emergence of a democratic, prosperous and multicultural modern nation in which the egalitarian ethos flourishes and where racism is the refuge of the uneducated, the marginal and those bypassed by global modernity. The counter-narrative remembers Indigenous people as the First Nations people, acknowledges their lands were stolen and their peoples massacred and sees the racism that could

A. Jakubowicz et al., *Cyber Racism and Community Resilience*, Palgrave Hate Studies,
DOI 10.1007/978-3-319-64388-5_6

justify such dispossession and its denial as a continuing and relentless mark of shame in what remains an ethnocracy (Jakubowicz 2016).

Overlaying this narrative binary, the articulation of a White power and nationalist view of the past and present penetrates social media and the wider Internet. While racist speech is condemned in Australian public discourse, it is not prohibited by Australian law. Thus, racism on social media, as we have seen in previous chapters, has become almost all-pervasive: a maelstrom of opinion, abuse, truths and post-truths. A narrative frame of reference enables this chapter to explore more deeply the basis for the arguments that feed the growth of cyber racism. The role of narrative may be to assist people to make sense of the world around them (Mattingly et al. 2008). Narratives both reflect and influence this understanding or world view. Such an understanding is therefore not static, but rather constantly changes and occurs at both individual and societal levels. A narrative analysis can focus on the content and context of cyber racism arguments and explore how they attempt to influence others.

The first part of this chapter uses the Australian example to demonstrate the development of contrasting national identity narratives and their interpretation on social media. This includes contextualising the narratives in terms of the historical and political development of Australia as a multicultural nation. The second part of this chapter uses examples from Australia and around the world to explore the discursive strategies employed by exponents of cyber racism to promote their version of national identity narratives. These complementary approaches, drawing from social psychology, sociology and anthropology, aim to give insight into the dynamic through which cyber racism can be legitimised through narratives about national identity.

National Identity Narrative Development and Context

The online contestation of national identity occurs within existing narratives of inclusion and exclusion (Connelly 2016). Such narratives are primarily a debate over who has the power to make decisions about which races, cultures and religions should be included in the national identity. Social actors use inclusion and exclusion arguments to justify and advance particular ideologies that support their claim to power. These narratives are also contextual and evolving, depending on the history and politics of a nation.

In Australia, the current claim to power began with the narrative of colonisation. Colonisation in itself is based on an overarching ideology that the White British colonisers had the right to invade Australia (based on the doctrine of terra nullius) and therefore claim power over decisions of inclusion and exclusion (Council for Aboriginal Reconciliation 2001). The colonisers also brought a nationalistic view of Australian society that privileges people considered as fitting into the category of "White Australian." Indigenous Australians and non-White European migrants were originally judged as less civilised, regarded and treated as inferior and considered as not belonging in the category of White Australian (Hage 2002; Perera 2005). However, as migration increased, the category of White Australian was contested and subsequently expanded to include some migrants based on their willingness to adopt the dominant White Australian culture (Perera 2005).

The definition of dominant White Australian culture is also contested. Initially, Australian culture primarily referred to inherited (White British) nationalist ideological values such as democracy, freedom and the Christian religion (Hage 2002). However, over time, values such as egalitarianism and inclusiveness are seen by some as potential additions to the definition of Australian cultural values. This view appears to resemble ideologies of cosmopolitanism, which recognises the value in engaging with migrants and Indigenous Australians and their cultural differences (Jakubowicz 2011; Skrbis et al. 2004; Werbner 2008). However, while many Australians may appear to hold these values, their view of inclusiveness of other cultures is not straightforward.

A recent survey showed that over 83% of Australians agree that "multiculturalism has been good for Australia" (Markus 2016, p. 50). This survey also found that a majority of Australians believe in an approach to multiculturalism that more closely resembles cosmopolitanism as a two-way process involving both Australians and migrants productively engaging with cultural differences. In reality, people with different cultural backgrounds engage on a daily basis in Australia, through everyday interactions in their neighbourhoods, at schools and so forth (Ho and Jakubowicz 2013). In this way, cultural difference is regarded as an unremarkable part of everyday life, and the potential for cultural inclusiveness is increased.

However, Markus (2016, p. 51) also found 28% of Australians believe migrants should "forget their ethnic and cultural backgrounds as soon as possible." This aligns more with assimilationist views that argue there should be one (White) culture in Australia and that cultural diversity is dangerous (Salter 2012). For Australians who believe in assimilation, migrants and Indigenous Australians who appear unwilling or unable to adapt to Australian culture may be considered as a threat to the dominant White Australian culture. This supposed threat may be used to support nationalistic ideologies and a claim to power that legitimises exclusion and racism.

Fantasising Australian Identity Narratives on Social Media

The struggle over what best represents a "true" Australian national identity is becoming increasingly visible in the digital realm. The spread of Facebook's social media pages has provided ideal locations for creative interpretations of Australia's history and identity to be constructed, refined and circulated. The authors of one nationalistic Australian Facebook page claim authority based on the belief that they represent Australians who wish to preserve an identity based on a specific White version of the "Australian way of life" (Facebook 2017a). The narrative on this page is based on an underlying ideology that not all migrants (or Indigenous Australians) should be included in their definition of Australian identity. The authors argue that people with desirable Australian cultural values such as patriotism should be included in their version of the Australian way of life, while some migrants and others with different cultural values are portrayed as threatening and are excluded. In particular, refugees and Muslims are the target of racist vilification and ridicule on this page on the basis that they pose an unacceptable threat to Australian culture (Connelly 2016).

This narrative is also spreading beyond followers of this Facebook page to the wider community. The ease of interaction on Facebook means that posts that are liked, shared and commented on by the followers of this page not only show up in their news feed but may show up in their friends' news feeds as well (depending on the privacy options selected), and friends of friends, and so on. This potentially increases the number of people who are exposed to this ideology exponentially. One post that made a link between a Muslim that was accused of planning a terrorist attack in Australia and all Muslims attracted 120,000 likes despite the number of followers of the page (in January 2017) being 84,000. This demonstrates that Facebook users are becoming more exposed to exclusion narratives that legitimise cyber racism.

In contrast to this exclusion narrative, other Facebook pages put forward ideologies that are more culturally inclusive. The aim of one Facebook page with an alternative view is to welcome all refugees and migrants to Australia regardless of the country of origin or religion (Facebook 2017b). Posts on this page encourage engagement with cultural difference. For example, one post refers to a billboard showing two Muslim Australian girls celebrating Australia Day. This post encourages engagement with an alternative view of Muslim Australians as non-threatening and worthy of inclusion in the Australian identity. One comment on this post referred to the lines of the song *I am Australian* by The Seekers: "We are one, but we are many and from all the lands on earth we come" to make their point. Although this post received only just over 1500 likes, the total likes for the page is almost 130,000, showing that this narrative is also gaining momentum.

The current narrative about who is included or excluded in Australia continues as a contested domain. Both of these Facebook pages have the power to delete comments that do not align with their version of the inclusion/exclusion narrative (and often do so). Their pages therefore stand as representative of two delineated narratives about Australian identity. On one side of the argument are people who align more with a White nationalist assimilationist view that regard their version of White cultural values as superior and call on the exclusion of those who do not align with those values, while on the other side are people who align more with the inclusiveness and cultural engagement of cosmopolitanism. As the research by Markus (2016) shows, while most Australians support multiculturalism, some are unsure about cultural inclusiveness and engagement. These Australians may be influenced by White nationalistic ideologies that use cyber racism to assert their claim to authority about who should be included in the narrative about Australian identity.

National Identity Narrative Construction and Cyber Racism

The narrative construction of national identity on social media occurs against the backdrop of the growing influence of cyber racism. At its most general, cyber racism has been defined in the literature as racism expressed or communicated via the Internet (Bliuc et al. 2016). Cyber racism is sometimes conceptualised in more strategic terms as "the use of the World Wide Web by white supremacist groups in a manner directed towards identity building, grievance framing and efficacy" (Adams and Roscigno 2005, p. 763).

Racist narratives are often generated by organised groups aiming to gain socio-political leverage or popular support by taking advantage of the affordances made possible by the widespread use and development of new communication technologies. Indeed, research suggests that racist, extremist groups were always early adopters of technologies (Back 2002), quickly becoming adept in using them to advance their group's goals (e.g., expanding their support base, disseminating their radical ideologies and increasing visibility and gaining economic benefits through selling their branded merchandise, see Gerstenfeld et al. 2003).

However, it is not only the White supremacist groups that are known to propagate racist discourses in the online world. "Lone wolves" or isolated individuals with racist views can be also prolific in disseminating racist discourse on the Internet. These are often disguised as patriotic defenders of a particular nation's values who express "moderate" views on forums and news commentary websites. Our recent research on racism in

Australian online forums shows that in many cases racists can develop complex arguments—built, for instance, around moral disengagement strategies—that they use to legitimise and mainstream their views (Faulkner and Bliuc 2016).

Narratives of racism are more often than not hotly contested, so for researchers, one benefit of the widespread use of the Internet and social media is that they can explore how arguments are developed across both sides of the debate. That is, online debates around racism represent rich and reliable sources of spontaneously generated data that reflect current societal issues and divisions. Researchers in many disciplines are increasingly using these large and readily available data sources to analyse inter and intragroup dynamics in many domains including cyber racism (Lazer et al. 2009).

Categorising Racist Discourse in Cyber Racism Narratives

Cyber racism narratives can be better understood by delving more deeply into the source of the communication and the nature of the discourse. If the source is a "White Power" or "alt right" site, where racist views are expressed alongside other anti-progressive, more general-type messages, the type of racist discourse most likely to be encountered is deliberate or **blatant**. Sometimes more sophisticated or "socially creative" arguments are made on such sites, and there is a tactical dimension to their rhetoric. Douglas et al. (2005) show that, while their (exclusive) attitudes are blatant, the arguments supporting these attitudes can be masterfully crafted to deceive (e.g., by creating the impression they are socially justified and broadly shared by many people). This type of racist discourse is likely the predominant one in the current context of widespread technology use and highly tensioned intergroup relationships. These sites are currently proliferating so fast, it is virtually impossible to estimate their numbers (Perry and Olsson 2009).

Alternatively, when racist discourse is communicated by a single individual (as, for example, on a forum or a news commentary website), the type of racism expressed is often of a **subtle** nature, possibly due to fear of exposure or retaliation in a public forum. There is an opportunistic element to this type of communication made by a single individual.

The nature and expression of contemporary racism has been shaped by both socio-economical and historical factors (societal forces), and individual factors (Dovidio 2001). Social psychologists talk about racial prejudice as shaped by historical factors, that is, contemporary racism is seen as developing from normal social process, and it is described as a multidimensional construct that includes implicit (unconscious) elements as well as subtle expressions, which are regulated by the social context. The goals of racist discourse can be also mapped along the same dimensions to include goals that are:

- Intentionally racist—where the message is communicated with the explicit goal to hurt, harm or discredit a member of a different ethnic group or a whole group;
- Strategically racist—where racist discourse is used to make a point that is part of a broader argument designed to persuade the public, or gain political support and
- Unintentionally racist—where the communicators are not aware or are uncaring that their arguments have a racist impact.

Cyber racism narratives may be categorised as blatant or subtle, intentional, strategic or unintentional depending on the source of the communication and the aims of the authors. The following sections will further explore the discursive strategies used by racists to achieve their goals.

The Use of Identity Rhetoric to Target Minority Groups

Conceptually, cyber racism integrates:

a) the communication medium (the Internet);
b) the content of this communication (racial hatred);

c) the sources of the communication/the groups or individuals that generate the content (White supremacist and others) (see Chap. 2) and
d) the targets of the communication (see also Chap. 5).

In terms of the targets of racist communication, racism occurs against almost all existing ethnic groups around the world, including, for example, anti-Indigenous (in Australia and Canada, etc.), antisemitic, anti-African American (mostly in the USA), anti-Indian and Pakistani (in UK), anti-Hungarian (in Romania), anti-Arab and anti-Muslim (in much of Europe). In Australia, particular minority groups have traditionally been more targeted than others, including Asians and Indigenous Australians. However, the current socio-political climate contributes to an increasingly strong wave of anti-Muslim sentiment. Middle Easterners, North Africans, South East Asians and others are now most frequently attacked on religious grounds (see Chap. 5).

The justification for the targeting of these groups is based on "identity rhetoric." Analyses of both racist and anti-racist identity rhetoric (Bliuc et al. 2012; Faulkner and Bliuc 2016) have documented the emergence of specific narratives designed to align contrasting identities (e.g., White Australian/racist vs multicultural Australian/anti-racist) to an "ideal," representative Australian identity. Narratives on both sides are underpinned and shaped by particular key dimensions, including identity content, in-group position, norms, values and behaviours that are all portrayed as being representative of a true Australian identity.

Research by Bliuc and colleagues (Bliuc et al. 2012) on the role of contrasting opinion-based groups in the context of the 2005 Cronulla riots in Sydney illustrates this point well. The Cronulla riots were instigated by a crowd of approximately 5000 mostly white, English-speaking young men who were initially protesting against what they saw as the unacceptable behaviour of a group of young Lebanese men at Cronulla beach (Noble 2009). This protest turned into an ethnically motivated riot where the crowd went on a rampage, attacking anyone of Middle Eastern appearance. An analysis of online rhetoric around these riots from both supporters of the riots (promoting racist and exclusive views of Australian identity) and opponents of the riots (promoting anti-racist and inclusive views of Australian identity) shows how rhetoric strategies are used by opinion-based groups with the purpose of achieving dominance in society.

This argument is consistent with the concept that one way to achieve societal dominance and consensus about who should be included as part of the national identity is by imposition of the definition of one group's category over alternative definitions (Dixon and Wallwork 2004). This involves constructing and reconstructing the relationship between the nation and identity in order to establish "a consonance between the nature of the categories used by a speaker and the constituency that the speaker is attempting to mobilise" (Reicher and Hopkins 2000, p. 75). The use of identity rhetoric by cyber racists aims to situate particular identities as a preferable part of the national identity narrative while undermining identities in targeted groups.

Building Cyber Racist Communities

The promotion of identity narratives that use cyber racism to undermine targeted groups is reliant on the narrative being spread and adopted by members of the general public. Research by Bliuc et al. (2016) has identified two key strategies used to achieve this:

1. building online communities of support through intragroup interaction (i.e., engaging like-minded people and the broader public through the use of images, videos, emoticons and likes) and
2. developing and putting forward sophisticated ideological arguments through intergroup interaction (i.e., engaging the other side in debates about race in virtual locations such as blogs and discussion sites that enable the communication and debate of complex and lengthy arguments).

Online interactions seek to build consensus and create a sense of community for people who share racist views to varying degrees. They are often carefully designed to create the impression that they promote widely supported views: markers of support are used to legitimise racist views. This strategy is primarily used to disseminate simple but powerful messages to the online audience, effectively validating already existing views (therefore strengthening the commitment of people who are already racist).

For example, if we examine the Twitter account of Pauline Hanson, leader of the Australian political party One Nation, we can see how videos, images and slogans are effectively used to further develop and endorse her radical anti-Muslim stance that preceded her 2016 election to the Australian Parliament. The following simple and powerful message posted in mid-December 2016 (after the Berlin Christmas market attack) received 423 replies, 835 re-tweets and 1.7K likes: "You only have to look at the horrors being committed by Islamic terrorists today in Europe to understand why we must ban Muslim immigration." Although brief, this message is opportunistic and in line with Hanson's position. Hanson uses it to further strengthen her stance and sustain her elaborate anti-Muslim narrative. It represents an effective way to capitalise on the public's emotions elicited by this event (e.g., disbelief, grief and, in particular, fear and anxiety) to achieve sympathy for the anti-Muslim immigration cause. Hanson is strategic in engaging with existing and potential supporters at an affective rather than at a cognitive level (it is likely that an elaborated, logical and well-informed argument in support of her stance would have been far less effective here).

Developing Sophisticated Ideological Arguments

The second strategy entails the engagement of the public (both supporters and potentially opponents of racism) through much more well-developed and complex narratives. The use of this strategy can be seen as more concerning from a public interest point of view because it has the potential to influence public opinion by being more persuasive to those who feel threatened by the "other" but do not yet have well-formed racist views. The opportunity to present elaborate and detailed narratives that often exploit the very salient tensions in Western societies across the world is particularly daunting. The effectiveness of this strategy rests on its ability to feed on people's insecurities and fears and proposes alternatives to current crises—alternatives which might be impractical to achieve, or even utopian (e.g., Donald Trump's "Muslim ban" (Killalea 2017)).

Opportunities for social influence within the broader public arena can be seen on blogs that allow comments from both supporters and opponents of racism. For example, if we look at one of the alt-right blogs on BlogSpot, we discover carefully constructed arguments designed to convince readers that the position presented is not only legitimate, but also morally justified. In a blog post by Colin Liddell (a regular alt-right blog contributor) on the "Black Lives Matter" (BLM) movement, the use of suggestive visual imagery, such as caricaturised images of an African American wearing an ancient Egyptian headdress, and photos of a young man (who Liddell claims to be Trump supporter) being attacked by African American youths, serves only as an introduction to the topic. While the gist of the narrative is encapsulated in the image, the "meat" of the narrative is presented in the blog post. There, the author uses various rhetorical strategies to elaborate his viewpoint. Specifically, to support the point he makes that the BLM movement has double standards, commenting on the motto of BLM he counters that "all lives matter" suggesting this would not apply in the case of the young White man.

In this example, the author seeks to make use of sarcasm, which, in this case, fulfils two purposes: of communicating offensive and hostile content, and of winning approval and sympathy from his readers by making himself appear clever and possessing a sense of humour. Humour is often used online by White supremacists as a rhetorical device primarily because of its strong persuasive potential (Weaver 2010, 2011). It is a communicative device that supports concepts of "race truth" (Weaver 2013) and has been shown to increase tolerance to racism in the public arena (Ford and Ferguson 2004).

Using terminology from the disciplines of logic and rhetoric, so his arguments seem rational and well supported by evidence, Liddell defends his viewpoint by attacking the "false premises" that the BLM movement has at its basis (according to the author).

In his words:

> Because it is based on *two* false premises, BLM is incapable of generating anything else besides the kind of hatred that we see here. The first premise is the standard, biologically-illiterate Leftist one, namely that all races are essentially equal, identical, and interchangeable. With this as a foundational starting point, the only logical explanation for Black inequality—in terms of poverty, crime, and arrests—must be down to "imaginary White "racism." (…) The other false premise is the characteristically Black one of Black supremacism, which we in the Alt-Right deride as "We Wuz Kangz." Roland Martin (…) is a prime exponent of this ludicrous view that holds that Blacks founded a great civilization on the banks of the Nile, but were somehow incapable of founding one on the banks of the Limpopo, Vaal, Zambezi, Congo, or Niger.

While the authors of such posts (which are found not only on blogs, but also on other virtual spaces such as forums, chat rooms and Facebook) can present to the public elaborated expositions of their arguments, the public also has the opportunity to respond and therefore actively participate in the discussion. This is a strategic process of consensus building within the alt-right community that can create the impression to members and supporters that constructions of racism are based on a consensus-building collective process. However, these spaces are deeply hierarchical, so opposition or even slight disagreement can be swiftly silenced (through ridicule and abuse, if not through outright censorship by the blog owners). This strategy is effective because by allowing debate it creates the (false) impression of a fair dialogue and democratic participation. It is false because in reality, alternative or contrasting views are not present in these spaces through either self-exclusion or censorship.

This process of discursive, (apparently) participative construction of racism in virtual spaces has also been detailed by Atton (2006) in relation to the British National Party (BNP). The ideological narrative of BNP is one characterised by anti-multicultural, anti-equality and anti-freedom views; it uses the "language of progressive media" to claim it engages and encourages its members to engage in cyber activism. However, it is what

Atton calls "authoritarian populism" (p. 583) that is exercised to ensure their views are not contested in online spaces. What BNP proposes:

> is very different from the cyber activism found within progressive social movements. The activism promoted by the BNP is centralized and party-based. It does not encourage its members to use the Internet as a tool for protest, to make direct contact with its opponents (whether construed as the "politically correct" ideologues of New Labour, or the Othered ethnic minorities against which the BNP sees itself as struggling). (p. 583)

Moral Disengagement Strategies

Increasingly subtle expressions of racism are often found in online open spaces of debate such as news commentary websites. In such places, blatant racist views are likely to be censored by moderators, but also due to the likelihood of a mixed audience it might seem more effective for racists to present more moderate or "mainstream views." (If strong blatant racism is expressed in moderate forums, there is a possibility of alienating "emerging radicals," or people who currently hold moderate views but have the potential of developing into fully fledged racists).

"Moral disengagement" has been described as a key strategy used by online racists in subtle cyber racism narratives (Faulkner and Bliuc 2016). Moral disengagement is a concept that was initially proposed by Bandura (1999, 2002) who distinguishes between several main categories of moral disengagement:

1. *Framing the harmful action or behaviour to appear moral*—this strategy can manifest as moral justification when the "pernicious conduct is made personally and socially acceptable by portraying it as serving socially worthy or moral purposes" (Bandura 2002, p. 103); advantageous comparison when harmful behaviour is compared favourably to some other unethical example; or euphemistic labelling, when the harmful behaviour is reframed using "sanitizing language."
2. *Minimising the perpetrators' role in causing harm* (displacement of responsibility)—this strategy operates by obscuring or minimising the

perpetrators' role in committing the harmful behaviour. For instance, claiming the harmful behaviour was imposed by an authority or dictated by a particular situation (so taking the responsibility away from the person engaging in the harmful behaviour—"I had to do it, it was not my choice"). In addition, the responsibility for the harmful behaviour can be diffused by highlighting how other people contributed to, or were complicit in, the harm (moving from personal to collective responsibility).

3. *Reframing the victim as deserving of the harmful action (assigning blame to the victim, denigrating the victim)*—this strategy operates by shifting the focus from the perpetrator of a harmful behaviour to the victim of that behaviour. Specifically, in using this strategy to avoid a sense of moral guilt, perpetrators of harmful behaviours may dehumanise their victims, or blame them for being the ones causing the harm to themselves.

This framework of analysis can be effectively applied to understand how subtle discursive strategies are used in online spaces that allow for debate and elaboration of arguments to occur. For example, moral disengagement has been used extensively in racist arguments in relation to an incident where Indigenous AFL player Adam Goodes was racially abused by a young girl during a game. The "Adam Goodes incident" occurred during a 2013 AFL game in which he was called an "ape" by a 13-year-old girl spectator. Goodes overheard the slur while playing and pointed the girl out to security; she was subsequently escorted from the field by police. Goodes described the racial slur as "shattering" and very hurtful (Windley 2013). Our analysis of comments about the Adam Goodes incident on the websites of two Australian newspapers reveals that when arguing their viewpoints, commentators who promoted racist views were more likely to use moral disengagement strategies (in comparison with those commentators who promoted anti-racist views).

For example, racists were more likely to reframe the harmful action (of insulting Adam Goodes) to appear moral compared to anti-racists. Their strategies included euphemistic labelling, that is, attempting to make the racist act appear moral by using sanitising language that relabelled the act as "name-calling," and not racism. Some extracts from online comments illustrate this point well:

> "Call it racism or just call it name calling what the heck…it's just name calling. It's been going on since time immemorial."
>
> "Don't condone what the girl said at all, but I do query whether it is verbal or racial abuse."
>
> "Did you ever say anything foolish as a teenager? No? Liar! Is this the way we want to treat every teenager who calls someone a name? Seriously?"

Attempts to make the racist (harmful) incident seem less morally dubious were made by comparing the incident to other incidents that could be classed as even less moral, such as drug use in the sport:

> The AFL, like the media will make a mountain out of a mole hill. You go on about silly things like this. What about the pathetic excuse for a drug policy the AFL has. Young kids will use drugs because their hero players do and they get 3 strikes. Come on AFL, prove you care on all fronts not just racial vilification.

In some cases, racist commentators used examples of more extreme harmful and morally problematic behaviour to make their point:

> I also note with interest that there are some on this forum that condemn this little girl and apparently the culture from which she comes, but are quick to excuse the ideology of two murders in London who CUT A SOLDIER'S HEAD OFF (spinning my finger around my ear with abandon right now).

Other examples of using moral disengagement strategies in discourse around the Adam Goodes incident included:

A) *Minimising the perpetrator's role:*

> It's a kid for goodness sakes, let's not lose sight of that, yes it was wrong, yes she has been escorted from the ground blah blah blah, but keep it in perspective people, she's reportedly 14 years old I'm sure she's learnt her lesson by now.

B) *Assigning blame to the victim:*

> if Adam Goodes chooses to grow a beard that makes him look like an ape. Then what does he expect. Since when is ape a racist word? I guess we all need a racist word book. Adam Goodes is nothing but a big bully (hope that is not a racist word) picking on a child, (…) picking on children in my book is a big no no.

C) *Denigrating the victim:*

> Unbelievable over-reaction. I don't believe people should be racist but surely Goodes should harden up a bit!! Give me a break!!

Opportunities provided on platforms where racists can strategically communicate in highly visible online forums allow massive public exposure of cyber racism discourses and potential for influence. Their arguments can enter the mainstream through mere exposure to vast audiences. Moreover, this liability is enhanced when these discursive strategies are successful not only in making the perpetrator feel better, but also in persuading the audience that particular racist acts are morally acceptable.

Racial Microaggression and Hidden Harmful Effects

The casual nature of cyber racism such as that expressed towards Adam Goodes hides the potential harmful effects in people who experience it. In studying casual racism in psychology, Sue and colleagues (Sue 2007, 2017) have proposed a very useful taxonomy of racial microaggressions (defined as "brief and commonplace daily verbal, behavioural, or environmental indignities, whether intentional or unintentional, that communicate hostile, derogatory, or negative racial slights and insults toward to the target person or group" Sue et al. 2008, p. 273). The term "racial microaggressions" was first introduced by Pierce in the 1970s, to refer to those "subtle, stunning, often automatic, and non-verbal exchanges which are 'put downs'" (Pierce et al. 1978, p. 66). According to Sue et al. (2008), racial microaggressions can take the form of:

1. Microassaults (i.e., explicit racial derogation meant to hurt the victim through name-calling, avoidant behaviour or intentional discrimination)—the closest form to old-fashioned racism;
2. Microinsults (i.e., communications that convey rudeness and insensitivity, and demean a person's racial heritage or identity) and
3. Microinvalidations (i.e., communications that exclude, negate or nullify the thoughts, feelings and experiences of a person from a diverse cultural background).

In analysing those different forms of microaggressions in the Australian context, Sue and colleagues (2008) identified a number of distinct themes, some of which can also be found in cyberspace:

1. "Alien in one's own land"—the underpinning message is that people who are not white Australian are not "true blue, real Australians"; they will always be foreigners—even Aborigines can be invalidated in this way;
2. "Ascription of intelligence" or other stereotypical characteristics—this type of microaggression represents the enacting of a particular stereotype as, for example, being surprised of the achievements of a person of Aboriginal background;
3. "Colour blindness"—this relates to the minimalising or denying the right of cultural difference as, for example, when referring to immigrants from Muslim countries as incapable of integrating and appreciating the "Aussie way of life";
4. "Denial of individual racism"—this form is often found as a preamble of openly racist statements as, for example, "I am not racist, I am even married to an Asian, but I have participated in the Cronulla riots" and
5. "Myth of meritocracy"—the underpinning message is, for example, that immigrants are coming to Australia because they want to get a free ride (the same type of argument is often made in the context of the Syrian refugee crisis in Europe).

In terms of the effect on their targets, these types of subversive racist discourse are likely to be, in some instances, even more harmful than blatant racism (Chakraborty and McKenzie 2002; Clark 1999; Collins and Williams 1999; Williams and Collins 1995; Williams and Wyatt 2015). There is strong support that microaggressions are particularly harmful when experienced on a daily basis—which is further enabled by living in an increasingly connected and online world. For instance, a study by Solorzano (2000) on the experiences of African Americans targeted by microaggressions found the cumulative effects of those experiences can be very damaging (Sue et al. 2008). They found that experiences of microaggressions lead to an increase in self-doubt, frustration and feelings of isolation, all indicators of decreased well-being.

Conclusion

This chapter has demonstrated that a key aim of the authors of cyber racism is to influence national identity narratives, both by reinforcing racially focused national identities, and fragmenting the identities and communities of those they detest. By developing their arguments around race, culture and attitudes to diversity, different segments of, for instance, Australian society are attempting to align themselves to the valued social category of national identity, so they come to constitute the "truest" representation of Australian identity. These contested narratives are constructed in nation-states in the context of the history and politics of a nation. They are primarily about decisions of inclusion or exclusion of different races, cultures and religions in the national identity narrative. The perpetrators of cyber racism put forward arguments based on White nationalist views that aim to strengthen their position in relation to national identity narratives and legitimise cyber racism.

Cyber racism can be perpetrated by members of White power or alt-right groups who use more blatant forms of cyber racism with complex and strategic arguments, or individuals who may use increasingly subtle forms of cyber racism to express their exclusionary and racist attitudes against particular targeted groups and individuals. Online communities of racists are built and sustained through the use of a variety of discursive techniques including identity rhetoric, development of sophisticated ideological arguments and moral disengagement strategies. Subversive racist discourse (microaggression) in particular has been shown to have a severe negative impact on targets.

The Internet and social media have central roles in enabling cyber racism narratives and the proliferation of the accompanying messages. Racists (particularly those who identify themselves as White supremacists) have taken full advantage of the affordances that the Internet and social media platforms provide. They are adept at using new and emerging technologies and effective at getting their message across to the general public. Their attempts to influence people who may support multiculturalism but are fearful of migrants (because of contextual factors like terrorist attacks) are particularly strategic. They are likely to prove successful in some cases, and this represents a threat to those versions of identity narratives that aim to promote the values of diversity, fairness and integration.

Given how peoples' world views are constituted through the narratives and discourses they imbibe and reproduce, understanding cyber racism is crucial to countering its proliferation. In the following chapters, we will further explore how cyber racism narratives can be countered, including through building communities of resistance and solidarity, enabling a more civil discourse about national identity.

References

Adams, J., & Roscigno, V. J. (2005). White Supremacists, Oppositional Culture and the World Wide Web. *Social Forces, 84*, 759–778.

Atton, C. (2006). Far-Right Media on the Internet: Culture, Discourse and Power. *New Media and Society, 8*, 573–587.

Back, L. (2002). Aryans Reading Adorno: Cyber-Culture and Twenty-First Century Racism. *Ethnic and Racial Studies, 25*, 628–651.

Bandura, A. (1999). Moral Disengagement in the Perpetration of Inhumanities. *Personality and Social Psychology Review, 3*, 193–209.

Bandura, A. (2002). Selective Moral Disengagement in the Exercise of Moral Agency. *Journal of Moral Education, 31*(2), 101–119.

Bliuc, A.-M., McGarty, C., Hartley, L., & Muntele Hendres, D. (2012). Manipulating National Identity: The Strategic Use of Rhetoric by Supporters and Opponents of the 'Cronulla Riots' in Australia. *Ethnic and Racial Studies, 35*, 2174–2194.

Bliuc, A.M., Betts, J., Faulkner, N., Vergani, M., Deakin University, Chow, R.J., & Iqbal, M. (2016). *Group Cohesion in Online White Supremacist Communities – Integrating Social Network and Computerised Linguistic Analyses*. Paper presented at the 'Taking Stock of Research on Violent Online Political Extremism' Vox-Pol Mid-Project Conference, Dublin, Ireland. http://www.voxpol.eu/vox-pol-mid-project-conference-taking-stock-of-research-onviolent-online-political-extremism

Chakraborty, A., & McKenzie, K. (2002). Does Racial Discrimination Cause Mental Illness? *The British Journal of Psychiatry, 180*, 475–477.

Clark, R. (1999). Racism as a Stressor for African Americans: A Biopsychosocial Model. *American Psychologist, 54*, 805–816.

Collins, C. A., & Williams, D. R. (1999). Segregation and Mortality: The Deadly Effects of Racism? *Sociological Forum, 14*(3), 495–523.

Connelly, K. (2016). Undermining Cosmopolitanism: Cyber-Racism Narratives in Australia. *Sites of Social Anthropology and Cultural Studies, 13*(1), 156–176.

Council for Aboriginal Reconciliation. (2001). *Terra Nullius and Sovereignty*. (online) Available at: http://www.austlii.edu.au/au/orgs/car/docrec/policy/brief/terran.htm. Accessed 9 Feb 2017.

Douglas, K. M. (2005). Understanding Cyberhate Social Competition and Social Creativity in Online White Supremacist Groups. *Social Science Computer Review, 23*, 68–76.

Dovidio, J. F. (2001). On the Nature of Contemporary Prejudice: The Third Wave. *Journal of Social Issues, 57*(4), 829–849.

Facebook. (2017a). *Take Back Australia*. https://www.facebook.com/fightback-aussie/. Accessed 11 Feb 2017.

Facebook. (2017b). *Welcome to Australia*. https://www.facebook.com/welcomet oaustralia/?fref=ts. Accessed 11 Feb 2017.

Faulkner, N., & Bliuc, A.-M. (2016). It's Okay to Be Racist': Moral Disengagement in Online Discussions of Racist Incidents in Australia. *Ethnic and Racial Studies, 39*, 1–19.

Ford, T., & Ferguson, M. (2004). Ford and Ferguson and Social Consequences of Disparagement Humour: A Prejudiced Norm Theory. *Personality and Social Psychology Review, 8*, 79–94.

Fozdar, F., Spittles, B., & Hartley, L. K. (2015). Australia Day, Flags on Cars and Australian Nationalism. *Journal of Sociology, 51*(2), 317–336.

Gerstenfeld, P. B., Grant, D. R., & Chiang, C.-P. (2003). Hate Online: A Content Analysis of Extremist Internet Sites. *Analyses of Social Issues and Public Policy, 3*(1), 29–44.

Hage, G. (2002). Multiculturalism and White Paranoia in Australia. *Journal of International Migration and Integration, 3*(3–4), 417–437.

Ho, C., & Jakubowicz, A. (2013). The Realities of Australian Multiculturalism. In A. Jakubowicz & C. Ho (Eds.), *'For Those Who've Come Across the Seas': Australian Multicultural Theory, Policy and Practice*. Melbourne: Australian Scholarly Publishing.

Jakubowicz, A. (2011). Playing the Triangle: Cosmopolitanism, Cultural and Social Capital as Intersecting Scholarly Discourses About Social Inclusion and Marginalisation in Australian Public Policy Debates. *Cosmopolitan Civil Societies Journal, 3*(3), 68–91.

Jakubowicz, A. H. (2016). Once Upon a Time In… Ethnocratic Australia: Migration, Refugees, Diversity and Contested Discourses of Inclusion. *Cosmopolitan Civil Societies: An Interdisciplinary Journal, 8*(3), 144–167.

Killalea, D. (2017). *Donald Trump's Muslim Ban: Why Weren't These Countries Included*. www.news.com.au. Accessed 14 Feb 2017.

Lazer, D., Pentland, A. S., Adamic, L., Aral, S., Barabasi, A. L., Brewer, D., Christakis, N., Contractor, N., Fowler, J., & Gutmann, M. (2009). Life in the Network: The Coming Age of Computational Social Science. *Science, 323*(5915), 721.

Markus, A. (2016). *Mapping Social Cohesion. The Scanlon Foundation Surveys 2016*. http://monash.edu/mapping-population/public-opinion/social-cohesion-report. Accessed 14 Feb 2017.

Mattingly, C., Lutkehaus, N. C., & Throop, C. J. (2008). Bruner's Search for Meaning: A Conversation Between Psychology and Anthropology. *Ethos, 36*(1), 1–28.

Noble, G. (2009). *Lines in the Sand: The Cronulla Riots, Multiculturalism and National Belonging*. Sydney: Institute of Criminology Press.

Perera, S. (2005). Who Will the Multiple Formations of Australian Whiteness. *Australian Critical Race and Whiteness Studies Association Journal, 1*, 30–39.

Perry, B., & Olsson, P. (2009). Cyberhate: The Globalization of Hate. *Information and Communications Technology Law, 18*, 185–199.

Pierce, C., Carew, J., Pierce-Gonzalez, D., & Wills, D. (1978). An Experiment in Racism: TV Commercials. *Education and Urban Society, 10*(1), 61–87.

Reicher, S., & Hopkins, N. (2000). *Self and Nation*. London: Sage.

Salter, F. (2012). *The War Against Human Nature III.* http://quatrant.org.au/magazine/2012/11/the-war-against-human-nature-iii/: Quadrant Online. Accessed 14 Feb 2017.

Skrbis, Z., Kendall, G., & Woodward, I. (2004). Locating Cosmopolitanism: Between Humanist Ideal and Grounded Social Category. *Theory, Culture & Society, 21*(6), 115–136.

Solorzano, D. (2000). Critical Race Theory, Racial Microaggressions, and Campus Racial Climate: The Experiences of African American College Students. *Journal of Negro Education, 69*, 60–73.

Sue, D. W. (2007). Racial Microaggressions in Everyday Life: Implications for Clinical Practice. *American Psychologist, 271*, 62.

Sue, D. W. (2017). Microaggressions and "Evidence" Empirical or Experiential Reality? *Perspectives on Psychological Science, 12*(1), 170–172.

Sue, D., Capodilupo, C., Nadal, K., & Torino, G. (2008). Racial Microaggressions and the Power to Define Reality. *American Psychologist, 63*(4), 277–279.

Wallwork, J., & Dixon, J. A. (2004). Foxes, Green Fields and Britishness: On the Rhetorical Construction of Place and National Identity. *British Journal of Social Psychology, 43*(1), 21–39.

Weaver, S. (2010). Developing a Rhetorical Analysis of Racist Humour: Examining Anti-Black Jokes on the Internet. *Social Semiotics, 20*, 537–555.

Weaver, S. (2011). Jokes, Rhetoric and Embodied Racism: A Rhetorical Discourse Analysis of the Logics of Racist Jokes on the Internet. *Ethnicities, 11*, 413–435.

Weaver, S. (2013). A Rhetorical Discourse Analysis of Online Anti-Muslim and Anti-Semitic Jokes. *Ethnic and Racial Studies, 36*(3), 483–499.

Werbner, P. (2008). *Anthropology and the New Cosmopolitanism* (1st ed.). New York: Berg Publishers.

Williams, D., & Collins, C. (1995). US Socioeconomic and Racial Differences in Health: Patterns and Explanations. *Annual Review of Sociology, 21*(1), 349–386.

Williams, D. R., & Wyatt, R. (2015). Racial Bias in Health Care and Health: Challenges and Opportunities. *JAMA, 314*(6), 555–556.

Windley, M. (2013, May 25). Adam Goodes 'Gutted' After 13-Year-old Girl's Racial Slur, Who Called the Sydney Champion Today to Apologise. *Herald Sun.* http://www.heraldsun.com.au/sport/afl/adam-goodes-gutted-after-13-year-old-girls-racial-slur-who-called-the-sydney-champion-today-to-apologise/story-fni5fan7-1226650256245.

7

Building Online Communities of Resistance and Solidarity

Introduction

While there has been considerable discussion on the growth of online racism and debate on how it might be curtailed, its continuing expansion suggests a failure of the policy sphere to adequately define and engage with the issues involved. Over the last decade, technological progress has dramatically altered the way we interact, consume news and entertainment, form communities and engage in society (Daniels 2012). It has also redistributed influence (Jakubowicz 2012). The changes have enhanced the capacity of hate advocacy groups to spread their beliefs but, importantly, also transformed the potential for responses that successfully address the wider public on issues of social justice (Jakubowicz 2012; Connelly 2016).

The technology shift has taken individuals from being mere consumers of content to being producers, curators and broadcasters (Gehl 2010). Each person is part of multiple online communities, and through their posts, likes, shares, tweets, retweets, comments and e-mails, they play a key role in determining which narratives are spread throughout their communities (Miller 2011). Building effective online communities of

A. Jakubowicz et al., *Cyber Racism and Community Resilience*, Palgrave Hate Studies, DOI 10.1007/978-3-319-64388-5_7

resistance and solidarity requires new approaches and policies, which take into account these new relationships with technology. This will involve working with a wider range of stakeholders from civil society and doing so in new ways.

Traditionally, the policy focus has been on formal channels, and engagement with communities has largely focused on local government, community representative bodies and service provision organisations (Paradies et al. 2009; Office for the Community Sector 2013; Victorian Auditor-General's Office 2015). The problem of online racism, if addressed at all, was seen as an adjunct to racism generally, and the engagement with new technology was focused on marketing (Paradies et al. 2009). There has been a gap in how to deal with online racism, and into this gap have stepped new actors with innovative approaches, reactions, responses and initiatives to address this problem specifically (e.g., the Online Hate Prevention Institute [OHPI] and IndigenousX). Some have taken place within formal structures, while others, driven by grassroots activism, have stepped outside organised structures. These new approaches are enabled by new technologies and centred on the creation of online communities, either as explicit groups, or as ad hoc communities that form and disperse based on peoples' networks and interactions.

Online communities of resistance and solidarity need the flexibility and energy of activists operating through purpose-driven civil society organisations or leading grassroots action. They also need the formal authority given through the endorsement and support of government agencies, local governments, community service providers and community peak bodies. New policy responses are needed to facilitate the creation of online communities of resistance to cyber racism and solidarity in the face of cyber racism through synergies between different stakeholders and approaches.

Some of the key issues in building resilience have already been outlined in Chap. 1. In this chapter, we provide a framework for building successful online communities that offer solidarity to their members in the face of online racism. It is worth nothing that aspects of this framework work equally well for online hate in general, a term which extends to online homophobia, online misogyny and other forms of online bigotry. The framework looks at:

- The eight different types of online communities;
- The types of stakeholders driving or empowering them and
- A range of proactive and reactive strategies they can adopt to tackle cyber racism.

In addition, the framework is illustrated with examples of each type of community, and two approaches to building online communities of resistance and solidarity from digitally based organisations are presented in depth: one from a civil society organisation focused on the digital space and the other from a grassroots online initiative. This is followed by a look at a civil society organisation, a government agency and a peak body representing a range of communities impacted by cyber racism. We examine how each of these three either is, or has the potential to, contribute to the creation of online communities of solidarity and resistance.

Online Communities

The Oxford Dictionary defines an online community as "a group of people who regularly interact with each other online, especially to share information and opinions on a common interest" (Oxford Dictionaries 2017). Online communities are more than just channels through which content can be broadcast to an audience. They require a sense of belonging and enough active engagement from community members to make it a shared space and not just a publishing platform for the community manager. The technology itself also has an impact. Members of a community can only engage when they see the content. Yet communities can also appear when their members unknowingly share characteristics that make them vulnerable to cyberattacks.

Facebook artificially limits the number of fans who see a post, but allows the number of viewers to be increased when the page owner pays to "boost" the post (Facebook 2017a). This revenue-raising exercise causes page owners to be selective about the content they push to their audience. Posts can also be boosted to targeted audiences, identified by location, age, gender and interests, who are not existing fans of the page. This can create an *ad hoc* community with both community members and those outside the community actively engaging in discussion on a particular post as well as engaging through likes and shares.

On Twitter, visibility is limited by time. Most people will only see a tweet if they are logged in and paying attention to Twitter or to notifications from Twitter and see the new tweet in real time. The engagement can involve replies, retweets and likes. Twitter also creates *ad hoc* communities around hashtags with people filtering by hashtag, then reading and engaging in the discussion that is linked through the use of that hashtag.

News websites and blogs which allow user comments can also develop into communities as people comment not only on the article but on each other's comments. Platforms which send an e-mail notification when a comment is replied to further facilitate this. E-mail itself can be used to create communities when the same people are included in multiple discussions. This can be achieved by addressing the e-mail to all, or through the use of a listserve. The use of e-mails with a blind carbon copy (BCC) or a listserve that only allows the list owner to message the list is an example of a channel which is prevented from becoming a community, as each recipient can only communicate back to the original sender.

As can be seen, the idea of "regular interaction" stretches from a permanent community, which members feel they are part of; through to an *ad hoc* community appearing around an active online discussion thread on a news article or surrounding a hashtag. In this case, participants may engage in a series of interactions while others look on, and then the *ad hoc* community disbands.

Types of Online Communities Tackling Racism

Online communities that support resistance to racism and solidarity in the face of racism can be categorised across three dimensions:

1. The **focus** of the community, whether it is *general* and works against all forms of racism or *specific* to racism against a particular target group;

2. The **scope** of the community's work, whether it is *dedicated* to anti-racism work or the anti-racism work is more *casual* with the community usually focusing on another activity
3. The **permanence** of the community, whether it is *ad hoc* in nature or part of an organised structure that intends to be *permanent.*

There are in total eight variations across these three factors as illustrated in Table 7.1:

The Community Builders

The examples highlighted in the table demonstrate the variety of actors that are running communities supporting resistance to racism and demonstrating solidarity with the victims of racism. These actors can be grouped into the following five categories:

1. Civil Society Organisations Affiliated with a Targeted Group

Civil society organisations that are affiliated with a particular target group bring an element of authenticity to the fight against racism when they share their experience as well as the impact racism is having on individuals and their community. For these organisations, emotional involvement is a strength that adds power to their narrative. Their arguments rely on empathy and can sway the disengaged bystander to take a stand against racism in solidarity with the victims and other allies in the community. These organisations are usually seen by members of the targeted community as having the community's interest at heart, which increases the trust members of that community place in the organisation. A special case of this kind of civil society organisations are the peak bodies representing homogeneous ethnic or cultural communities. Examples include state bodies such as the Jewish Community Council of Victoria, the United Indian Association and the Islamic Council of Victoria, as well as national bodies such as the Australian Federation of Islamic Councils and the

Table 7.1 Typology of resilience groups

Focus	Scope	Permanence	Examples
Specific	Dedicated	Permanent	Islamophobia register;[a] TheyCan't[b]
Specific	Dedicated	*Ad hoc*	Aboriginal Memes Petition;[c] Hashtag campaign #RideWithMe[d]
Specific	Casual	Permanent	IndigenousX[e]
Specific	Casual	*Ad hoc*	News article about racism against a particular community with an open comments section[f]
General	Dedicated	Permanent	Online Hate Prevention Institute;[g] All Together Now;[h] Believe in Bendigo;[i] No Hate Speech Movement;[j] INACH[k] UN Human Rights Commissioner[l]
General	Dedicated	*Ad hoc*	No Room for Racism Coalition which sprang up to counter rallies by the UPF[m]
General	Casual	Permanent	FECCA;[n] UNAOC[o]
General	Casual	*Ad hoc*	News article with comments against racism—S18C[p]

[a]https://www.facebook.com/islamophobiaregisteraustralia/
[b]https://www.facebook.com/TheyCant
[c]https://www.change.org/p/facebook-headquarters-immediately-remove-the-racist-page-called-aboriginal-memes
[d]http://www.bbc.com/news/blogs-trending-30479306
[e]https://twitter.com/IndigenousX?ref_src=twsrc%5Egoogle%7Ctwcamp%5Eserp%7Ctwgr%5Eauthor
[f]https://theconversation.com/booing-adam-goodes-racism-is-in-the-stitching-of-the-afl-45316
[g]https://www.facebook.com/onlinehate/
[h]https://www.facebook.com/alltogethernow.org.au/
[i]https://www.facebook.com/believeinbendigo/
[j]https://www.facebook.com/nohatespeech/
[k]https://www.facebook.com/pg/INACHnet/likes/?ref=page_internal
[l]https://www.facebook.com/unitednationshumanrights
[m]https://www.facebook.com/No-Room-for-Racism-357685031103855/?fref=ts
[n]https://www.facebook.com/iFECCA/
[o]https://www.facebook.com/unaoc.org/
[p]https://www.theguardian.com/commentisfree/2016/sep/01/the-debate-about-18c-doesnt-have-to-be-a-left-right-slanging-match

Executive Council of Australian Jewry. These umbrella bodies can speak with a united voice on behalf of their affiliated organisations, increasing the impact of their message.

2. Unaffiliated Civil Society Groups

Unaffiliated civil society groups bring an element of authority to discussions of racism through their objective and often non-emotional assessment of racism and its impact. Their authority can be applied to calling out racism and bringing it to the attention of supporters and the wider public. Their online communities provide a ready audience able to spread the message through social media to a wider audience. These civil society groups may be able to dedicate greater resources to empower their supporters and enable them to take positive action. A special form of civil society organisation is the association or peak body that brings different targeted group organisations together. Examples include the Federation of Ethnic Community Councils of Australia (FECCA) and the Jewish, Christian, Muslim Association (JCMA). These broader groups can run their own more general communities, or act as a converter, allowing communities affiliated with one targeted group to more easily stand in solidarity with other targeted groups. This is based on an explicit assurance of mutual solidarity coordinated through the peak body.

3. Government Agencies

Government agencies can bring formal authority to an online community but may be constrained in what they can do as a result of their remit and need to remain impartial. They are in a strong position to provide factual information to a community, particularly on laws which may apply, and the legal rights of individuals and targeted communities. Government agencies are also able to bestow authority on civil society organisations through the provision of grants. This provides valuable practical support to the organisation, but also demonstrates the government's confidence in their work.

4. Grassroots Activists

Grassroots activists are members of the public who take grassroots action. They are able to swiftly create large and effective *ad hoc* communities in social media, or through online petitions and crowdfunding (e.g., in 2017, Go Fund Me raised $100,000 within 24 hours to fund a new ad campaign to replace billboards of Muslim girls celebrating Australia Day, which had been taken down after the operators received death threats from far-right anti-Islamic group (Go Fund Me 2017)). These *ad hoc* communities may last, and in time become new civil society organisations, or, they may be short-lived, either achieving their purpose or not. The story of an everyday person taking action against racism has a high degree of human interest, particularly if there is evident success or novelty involved, and may lead to media coverage, which can grow the community.

5. Academics

Generally speaking, academics do not build communities, except for academic communities, but they can provide a source of content, insight and information that support active communities. This can be through the direct provision of information to be shared, or through the provision of evidence that can help other actors focus their strategic efforts more effectively.

Community Strategies for Resistance and Solidarity

Online communities of resistance to racism and solidarity in the face of racism can act both reactively and proactively.

Reactive responses include online community efforts to:

- Call out racism online or in daily life;
- Work collaboratively online to have racist material removed and

- Promote counter speech to racist narratives through online campaigns disproving racist claims with facts.

Proactive responses include community-based campaigns to:

- Promote positive values such as multiculturalism;
- Share knowledge about the culture and traditions of targeted groups;
- Promote narratives highlighting the harm in racism, both to individuals and society;
- Help people learn to recognise what other groups in society may consider racist and why;
- Countering historical and other narratives that reflect bias by the dominant cultural group to the detriment of minorities and
- Promote narratives that normalise positive intergroup relations.

Bringing It Together

The stakeholders running a community and the type of community will have a significant impact on which strategies will be possible. A stakeholder wishing to advance multiple strategies may use multiple online communities, possibly with overlapping membership, to achieve their goal. Some strategies work best when multiple stakeholders assist the community in different ways.

There are two support roles in which a stakeholder can support another actor's community:

1. **As a source**: providing content which keeps the community engaged. Organisations affiliated with a targeted group can provide their experiences and those of their community. Government agencies can provide information. Politicians can provide statements of policy. Researchers, whether from academia or other civil society organisations, can provide facts and opinions.
2. **As an amplifier**: an amplifier expands an online community. This usually occurs when one community shares the campaign of another. The organisation receiving content from a source will often be acting

as an amplifier. Online communities run by civil society organisations or government can amplify the impact of narratives from target groups or findings from research. This may involve more than simply sharing a link to the content. It can involve substantially repackaging the narrative or key research findings to make it work more effectively within the amplifying community. An amplifier can take a message out of the echo chamber of the targeted community or the ivory towers of academia, into the wider community.

Civil society organisations affiliated with a targeted group can be amplifiers, spreading calls to action from more general civil society organisations and government agencies to their community. They can also act as amplifiers for other target groups when those groups come under attack. Associations or peak bodies that bring different targeted group organisations together can be particularly effective in this role. By acting as amplifiers for the work of their member organisations and adding their own commentary, these groups can effectively spread messages between targeted groups, helping build cross-community empathy, mutual support and understanding as part of a network of resistance and solidarity.

The following tables show which strategies, both proactive and reactive, are most likely to work for each stakeholder group (Tables 7.2 and 7.3):

Case Study: The Online Hate Prevention Institute (Australia-Based, Globally Accessible)

Who: The Online Hate Prevention Institute (OHPI).

What: Founded in January 2012, the OHPI is Australia's only unaffiliated civil society organisation entirely dedicated to the problem of online hate speech.

How: Due to its strong focus on technology expertise, OHPI is unique among its peers, being as much a high-tech startup as a civil society organisation. This influences not only their deep understanding of the problem of online racism, but also the

Table 7.2 Groups and proactive modes of resilience

Proactive					
Mode of resilience	Affiliated civil society group	General civil society group	Government	Academics	Grassroots
Promoting positive values	Yes	Yes	Yes	No	Unlikely
Strengthening intergroup relations	Yes	Yes	Yes. Through event funding.	No	Unlikely
Highlighting harm in racism	Yes. Through personal narratives.	Yes. Through research and campaigns.	A little. Through statements.	Yes. Through research.	Unlikely
Education to recognise racism	Yes. Through introducing themselves.	Yes. Through publications.	A little. Through online resources.	A little. Through research (but needs conversion for public impact).	Unlikely
Countering historical and other narratives	Yes	Possibly	No	Yes	Yes
Narratives normalising positive intergroup relations	Yes	Yes	A little. Through promoting government support.	No	Yes

Table 7.3 Groups and reactive modes of resilience

Reactive					
Mode of resilience	Affiliated civil society group	General civil society group	Government	Academics	Grassroots
Calling out racism	Yes. Through public statements.	Yes. Through public statements and educational campaigns.	A little. Through statements.	No	Yes
Working to remove racist content	Yes. Through public statements and campaigns.	Yes. Through public statements and campaigns.	Yes in limited circumstances. Through the media and approaches to companies in extreme cases.	No	Yes
Disproving racist narratives	No	Yes. Through public campaigns.	A little. Through statements and speeches.	Yes	Yes

strategies they can use to counter it. OHPI is a registered charity, with an annual budget of under $200,000 in 2016 (Australian Charities and Not For Profits Commission 2017).

Why: OHPI seeks to empower and give voice to its stakeholders by removing the barriers that prevent the reporting of cyber racism. OHPI's approach aims to facilitate the creation of resilient online communities of resistance and solidarity to cyber racism.

OHPI grew out of a project focused exclusively on online antisemitism; however, the formation of the Institute signalled a shift to a wider anti-hate engagement tackling issues such as Islamophobia, anti-Aboriginal hate, homophobia, misogyny, cyberbullying, trolling and other forms of online hate. OHPI is now recognised globally as a key innovator in developing online strategies to address systemic governance failures on the Internet, which have allowed hate to go viral.

For regular social media users, the key barrier is an inability to meaningfully respond when they encounter racism online. Social media companies do provide systems for reporting online racism, but most reports are rejected by them, sending a negative and disempowering message to affected users. Real change is hampered due to the following:

- Those within social media companies who wish to see change are often blocked, as handling racism is not a business priority, and technical expertise is more focused on generating revenue;
- Researchers are blocked in their investigations of online racism due to the difficulty in acquiring sufficiently large and representative samples of online racism and
- Law makers are blocked due to both insufficient technical understanding and a lack of research data to provide the basis for evidence based policy.

Sourcing Hate Content from Their Online Community

OHPI's approach is centred on crowdsourcing hate speech data from the public. This is facilitated through their online community of over 24,000 fans on Facebook (Facebook 2017b) and their online tool at FightAgainstHate.com (Fight Against Hate 2015a). This tool enables users to report instances of cyber racism. OHPI issues regular campaigns, called briefings, which are published on the OHPI website and include calls to action to empower members to report specific instances of online hate. In the 2016 financial reporting year, OHPI published 84 briefings, which were collectively liked or shared 39,217 times (OHPI 2016a). OHPI also operates a Twitter account and mailing list, which provide secondary avenues of communication.

Each briefing on the OHPI website is announced with a Facebook post, and community members often respond with a comment when they have taken action. Some also tag friends in their comment, drawing them into the reactive campaign created by the briefing. Administrators of other Facebook pages act as key nodes to a wider network, often sharing content relevant to their community on their own pages. Some of these pages relate to civil society organisations, and others are purely based in social media. The sharing of OHPI's briefings into other online communities increases resilience by linking these communities into a wider action of solidarity against the cyber racism and active resistance against the examples explored in the briefings.

The FightAgainstHate.com reporting tool

The core of OHPI's system is FightAgainstHate.com, an online tool which allows the public to report the hate speech they see on Facebook, YouTube and Twitter. The system allows reporting of racism as well as cyberbullying, homophobia and misogyny. It also gives users the option to lodge a report and assign their own category label if none of those presented appear appropriate. Through the tool, a large sample of hate content is gathered, and its continued availability online can then be monitored. This not only provides data on the nature of online hate, but also allows monitoring of the effectiveness of self-regulation by social media companies in removing the hate. Critically for researchers, it converts the problem of finding the hate into the more manageable problem of quality control over reports made by the public (Oboler and Connelly 2014).

To report a hate content item, the user copies the Web address from the social media platform and pastes it into FightAgainstHate.com. The system then parses the address and extracts a unique identifier for the underlying content. This helps to merge duplicate reports even when the incoming addresses may differ. The system then invites the reporter to classify the type of hate they are reporting. Some categories, like antisemitism and anti-Muslim hate, have subcategories. A user selecting antisemitism, for example, would be able to further specify if the content was incitement to violence, Holocaust denial, traditional antisemitism or new antisemitism.

Users who report to FightAgainstHate.com know that, unlike most reports made directly to platform providers, their report will not be ignored and their effort wasted. They know their contribution is likely to make a real difference. This provides a further incentive for users to continue reporting; it strengthens their resilience and helps create a sense of resistance as users are able to share what they have reported and encourage others to take action. OHPI's briefings are often based on items of hate reported through the FightAgainstHate.com system.

Essentially, FightAgainstHate.com empowers members of the community, increases their individual resilience and makes the community as a whole a more dynamic, powerful and flexible community of resistance. The data it generates are further enhanced by reports made through Facebook messages and direct e-mails to OHPI—which may also be used in briefings for future collective action.

The Unity in Diversity on Australia Day Campaign

The OHPI community also promotes proactive campaigns. The largest was the "Unity in Diversity on Australia Day" campaign in 2016, which posted messages and images of positive Australian values such as multiculturalism, mateship, diversity and religious pluralism, which could be shared by the community on Facebook (OHPI 2016b).

Though proactive in the sense that the campaign didn't confront messages of cyber racism head on, it was a response to a hate campaign by far-right groups, which sought to declare everyone born in Australia as an "Indigenous Australian," both removing the special status of Australia's first peoples and seeking to exclude from the community of "real Australians" those who immigrated to Australia and have since become naturalised Australians, often through citizenship ceremonies held on Australia Day (OHPI 2016c). OHPI's briefing, explaining the reactive need for the campaign, was liked and shared over 3000 times. A second briefing was released ahead of the campaign launch after a backlash from some Indigenous community members who were planning a boycott of Australia Day, instead promoting it as Invasion Day. Due to the backlash, OHPI published a second briefing, highlighting the different meanings of Australia Day to different segments of the Australian community, including those in the Indigenous community who were promoting the Invasion Day campaign (OHPI 2016d).

On Australia Day itself, different images and messages about the campaign were posted on OHPI's Facebook page. These posts reached over 160,000 Australian Facebook users and gathered over 7200 likes, shares and comments.

The Spotlight on Anti-Muslim Internet Hate Campaign

The Spotlight on Anti-Muslim Internet Hate (SAMIH) campaign was created by OHPI to:

- Raise awareness about anti-Muslim hate in social media;
- Encourage active reporting by the public;
- Gather a large sample of anti-Muslim hate data and
- Track how social media companies were responding to reports about anti-Muslim hate (Fight Against Hate 2015b).

The campaign mirrored the "Spotlight on Antisemitism" campaign which had gathered over 2000 items of antisemitic social media content in a two-month campaign period (The Big Smoke 2015). The campaign also built on OHPI's 2013 report "Islamophobia on the Internet" (Oboler 2013) and a refined scheme of anti-Muslim harassment from a draft chapter later published in the book "Islamophobia in Cyberspace" (Oboler 2016a).

SAMIH was formally launched as part of an event run by La Trobe University whose students interned with OHPI and helped to produce some of the communications material and messaging for the campaign (OHPI 2015a). In addition to the support from the university, the campaign also received:

- A letter of support from the Victorian Minister for Multicultural Affairs on behalf of the state government (OHPI 2015b);

- Financial sponsorship from the Australian Federal Police and the Islamic Council of Victoria and
- Non-financial support from a range of other organisations including the Federation of Ethnic Communities' Councils of Australia (FECCA), the Multicultural Communities Council of Illawarra and All Together Now (Fight Against Hate 2015c).

The campaign was built with three possible levels of implementation. The cost ranged from $41,600 to $86,700, with OHPI planning to cover over half from existing resources and seeking the rest from sponsorships and donations through a crowdfunding campaign (Fight Against Hate 2015d). The crowdfunding campaign on the Indiegogo platform raised $3400 from 43 people and also helped raise awareness of the campaign (Indiegogo 2015). There were 48 Facebook posts shared with OHPI's Facebook community about the campaign—some linking to the campaign mini-site with a counter for the number of reports gathered so far, while others linking to the crowdfunding campaign. Some of these were boosted with advertising to the wider Australia community. In total, the posts had a reach of over 224,300 and were liked, shared or commented on over 7600 times. The online community was kept up-to-date on progress towards the funding and reporting goals.

The campaign gathered over 1000 items of anti-Muslim hate, and the resulting breakdown, as well as the social media platforms' initial responses in removing the hate, was published in an interim report on International Human Rights Day in December 2015 (Oboler 2015). The collected data were shared with multiple experts for manual vetting and inclusion in a final report. Unfortunately, this work was never completed due to the funding shortfall.

OHPI Data: Feeding into Research and Policy

OHPI seeks to be a bridge between the online community wanting to take action against cyber racism, and the researchers, policymakers and government agencies working to address the problem of cyber racism. This work started in the policy sphere, with the plans to develop

FightAgainstHate.com as a solution to the difficulty in gathering data about online hate in social media. OHPI, while engaging in many other activities, was created as the institutional vehicle to drive forward work on the FightAgainstHate.com platform.

The origins of FightAgainstHate.com are in a software specification prepared in consultation with international anti-racism experts at the Global Forum to Combat Antisemitism in 2011 (Abitbol 2011). It was launched in Sydney by The Hon Paul Fletcher on behalf of the then Minister for Communications, the Hon Malcolm Turnbull, in December 2014 (Roth 2014). In 2015, the Global Forum to Combat Antisemitism endorsed the tool and circulated the first interim report created using FightAgainstHate.com. The completed report, which included additional details of the takedown rates for different types of antisemitism across Facebook, YouTube and Twitter, was published in early 2016 (Oboler 2016b). This provided the template later used for the "Spotlight on Anti-Muslim Internet Hate Interim Report" in the SAMIH campaign. The reports demonstrated the impact of FightAgainstHate.com, which was described in two UNESCO reports as one of the "innovative tools for keeping track of hate speech across social networks and how different companies regulate it" (Gargliardone et al. 2015, p. 4; Pollack 2015, p. 46). Data from the interim reports were also presented before diplomats at the United Nations in New York as part of a United Nations Alliance of Civilizations conference "Tracking Hatred: An International Dialogue on Hate Speech in the Media" (OHPI 2015c).

These reports also highlight the types of hate which platforms are less willing to remove. They provide a baseline on how effective social media companies are with self-regulation to remove hate. Evidence of the failure of self-regulation has provided the basis for introducing co-regulation, first in Germany and then in Europe. In May 2016, for example, the European Commission and technology companies reached an agreement that 50% of hate speech would be removed within 24 hours (European Commission 2016). Data from another study demonstrating that in a sample of 600 items only 25% of the hate was removed within 24 hours led to calls by European officials for technology companies to do more to tackle the problem (*The New York Times* 2016). Software like

FightAgainstHate.com could facilitate such monitoring to occur on a daily basis and using significantly larger samples. The one difficulty is that, at present, OHPI needs to extract the data from the FightAgainstHate.com system and manually create the summary statistics and graphs used in the reports.

The CSI-CHAT Analysis Tool

A prototype tool called CSI-CHAT (Crowd Sourced Intelligence—Cyber-Hate and Threats) was completed by OHPI in late 2016 to allow organisations from key stakeholder groups, such as academics, government agencies, civil society organisations and community organisations, to:

- Create their own schema of hate types;
- Review the data reported through FightAgainstHate.com;
- File it according to their organisations schema and
- Produce summary statistics.

CSI-CHAT not only provides data collated through FightAgainstHate.com, such as the total number of people who have reported a particular social media item, the date it was first reported and the date it was most recently reported, but also:

- allows users from stakeholder organisations to add their own annotations to reported items;
- allows users from stakeholder organisations to create incident reports which carry a description as well as embedding related items which have been reported and
- provides summary statistics by generating tables and graphs, giving a live summary of the organisation's sorted data as well as trend analysis based on that data for any designated time period.

Essentially, CSI-CHAT takes the power of FightAgainstHate.com and makes it available to a wider audience of stakeholders. This makes it easier for experts in anti-racism to become part of the expert community tackling online hate. It also increases the impact and value of the contributions of the online community reporting content through FightAgainstHate.com.

Case Study: IndigenousX

Who: IndigenousX

What: A permanent and growing affiliated grassroots Twitter community made up of Indigenous Australian guest tweeters and followers who are either Indigenous Australians or supporters of IndigenousX (IndigenousX 2017a).

How: IndigenousX is a private company that runs on a modest budget plus significant voluntary time.

Why: The primary aim of IndigenousX is to give Indigenous Australians a stronger voice in Australian society. As part of this, IndigenousX aims to increase awareness about Indigenous Australian issues, including but not limited to racism, which impact on the Indigenous Australian community.

An examination of 600 tweets by four guest tweeters during January and February 2016 focused specifically on how IndigenousX achieves its aims and how this contributes to community resilience. Findings show that the narrative developed over that period counteracts the racist narrative directed at the Australian Indigenous community. This racist narrative is the manifestation of attitudes that have developed in Australia since colonisation (Connelly 2016). These attitudes are based on a version of events and experiences that privileges the dominant (White) perspective. This perspective ignores or downplays the Indigenous Australian perspective and negatively stereotypes Indigenous Australian people and their culture. Indigenous Australians are often portrayed as hopeless victims without a viable culture; lazy, drunk or affected by drugs; dependent on welfare; and without the ability to determine their own future (Kowal 2015; Graham 2015).

The IndigenousX counter-narrative that developed over the four-week period of this research contrasts with Australian racist narratives. The topics tweeted about are those that are important to Indigenous Australians and include:

- Invasion Day;
- The National Apology to the Stolen Generation;
- Indigenous health;
- Direct racism and
- Maintenance of language and culture.

The counter-narrative challenges the dominant racist perspective and also focuses on the diversity of Indigenous Australian people, their perspectives and their performance of excellence, thereby challenging negative stereotypes. They also show, through their emotional expression, that they are hopeful about their future. This powerful counter-narrative acts as a form of resistance to racism.

The IndigenousX Community

The founder of IndigenousX, Luke Pearson, started tweeting about Indigenous issues because he felt "the national dialogue was characterised by a consistent lack of awareness, understanding and respect for Indigenous people" (IndigenousX 2017a). In 2012, this expanded to the IndigenousX account, the aim of which is to enhance the platform and profile he had created by providing an opportunity for 52 other Indigenous people per year "to share their knowledge, opinions and experiences with a wide audience of interested Tweeps" (IndigenousX 2017a). Each week, a different Indigenous Australian person guest hosts the account, tweeting about topics that are important to them.

The IndigenousX Twitter community is supported through a variety of online resources, which work to raise the profile of the Twitter account and grow the community, including:

- A website that gives an overview of IndigenousX (IndigenousX 2017a);

- A Facebook page that posts news articles and refers back to the Twitter account (Facebook 2017c);
- A blog connected to the website that posts educational and news articles (IndigenousX 2017b);
- A YouTube channel that presents talks, interviews and campaign information about IndigenousX and provides music and other playlists (YouTube 2017);
- A dedicated section in online news site Guardian Australia publishing an online profile of the guest tweeter on a weekly basis (Guardian Australia 2017). The section, entitled "Our stories, our way," provides an opportunity for the guest tweeters to explain who they are and what they are passionate about. This gives readers and followers of IndigenousX insights into the lives of IndigenousX tweeters. It also gives a preview of what they might tweet about when they are a guest on the IndigenousX Twitter account.

Australia Day, the National Apology Day and Mabo Day

Two major events, Australia Day and the anniversary of the National Apology (to the stolen generation), fell in the research period. Originally, the focus of Australia Day was about the arrival of the First Fleet and the raising of the British Flag. However, Australia Day has evolved to be a national holiday put aside to celebrate "What's great about Australia and being Australian" (NADC 2017). Australia is now recognised and embraced by most Australians as a multicultural nation (Markus 2016). As such, contemporary Australia Day celebrations also focus on promoting diversity and positive Australian values such as mateship. Campaigns such as OHPI's 2016 "Unity in Diversity on Australia Day" campaign, described earlier, reinforce these ideas.

However, the current date of Australia Day falls on the anniversary of the day Australia was colonised by the British in 1788. For this reason, many Indigenous Australians believe the narrative about Australia Day represents the invasion of their country by the British, and therefore an ongoing series of traumas. In response to this narrative, the number one hashtag on the IndigenousX Twitter account around Australia Day was #Invasion Day 2016. Other common words used in the tweets to describe Invasion Day include "massacre" and "genocide." These words represent a part of the IndigenousX narrative that is ignored or somewhat hidden from the Australian narrative. For example, one tweet included a link to a list of Indigenous Australian massacres that occurred after colonisation, but this part of Indigenous history is generally not known or is unacknowledged by many White Australians in their celebration of Australia day.

The hashtags #Apology and #Apology2016 were also prominent during this time. The anniversary of the Australian Government Apology to Indigenous Australians (February 13, 2008) for the forced removal of their children ("The Stolen Generation") is an important part of the Australian narrative that goes some way to recognising the past trauma of Indigenous Australians. The tweets suggest that while nothing can make up for the pain and suffering caused by the policy of forced removal of Indigenous children enacted by previous governments, the Apology at least recognises that this policy was wrong and hurtful. The prominence of these tweets as part of the IndigenousX narrative highlights the importance of ensuring the Indigenous Australian perspective is included as part of Australian history. The #Maboday (named after the late Eddie Mabo, a modern advocate for the recognition of native title, and the return of Indigenous lands) was also one of the top hashtags and represents a call for a day recognising Indigenous Australians (replacing the Queen's Birthday) that is not tied up with the White version of events.

Indigenous Health

The failure of government policies to "close the gap" with regard to Indigenous health was another prominently tweeted topic. Tweets were about the waste of government funds, continued evidence of family violence and prescription drug misuse. It was also pointed out that the "Northern Territory Intervention" ("'a national emergency response to protect Aboriginal children in the Northern Territory' from sexual abuse and family violence" (Australian Human Rights Commission 2007)) has resulted in more Indigenous children being taken away from their families, creating what has been referred to as a "new stolen generation". A comment from radio broadcaster Alan Jones that implied this new stolen generation was necessary to fix Indigenous issues "for their own good" engendered an angry response. This is an example of the subjugation in the racist narrative that IndigenousX tweeters are affected by and respond to.

Direct Racism

Bringing topics into the limelight that would otherwise be missed or whitewashed by the mainstream media is also part of the IndigenousX narrative. Direct racism against Indigenous Australians is one such topic. Some tweets highlighted an incident where prominent sports people used blackface (representing Indigenous Australians) at a fancy dress party. Blackface is widely recognised as a form of racism because of its past use in demeaning people of colour. Other tweets focused on a video game called *Survival Island 3* that was available for sale on Google Play and iTunes. The game called for users to kill Indigenous Australians. IndigenousX tweets highlighted these racist incidents from the perspective of Indigenous Australians.

Cultural Revival

Cultural revival through education is also part of the IndigenousX narrative. The recognition of Aboriginal English (#AbE) as a language was highlighted as an important issue by a teacher who was a guest tweeter. Many other tweets by this tweeter recognised the importance of literacy for Indigenous children, not just in English but in Indigenous languages as well. Other tweets called for the teaching of Indigenous culture and history as part of the mainstream Australian school curriculum and pointed out the significance of language revival in maintaining Indigenous Australian culture. While not discussed as part of the mainstream White narrative, the maintenance of language and culture through education is put forward as a positive step in improving Indigenous Australians lives.

IndigenousX Optimism and Excellence

The perception that Indigenous Australians are hopeless victims is also challenged through the emotion words used in the IndigenousX narrative. Mohammad and Kiritchenko (2015) argue that all language, including tweets, is used to convey a range of emotions. An examination of IndigenousX tweets, exploring some common emotion words, shows 87.5% were positive, with words like "amazing" and the #awesome featuring. The most common emotion words expressed were "love" and "hope," showing that these are important among the IndigenousX tweeters in this research. Despite many of the tweets highlighting negative aspects of Indigenous Australian culture in Australia, the tweets show an overwhelmingly optimistic overtone through the emotion words used.

As well as highlighting events, experiences and emotions from the perspective of Indigenous Australians, IndigenousX proactively challenges negative stereotypes of Indigenous Australians through the promotion of Indigenous excellence. The Guardian Australia profiles show that IndigenousX guest tweeters have different life experiences and come from a variety of backgrounds. The four guest tweeters tweeting during the period of this research included a musician, a teacher, a journalist and a solicitor, and their backgrounds were both urban and rural. Other IndigenousX guest tweeters (who were not part of this research) also vary greatly in their life experiences and backgrounds, reflecting the general diversity of Indigenous Australians. In addition, the "X" in IndigenousX stands for excellence. IndigenousX demonstrates that excellence can be "performed" through the diversity of the guest tweeters, their different ways of experiencing and communicating their worlds and their powerful individuality.

Resistance, Solidarity and Community Resilience

The individual IndigenousX guest tweeters are drawn together as a community with a common goal of sharing their knowledge, opinions and experiences. The authenticity of the IndigenousX tweeters also attracts followers who feel empathy with them and/or support their tweets. These followers act in solidarity, with the guest tweeters increasing the visibility and strength of this community and the counter-narrative. At the time of writing, the IndigenousX community was made up of over 250 Indigenous Australian guest tweeters and 29,300 followers. In addition, the IndigenousX counter-narrative has a high level of human interest and has led to mainstream media coverage potentially also growing the community. As the IndigenousX community grows, so too does the spread and power of the counter-narrative of resistance to racism and the solidarity with others who have the same views.

There is growing evidence that this form of resistance and solidarity may also contribute to Australian Indigenous community resilience. An Australian study (Bodkin-Andrews 2013) identified that acknowledging and challenging racism are important factors in individual Indigenous Australian's resilience to racism. An international study on community resilience using Twitter (Rodgers and Scobie 2015) also identified that working towards a common goal against racism strengthened a community's common bond and belief in themselves. In this case, there is further potential for members of other Australian Indigenous communities and the general public to feel empowered to speak up in solidarity with the IndigenousX Twitter community, making the IndigenousX counter-narrative a more visible part of the Australian narrative and contributing to Australian Indigenous community resilience.

Additional Approaches

An Unaffiliated Civil Society Group

All Together Now is a purpose-driven civil society organisation focused generally on encouraging the embracement of cultural diversity and the eradication of racism (All Together Now 2017). Following recommendations in the *Building on our Strengths' Report* (Victorian Health Promotion Foundation 2009), All Together Now uses a combination of online and traditional educational and campaigning strategies to achieve its aims. Work so far has included:

- The development of the award-winning **Everyday Racism Mobile App** (based on giving individual users an immersive experience);
- The similarly inspired **Anti-racism App for primary school children** (promoting positive peer relations among students);
- The **CAPE project** (designed to counter White nationalist conspiracy theories);
- The **#Eraseracism campaign** (designed to counter racism in Football) and
- The award-winning **One Parramatta Project** (a film project targeting youth, that inspired self-reflection about racism and cultural inclusion).

All Together Now is a registered charity with a budget in 2016 of just over $100,000 (Australian Charities and Not For Profit Commission 2016).

The use of a social media is seen as a vital marketing tool to promote All Together Now's projects and campaigns, and to spread the message challenging racism in Australia and building a culture of solidarity against racism. Although not part of the original intended strategies, the use of social media has evolved to include building and maintaining online communities of resistance and solidarity to racism. The All Together Now Facebook page has 4000 fans (Facebook 2017d), and the Twitter account has over 2000 followers (Twitter 2017a). In both cases, All Together Now uses these channels not only to share their own campaigns, but also to share content from other sources such as news stories, videos and the social media campaigns of others.

Some social media content directly requests community members' engagement, for example, the Facebook post on December 16, 2016, which shared a video with the comment "We all know Australia is a culturally diversified country, but sometimes we may encounter some cultural stereotype conditions. Check out this interesting video and post comments about your encounters and feelings!" (Facebook 2017d). Through their engagement with All Together Now social media content, supporters contribute to the discussion and build the sense of community. They thus raise their own and each other's awareness of racism and issues surrounding cultural diversity, while joining together on campaigns in solidarity against racism.

A Government Agency

The AHRC is an independent statutory organisation that reports to the Australian Parliament through the Attorney-General (Australian Human Rights Commission 2017a). The Commission's mission is "leading the promotion and protection of human rights in Australia," and it includes a president who serves as the CEO and seven commissioners with specific portfolios for different areas of human rights, including a Race Discrimination Commissioner (Australian Human Rights Commission 2017a). The Commission employs over 100 staff and has an annual operating budget of over AU$22.5 million (Australian Government 2016). In August 2012, the Commission launched the "Racism. It Stops With Me" national anti-racism campaign as part of a National Anti-Racism Strategy (Australian Human Rights Commission 2013a).

The national anti-racism campaign allows businesses, sports bodies, educational providers, local governments and civil society organisations to become supporter organisations by agreeing to "endorse the campaign message, promote the campaign and identify specific activities they will undertake in the anti-racism space" (Australian Human Rights Commission 2013b). Supporters sign a legal agreement with the Commission, though there is no financial commitment from either side: supporting organisations do not pay to participate, nor does the Commission provide grants to support the activities of supporting organisations. Individuals were also able to become supporters by entering their name, e-mail address and their photograph. After its first year, the campaign had 160 organisational supporters and 900 individual supporters (Australian Human Rights Commission 2013c). By June 2015, the number of organisations had grown to over 360, and by February 2017, there were over 2400 individual supporters signed up (Australian Human Rights Commission 2015; Australian Human Rights Commission 2017b).

The national anti-racism campaign has a community based around its Twitter account @itstopswithme, which, in January 2017, had 8300 followers and had made 1575 tweets (Twitter 2017b). Followers engage by sharing their own anti-racism work through tweets to the campaign account and using the hashtag #itstopswithme, and by liking, retweeting and replying to the campaign account's tweets. The campaign account often likes or retweets posts directed to it, sharing them with the wider campaign community. In this way, the account acts as an amplifier for the anti-racism content from supporters. The campaign does not operate a Facebook page, a move which has allowed its name and logo to be appropriated by a group of students outside Australia who have created a small and largely inactive page combating racism (Facebook 2017e).

The AHRC also operates a large Facebook page, with over 85,000 fans (Facebook 2017f). The page:

- Promotes the events and activities of the Commission;
- Shares news articles about the Commission;
- Features commentary from the Commissioners;
- Posts links to new content on the Commission's website and
- Shares content from the individual pages operated by each Commissioner.

The page does not share social media content from other organisations, though an occasional exception is made, for example, for content from the United Nations. This makes it far more a broadcast mechanism than an online community. This is also reflected in interactions. The audience shares the content, but discussion is limited and is usually in the form of requests for additional information.

The Race Discrimination Commissioner's page has just under 3000 fans (Facebook 2017g). On his page, the Race Discrimination Commissioner shares his own articles in the media, relevant news stories and photographs and reports on his activities and engagements with the community. The Commissioner makes around 10–15 posts a month compared to around 60 a month made on the Commission's page. While the engagement is higher on the Commissioner's page, most of it is negative. Some comments are critical of the arguments made in the

Commissioner's articles, but much of it is opposition to anti-racism work in general, and to the role of the Commissioner and the Commission specifically. If the Commissioner's page constitutes an online community, that community is not one of solidarity and resistance to racism. The high number of likes and shares, however, shows the page does provide useful content for other communities. The decision to allow those opposed to the Commissioner's work to have their say, while perhaps appropriate for a Government Commissioner, has resulted in a lack of a safe space for those opposed to racism and as a result has significantly inhibited the formation of a community of solidarity and resistance to racism forming around the page.

A Peak Body for Impacted Communities

The Federation of Ethnic Communities' Councils of Australia (FECCA) is an example of a permanent national peak body that has a general focus on anti-racism as part of its work (FECCA 2017). FECCA's general priorities are "the promotion of multiculturalism, the protection of community harmony, the advocacy of social justice, and the rejection of racism and discrimination" (Facebook 2017h). FECCA's work focuses on advocating and promoting general issues on behalf of Australians from culturally and linguistically diverse (CALD) backgrounds to government, business and the broader community. FECCA runs meetings and conferences to canvas issues among the various communities and shares a number of resources including fact sheets, reports and the Australian Mosaic Magazine.

It is difficult for FECCA to run its own campaigns on specific issues related to cyber racism because of the limitations created by the competing priorities of the many communities that form part of FECCA and insufficient government funding. This was demonstrated recently when the FECCA Youth Group, which comprised representatives of CALD youth from around Australia, expressed an interest in creating a project that focused on positive ways to respond to cyber racism. However, research as part of the CRaCR project showed that despite many meetings and the recognition among participants that this was an important issue for Australian CALD youth, no one was able to take responsibility for this project, and funding was not specifically allocated to it. The FECCA Youth Cyber-Racism Project consequently has not yet proceeded.

FECCA's role in developing online communities of resistance and solidarity to cyber racism is more likely to be in its potential to amplify campaigns on behalf of other organisations. FECCA's social media presence is growing: its Facebook page has almost 1500 followers, and its Twitter account has over 2700 followers (Facebook 2017g; Twitter 2017c). There is already evidence that amplification is occurring. For example, on January 5, 2017, FECCA posted a link on their Twitter account to a grassroots kickstarter page that is trying to raise funds to support a campaign challenging anti-Muslim sentiment by promoting Muslim diversity. This would increase the visibility and effectiveness of that campaign. So far, most of the amplification through links and retweets or posts on FECCA's social media is from organisations promoting cultural diversity or highlighting excellence among migrants and refugees. A stronger focus on supporting existing online anti-racism communities would amplify the resistance and solidarity of those communities.

Conclusion

The case studies in this chapter represent some of the different approaches that can be taken to build online communities of resistance and solidarity. These approaches include:

- Grassroots initiatives;
- Civil society organisations affiliated with targeted communities;
- Civil society organisations addressing racism more broadly either online or in general;
- Civil society organisations representing communities or groups of communities and
- Initiatives by government agencies.

There are different challenges to creating a vibrant community depending on the type of stakeholder creating the community. In all cases, to be an effective community, there needs to be a level of interaction and engagement. This needs to go beyond the use of the online space as a channel for advertising the sponsoring organisation, or the use of the space as a news channel to gather and share stories by those interested in anti-racism work.

There are a wide variety of communities that can form, including:

- Permanent communities that bring together people interested in resisting racism and standing in solidarity against it;
- Temporary and *ad hoc* communities that coalesce around an incident, a campaign or through ongoing responses by the same group to a particular item of content such as a news article;
- Communities that are focused on one type of cyber racism;
- Communities that address cyber racism in general;
- Communities that are dedicated to combating cyber racism as a whole or
- Communities with a broader focus, which occasionally engage in activities of resistance to racism and solidarity with victims of racism.

Different approaches will be easier to implement, or have a greater chance of successful engagement when spearheaded by different types of online communities supported by different types of stakeholders. **Cooperation between organisations can significantly increase the effectiveness of the resulting community.**

Online communities run by civil society groups affiliated with a particular community impacted by cyber racism may see resistance to racism against their community as one of many competing aims for their community. Creating a community of solidarity against other forms of cyber racism may or may not be considered relevant in the ordinary course of events, but such communities can become communities of solidarity in times of crisis. The special case of community peak representative bodies also serves a vital function in speaking out on behalf of their members when the community comes under attack (as demonstrated in Chap. 5). This provides a narrative, which communities can gather around if the processes and relationships exist to support this.

Civil society organisations dedicated to anti-racism work in general may see building an online community of resistance and solidarity as one of their key goals. A community focused on multiple kinds of cyber racism benefits from greater reuse of approaches, tools and experience. These communities may entice people to initially engage as an act of resistance to an attack on their own community, but can also convert this support into solidarity for other communities that come under attack at a later stage.

Australian anti-racism policies have in the past primarily focused on the use of new technology as a channel for advertising. **Attempts to fully control online spaces and keep the discussion focused on an organisation or agencies own work run counter to the collaborative nature of social media.** The size, significance and impact of an online community are not primarily determined by the size, age or authority of the sponsoring organisations. Digitally focused civil society organisations and grassroots campaigns may be better able to create effective online communities than their larger, more established and better funded counterparts.

Collaborations involving material support and endorsements to organisations running effective online communities from organisations and agencies in a position to provide this support can be an effective path to creating stronger online communities of resistance and solidarity. Evidence of support for a community from those with formal authority and public profile can help energise and grow the community. **Endorsing and materially supporting the work of those running effective online communities is a quick and cost-effective path to becoming a significant member of the community.** The community then becomes a channel, promoting the relevance of the supporting organisation in tackling cyber racism. This can be far more cost-effective and impactful than seeking to build a new community, which may involve additional staff, project management and outsourcing of technical work, which a civil society organisation focused on digital advocacy already has in-house. There are also limitations through policies and public expectations when campaigns are run by larger organisations or government agencies. Supporting the communities of small organisations allows the same impact without the overheads and limitations.

This chapter has articulated approaches to creating effective online communities of resistance and solidarity, including detailing two successful case studies. The key is **encouraging meaningful engagement in a safe space**. In the case of OHPI, this engagement occurs in a Facebook based community which denies access to those associated with hate groups, and where active participation is encouraged through community campaigns, supported through the use of FightAgainstHate.com reporting tool, which draw people together to take action against different forms of racism in Australia. IndigenousX focuses on providing a safe space for Indigenous Australian tweeters to counter the racist narrative against their community. It not only draws IndigenousX tweeters together but also builds a community of supporters who counter racism against Indigenous Australians.

Communities can also be strengthened by engagement from the public and other stakeholders. Partnerships and collaborations can be formed between different organisations with similar aims. These organisations can also promote the good work of others, even in the absence of formal partnerships. Online communities are also not limited geographically and can expand to like-minded people and groups around the world. The recognition of the value of building successful online communities of resistance and solidarity and the support of grassroots civil society organisations leading the way are vital in the effort to combat cyber racism.

References

Abitbol, C. (2011, July 27). The Fight Against Online Hate. *The Australian Jewish News*. https://www.jewishnews.net.au/the-fight-against-online-hate/22046. Accessed 9 Feb 2017.

All Together Now. (2017). *All Together Now About Us*. http://alltogethernow.org.au/about-us-2/. Accessed 10 Feb 2017.

Australian Charities and Not For Profit Commission. (2016). *All Together Now*. https://www.acnc.gov.au/RN52B75Q?ID=21EBC005-6E2A-454F-B519-5A5DB6C842B5&noleft=1. Accessed 10 Feb 2017.

Australian Charities and Not For Profit Commission. (2017). *Online Hate Prevention Institute*. https://www.acnc.gov.au/RN52B75Q?ID=08FA27C4-3484-4CCD-89F0-D74243903476&noleft=1. Accessed 9 Feb 2017.

Australian Government. (2016). *Australian Human Rights Commission*. https://www.ag.gov.au/Publications/Budgets/Budget2016-17/Documents/Portfolio-budget-statements/PBS-AHRC-2016-17.pdf. Accessed 10 Feb 2017.

Australian Human Rights Commission. (2007). *Social Justice Report 2007—Chapter 3: The Northern Territory 'Emergency Response' Intervention*. https://www.humanrights.gov.au/publications/social-justice-report-2007-chapter-3-northern-territory-emergency-response-intervention. Accessed 10 Feb 2017.

Australian Human Rights Commission. (2013a). *Racism. It Stops with Me*. https://itstopswithme.humanrights.gov.au/. Accessed 10 Feb 2017.

Australian Human Rights Commission. (2013b). *Racism. It Stops with Me*. Support the Campaign: Organisations. https://itstopswithme.humanrights.gov.au/what-can-you-do/join-campaign/support-campaign-organisations. Accessed 10 Feb 2017.

Australian Human Rights Commission. (2013c). *Racism. It Stops with Me and the National Anti-Racism Strategy: One Year on, Australian Human Rights Commission*. https://itstopswithme.humanrights.gov.au/sites/default/files/RISWM%20One%20Year%20On%20Report_0.pdf. Accessed 10 Feb 2017.

Australian Human Rights Commission. (2015). *National Anti-Racism Strategy and Racism. It Stops with Me: Summary Evaluation and Future Directions*. https://www.humanrights.gov.au/our-work/race-discrimination/publications/national-anti-racism-strategy-and-racism-it-stops-me. Accessed 10 Feb 2017.

Australian Human Rights Commission. (2017a). *About the Commission*. https://www.humanrights.gov.au/about-commission-0. Accessed 10 Feb 2017.

Australian Human Rights Commission. (2017b). *Individual Supporters*. https://itstopswithme.humanrights.gov.au/what-can-you-do/join-campaign/individual-supporters. Accessed 13 Feb 2017.

Bodkin-Andrews, G. (2013). Promoting Resiliency to Counter Racism: The Lived Wisdom Within Aboriginal Voices. *InPsych, 35*(4), 14–15. www.psychology.org.au/publications/inpsych/2013/august/bodkin-andrews. Accessed 10 Feb 2017.

Connelly, K. (2016). Undermining Cosmopolitanism: Cyber-Racism Narratives in Australia. *Sites, 13*(1), 156–176.

Daniels, J. (2012). Race and Racism in Internet Studies: A Review and Critique. *New Media & Society, 15*(5), 695–719. doi:10.1177/1461444812462849.

European Commission. (2016). European Commission—Press Release. http://europa.eu/rapid/press-release_IP-16-1937_en.htm. Accessed 9 Feb 2017.

Facebook. (2017a). *Facebook Business—Boost Your Posts*. https://www.facebook.com/business/help/547448218658012. Accessed 14 Feb 2017.

Facebook. (2017b). Online Hate Prevention Institute. https://www.facebook.com/onlinehate/. Accessed 9 Feb 2017

Facebook. (2017c). *IndigenousX*. https://www.facebook.com/Indigenousx/. Accessed 10 Feb 2017.

Facebook. (2017d). *All Together Now*. https://www.facebook.com/alltogether-now.org.au/. Accessed 10 Feb 2017.

Facebook. (2017e). *Racism Stops with Me*. https://www.facebook.com/pg/Racism-Stops-With-Me-565692716841638/about/?ref=page_internal. Accessed 10 Feb 2017.

Facebook. (2017f). *Australian Human Rights Commission*. https://www.facebook.com/aushumanrights/. Accessed 10 Feb 2017.

Facebook. (2017g). *Australian Race Discrimination Commissioner Tim Soutphommasane*. https://www.facebook.com/Australian-Race-Discrimination-Commissioner-Tim-Soutphommasane-653778944684939/. Accessed 10 Feb 2017.

Facebook. (2017h). *FECCA*. About. https://www.facebook.com/pg/iFECCA/about/?ref=page_internal. Accessed 10 Feb 2017.

FECCA. (2017). *FECCA*. http://fecca.org.au. Accessed 10 Feb 2017.

Fight Against Hate. (2015a). *Welcome to Fight Against Hate*. https://fightagainsthate.com. Accessed 9 Feb 2017

Fight Against Hate. (2015b). *SAMIH: Spotlight on Anti-Muslim Internet Hate*. About. https://fightagainsthate.com/samih/about.php. Accessed 9 Feb 2017.

Fight Against Hate. (2015c). *SAMIH: Spotlight on Anti-Muslim Internet Hate*. Sponsors, Partners & Supporters. https://fightagainsthate.com/samih/partners.php. Accessed 9 Feb 2017.

Fight Against Hate. (2015d). The SAMIH Project. Budget. https://fightagainsthate.com/samih/resource/budget.pdf Accessed 9 Feb 2017.

Gargliardone, I., Gal, D., Alves, T., & Martinez, G. (2015). *Countering Online Hate Speech*. Paris: UNESCO.

Gehl, R. W. (2010). *A Cultural and Political Economy of Web 2.0* (PhD Thesis). Online at: http://www.robertwgehl.org/text/gehl_diss_full.pdf

Go Fund Me. (2017). Put Aus girls back in Aus day ad. https://www.gofundme.com/aus-day. Accessed 9 Feb 2017.

Graham, D. G. (2015). Lost to View: The Resilience of Indigenous Australians in the Face of Systematic Racism. In V. Pulla & M. Bharath Bhushan (Eds.), *Some Aspects of Community Empowerment and Resilience* (pp. 49–62). New Delhi: Allied Publishers.

Guardian Australia. (2017). *IndigenousX*. https://www.theguardian.com/commentisfree/series/indigenousx. Accessed 10 Feb 2017.

Indiegogo. (2015). *SAMIH: Spotlight on Anti-Muslim Internet Hate*. https://www.indiegogo.com/projects/samih-spotlight-on-anti-muslim-internet-hate#/. Accessed 9 Feb 2017

IndigenousX. (2017a). *IndigenousX Our Story*. http://indigenousx.com.au/about/#.WJzkQKOr1hc. Accessed 10 Feb 2017.

IndigenousX. (2017b). *IndigenousX BlogX*. http://indigenousx.com.au/blogx/#.WGWJ66Or1DY. Accessed 10 Feb 2017.

Jakubowicz, A. (2012). Cyber Racism. In H. Sykes (Ed.), *More or Less: Democracy and New Media*. Sydney: Future Leaders.

Kowal, E. (2015). Time, Indigeneity and White Anti-Racism in Australia. *The Australian Journal of Anthropology, 26*(1), 94–111.

Markus, A. (2016). *Mapping Social Cohesion. The Scanlon Foundation Surveys 2016*. Retrieved from http://www.monash.edu/mapping-population/public-opinion/social-cohesion-report. 10 Feb 2017.

Miller, D. (2011). *Tales from Facebook*. Cambridge: Polity Press.

Mohammad, S., & Kiritchenko, S. (2015). Using Hashtags to Capture Fine Emotion Categories from Tweets. *Computational Intelligence, 31*(2), 301–326.

NADC. (2017). *About Our National Day*. http://www.australiaday.org.au/australia-day/about-our-national-day/. Accessed 10 Feb 2017.

Oboler, A. (2013). *Islamophobia on the Internet: The Growth of Online Hate Targeting Muslims*. Online Hate Prevention Institute, Melbourne.

Oboler, A. (2015). *Spotlight on Anti-Muslim Internet Hate Interim Report*. Melbourne: Online Hate Prevention Institute. http://ohpi.org.au/anti-muslim-hate-interim-report/. Accessed 9 Feb 2017.

Oboler, A. (2016a). The Normalisation of Islamophobia Through Social Media: Facebook. In I. Awan (Ed.), *Islamophobia in Cyberspace* (pp. 41–62). London: Routledge.

Oboler, A. (2016b). *Measuring the Hate: The State of Antisemitism in Social Media*. Online Hate Prevention Institute. http://ohpi.org.au/measuring-antisemitism/. Accessed 9 Feb 2017.

Oboler, A., & Connelly, K. (2014,December 10–12). Hate Speech: A Quality of Service Challenge. In *Proceedings of the IEEE Conference on e-Learning, e-Services and e-Management* (pp. 117–121). Melbourne.

Office for the Community Sector. (2013). *Community Collaboration: The Changing Context of Local Government and Community Sector Partnerships*

(Report). Melbourne: Department of Health and Human Services. Retrieved from http://www.dhs.vic.gov.au/__data/assets/word_doc/0005/832172/Community-collaboration-The-changing-context-of-local-government-and-community-sector-partnerships-1-July-2013.doc. Accessed 30 June 2016.

OHPI. (2015a). *Launch of Spotlight on Anti-Muslim Internet Hate Campaign*. http://ohpi.org.au/launch-of-spotlight-on-anti-muslim-internet-hate-campaign/. Accessed 9 Feb 2017.

OHPI. (2015b). *Victorian Government Supports SAMIH*. http://ohpi.org.au/victorian-government-supports-samih/. Accessed 9 Feb 2017.

OHPI. (2015c). *Presentation in the UN (New York)*. http://ohpi.org.au/unaoc/. Accessed 9 Feb 2017.

OHPI. (2016a). *Annual Report 2016-Briefings*. http://ohpi.org.au/annual-report-2016-briefings/. Accessed 9 Feb 2017.

OHPI. (2016b). *Unity in Diversity on Australia Day*. http://ohpi.org.au/unity-in-diversity-on-australia-day/. Accessed 9 Feb 2017.

OHPI. (2016c). *Racist Groups Use Australia Day to Promote Bigotry*. http://ohpi.org.au/racist-groups-use-australia-day-to-promote-bigotry/. Accessed 9 Feb 2017.

OHPI. (2016d). *The Different Meanings of Australia Day*. http://ohpi.org.au/the-different-meanings-of-australia-day/. Accessed 9 Feb 2017.

Oxford Dictionaries. (2017). *Online Community*. https://en.oxforddictionaries.com/definition/online_community. Accessed 9 Feb 2017

Paradies, Y., Chandrakumar, L., Klocker, N., Frere, M., Webster, K., Burrell, M., & McLean, P. (2009). *Building on Our Strengths: A Framework to Reduce Race-Based Discrimination and Support Diversity in Victoria* (Report). Melbourne: Victorian Health Promotion Foundation. https://www.vichealth.vic.gov.au/~/media/ProgramsandProjects/Publications/Attachments/Building on our strengths-full report v2.ashx. Accessed 14 Feb 2017.

Pollack, I. R. (2015). *World Trends in Freedom of Expression and Media Development: Special Digital Focus 2015*. Paris: UNESCO.

Rodgers, K., & Scobie, W. (2015). Sealfies, Seals and Celebs: Expressions of Inuit Resilience in the Twitter Era. *Interface: A Journal on Social Movements, 7*(1), 70.

Roth, P. (2014). New Tool to Fight Online Hate. *The Australian Jewish News*. https://www.jewishnews.net.au/new-tool-to-fight-online-hate/38909. Accessed 9 Feb 2017.

The Big Smoke. (2015). Spotlight on Antisemitism—Illuminating the Nature of Hate. http://thebigsmoke.com.au/2015/03/31/spotlight-antisemitism-illuminating-nature-hate/. Accessed 9 Feb 2017.

The New York Times. (2016). Europe Presses American Tech Companies to Tackle Hate Speech. http://www.nytimes.com/2016/12/06/technology/europe-hate-speech-facebook-google-twitter.html?_r=0. Accessed 9 Feb 2017.

Twitter. (2017a). *All Together Now*. https://twitter.com/alltognow. Accessed 10 Feb 2017.

Twitter. (2017b). *Racism. It Stops with Me*. https://twitter.com/itstopswithme. Accessed 10 Feb 2017.

Twitter. (2017c). *FECCA*. https://twitter.com/ifecca. Accessed 10 Feb 2017.

Victorian Auditor-General's Office. (2015). *Public Participation in Government Decision-making: Better Practice Guide* (Report). Melbourne: Victorian Auditor-General's Office. http://www.audit.vic.gov.au/publications/20150130-Public-Participation-BPG/20150130-Public-Participation-BPG.pdf. Accessed 14 Feb 2017

Victorian Health Promotion Foundation. (2009). *Building on Our Strengths*. https://www.vichealth.vic.gov.au/media-and-resources/publications/building-on-our-strengths. Accessed 10 Feb 2017.

Youtube. (2017). *IndigenousX*. https://www.youtube.com/channel/UCTK6oqOtAgPONFsdVyf6fDw. Accessed 10 Feb 2017.

8

Promoting Resilience Through Regulation

Introduction

In 2016, an Australian Aboriginal woman, well-known sportswoman and former Australian Labor Party senator Nova Peris, was abused in a vicious Facebook post by a local Liberal Party official. The man, a formerly well-respected chiropractor, pleaded guilty and was sentenced for using a carrier service to menace, harass or offend pursuant to s 474.17 of the Criminal Code Act 1995 (Commonwealth). This case became an important test of the validity of the two competing narratives of Australian identity discussed in Chap. 6. As we have seen, a number of the Target groups in Chap. 5 also saw the case as an important example of racism online and what recourse its targets might have. The media reporting of the case (Fitzpatrick 2016; AAP 2016) pointed to two aspects of the outcome—firstly, that Nova Peris had bravely called out the racism while others, by implication, might have avoided it. Secondly, the state acted through a criminal prosecution using legislation, the use of a carriage service to harass, that had some years earlier been used to convict Man Haron Monis (the Sydney Lindt café terrorist bomber of 2014) for intimidation of the families of Australian soldiers who had died in Afghanistan (Wilson 2013).

A. Jakubowicz et al., *Cyber Racism and Community Resilience*, Palgrave Hate Studies,
DOI 10.1007/978-3-319-64388-5_8

While her harasser avoided a custodial sentence, Ms. Peris was satisfied that the suspended gaol sentence, monetary fine and good behaviour bond imposed on him were potent in sending a clear message "to others who use social media to disseminate such vile and racially abusive comments" (ABC News 2016). The online racism directed at Ms. Peris, as we have already shown, is one illustration of many in contemporary Australian society. What is interesting about this particular incident is that it provides an example of how criminal measures may be applied in reprimanding offenders. But, is this the optimal way to regulate online racism?

In this chapter, we examine how the law grapples with cyber racism, focusing on the complex layers of jurisdictions, initiatives and avoidances that have characterised the Australian situation. The regulation of online racism is often approached through well-rehearsed debates around freedom of speech and freedom from racism. We wish to add a further dimension to this debate: how can we promote respect for human rights on the Internet, but resist relying upon the criminal law as first resort? Moreover, how can we build a model of regulation that promotes community resilience? Drawing on our original data stream (described in Chap. 3), we begin by examining what online users experience and want in terms of balancing regulation with freedom of expression. To understand the complexities surrounding effective regulation, we delve into the challenges that face Internet Service Providers (ISPs), social media platforms, governments and law enforcement agencies globally. We find that there is a lack of consensus around the criteria for intervention, even the need for intervention, leading us to ask the following: how do we decide what should be regulated? How offensive does material and behaviour need to be to warrant censure? Each jurisdiction adopts its own approach to regulation. Here we take Australia as our example, providing an overview of the current legal and regulatory terrain. Within this context, we then focus on Facebook as a case study given its identification, in the Cyber Racism and Community Resilience (CRaCR) project's survey research, as a major site for online racism. With criminalisation at one end of the response spectrum and voluntary Codes of Conduct at the other, we find a gap in the regulation of cyber racism in Australia. We propose a multilayered response to cyber racism that:

(i) recognises voluntary community standards as providing a scaffold for an online regulatory framework;
(ii) advances a "light touch" civil penalty regime modelled on comparable approaches to regulating harmful material online and
(iii) allows for criminal law state intervention when other avenues are exhausted or where conduct is especially serious.

The overall goal of this approach is to empower Internet users to identify and resist cyber racism—backed by enforcement—not just punishment. We explore what form of regulation could operate most effectively to facilitate community resilience, that is, a strategy that sees regulation as a mode of setting civil standards while empowering communities in a manner that does not further exacerbate problems. Through this approach we seek to provide a clear illustration of why community-generated strategies are crucial and why they also need to be backed by legislative imperative. As racist and other forms of online hate continue to proliferate in the Internet's lawless and amoral realm, it is important to consider how we can raise and enforce standards in cyberspace.

Background

Before delving into the substance of this chapter, it is worthwhile revisiting the meaning of two terms that are critical to this book: resilience and racism. First of all, what is actually meant by resilience? We use resilience to describe a mode of enabling citizens to discern and name racism, and ultimately resist it (Bodkin-Andrews et al. 2013). It relates to developing skills and shoring up the capacity to unpack and challenge hostile online behaviour. Resilience is certainly *not* about victims of racism needing to "toughen up" or develop a "thick skin," nor is it any recognition of some perceived right of others to be bigots.[1]

What is meant by racism? In general terms, racism refers to unfavourable treatment, prejudice or discrimination and includes behaviour that is "likely to offend, insult, humiliate or intimidate" a person because of his or her "race, colour, national or ethnic origin" (Australian Human Rights Commission (AHRC) 2014b, p. 3). Racism is a combination of values, attitudes and behaviours that exclude people from society on the basis of their race, ethnicity, cultural practices, national origins, indigeneity and, in some instances, religious beliefs (Paradies et al. 2009).

Racism is essentially premised on heightened perceptions of difference—the binary opposites of superiority and inferiority. In this way, the target of the antagonism is regarded as "the other" merely because of his or her racial attributes. Racism and racial abuse occur in everyday public life: during situations of employment, education, housing, sport and public transport (AHRC 2014b), and adversely impact physical and mental well-being, undermine social cohesion and result in inequalities of opportunities between groups (Paradies 2006). Increasingly, racism is expressed and experienced online.

Racism may be understood from legal perspectives as well. For example, in Australia s 9 of the Racial Discrimination Act 1975 (C'th) seeks to regulate acts that involve "a distinction, exclusion, restriction or preference based on race, colour, descent or national or ethnic origin" which impair or seek to affect another person's human rights and freedoms in public life. Again, such a definition emphasises racism's harmful delineation of difference and dissimilitude. Racism is therefore contrary to ideologies of inclusion and social cohesion, and contrary to the notion that citizens should enjoy equality before the law. Such legislation operationalises international human rights measures. For example, the ICERD (1965) seeks to protect the universal dignity and equality of all human beings, without distinction or exclusion as to race, colour or ethnic origin. On this basis, doctrines of racial superiority or supremacy are recognised as morally repugnant, "socially unjust and dangerous" (ICERD 1965). However, Australia specifically filed a reservation on Article 4 of the Convention when it signed in 1967, thus freeing it from any requirement to criminalise race hate speech.

Boundaries are drawn around racism by international and domestic policy, legislation and conventions, yet, as we will see in this chapter, difficulties and ambiguities remain, raising regulatory problems, especially when it comes to racism on the Internet. As we have already seen in this book, racism is increasingly experienced online. Cyber racism refers to racism that manifests in the realm of the Internet, through words, images, videos and symbols. It is disseminated via channels as diverse as social media, online games, forums, messaging services and White supremacist and other "hate sites" (Tynes et al. 2012) Racist conduct online covers a wide spectrum ranging from racist material disguised as "humour" (Weaver 2011), to hateful comments on social media platforms (Oboler 2012), to direct threats and incitements, to violence targeting specific individuals or groups on the basis of race (Stephens 2015).

Much online hatred is spread via platforms where the content is often public, such as Facebook. Cyber racism represents a digital form of racism that is just as harmful as physical, "real world" racism. The Internet and social media can be seen as creating a new notion of "public" that opens up the private realm of users' domestic and work space, perhaps making online hatred particularly insidious in its infiltration of the target's private world. For instance, smart phones are increasingly our most constant companions in daily life, storing confidential details, close contacts, calendars and personal photographs, whilst also being a mode of online interaction, a conduit between our private and public lives. Much of what we do on our smart phones feels personal, almost an extension of our consciousness, yet, paradoxically, it is often very public, especially when we interact on social media. On this understanding, the infiltration of hateful content onto our personal devices, leading to it often being viewed while in our personal spaces, may seem a particularly potent form of violation and intimidation in its anonymity, speed and quantity from unseen and unknown sources. For instance, cyberbullying victims perceive online hostility to be the same as, if not worse than, face-to-face, real-world bullying (Dooley et al. 2009; Rohlfing 2015); cyber racism could be similarly understood.

Balancing Regulation and Freedom

How is cyber racism experienced? There is limited evidence on how experiencing "virtual" racism affects victims in comparison with direct, "in-person" or real-world experiences of racism, but it is apparent that online racism can negatively affect self-esteem and produce feelings of anger, frustration and hopelessness (Dunn et al. 2014). It is also significantly related to higher levels of depression and anxiety (Tynes et al. 2012). As detailed in Chap. 3, the "Encounters" stream of the *CRaCR Project* surveyed 2141 participants through an online platform, MyOpinion, targeting the general population as well as groups significantly at risk of racism. The purpose of the survey was to gauge:

- regular Internet users' encounters with and responses to cyber racism;
- the impacts of such behaviour;
- the effectiveness of existing regulation and
- users' experiences of counter-racism online.

The survey participants included authors of cyber racism as well as witnesses and targets of the online behaviour. Survey results indicated that the racist content was commonly encountered on social media, in particular Facebook. The differences in targets' and witnesses' responses were interesting. Targets of online racism were typically more (re)active than witnesses, with over two-thirds of targets making an active response to any racist content directed towards them. The most common forms of response for both witnesses and targets were "within platform," that is, using Facebook as an example, reporting the content and blocking or de-friending the author. Whilst significant numbers of users appeared to avail themselves of the reporting functions within platforms, the current regulations, as well as peer and community monitoring of Internet behaviour, were seen as insufficient in maintaining civil and healthy relations across the issues of cultural diversity.

The data showed that people want a level of regulation, but in a mode that does not impinge on their freedom of expression. For example, our "Encounters" survey of Internet users found that nearly half of those surveyed believed that freedom to speak your mind was more important than freedom from hate speech. Yet, in the same survey, 80% of people supported laws against racial vilification, and 69% supported laws against religious vilification, indicating that there is a high level of support for regulating the harms associated with those kinds of behaviours. This overwhelming support for laws that regulate racial/cultural/religious hatred is summarised in Table 8.1. In a similar vein, a 2014 Nielsen

Table 8.1 Results from the CRaCR Survey: Attitude on Action

	Agree %	Disagree %
It should be unlawful to offend someone on the basis of their race, culture or religion	65.8	9.9
It should be unlawful to humiliate someone on the basis of their race, culture or religion	74.1	6
It should be unlawful to insult someone on the basis of their race, culture or religion	71.5	7.3
It should be unlawful to intimidate someone on the basis of their race, culture or religion	79.3	5.3
There should be laws against racial vilification	79.5	3.6

survey shows that 88% of Australians believe it should be unlawful to offend, insult or humiliate on the basis of race (AHRC 2014a).

Looking specifically at online racism, our survey found that 73% of respondents believed that "Internet Services like Facebook and YouTube, etc should report all complaints about racism to the relevant authority," with 65% of participants believing that online commentary should be more controlled (Jakubowicz et al. 2014).

In light of these results, and a further study in 2017 (Jakubowicz et al. 2017) as part of another 2016 Parliamentary Inquiry into 18C (Jakubowicz et al. 2016), while it is evident that the majority of people do support the legal regulation of racial vilification, many online users still value freedom of speech over freedom from hate speech. This suggests that Internet users (and the broader public) want laws to be available when necessary, but do not want those laws to impinge on their online freedom of speech. It also suggests the need for a spectrum of regulatory options that enable people to address racist online speech in ways that are not too heavy-handed. Certainly, our results make a strong case for the need for more sophisticated regulatory measures.

The Major Challenges to Regulation

There are significant challenges to regulating cyber racism. Indeed, tensions "between virtual and physical perspectives" are a "recurring theme in the law of the internet" (Kerr 2008, p. 5). By its very nature, the

Internet is a lawless and non-governable space, presenting challenges to regulation stemming from the quantity of activity, the ease of anonymity, its borderless nature where "anything goes" (Rohlfing 2015, p. 297) and the lack of consensus regarding what content is acceptable and requirements for intervention. Because of these factors, the Web has presented as the perfect setting for unrestrained extremist activity (Corb 2015).

Volume and Anonymity

There is a vast amount of activity occurring continually online, for example, globally Facebook has over 1 billion daily active users (Facebook 2016c). With that volume of activity comes instantaneous content generation and sharing, the ability to disseminate material to large audiences (AHRC 2013) and the inability of enforcement agencies to effectively respond (Giannasi 2015). In addition to the sheer volume of activity, there is the issue of the perceived anonymity of the Internet, the lack of social cues and mob-mentality of hate sites that lead to disinhibited behaviour (Rohlfing 2015), and embolden racist authors, trolls and bullies (Bocij and McFarlane 2003; Citron 2014; Bakalis 2016). Anonymity "give[s] people the cover of darkness to say things that they wouldn't say publicly" (Clarke 2016). Moreover, anonymity makes it difficult for victims to seek redress (AHRC 2013; Mason and Czapski 2017). Rohlfing (2015) suggests that ISPs could better handle this situation. For instance, they could establish procedures for surveillance and detection of offensive Web content by requiring verifiable personal identification of users, thereby diminishing the possibilities of anonymity. This measure would assist in identifying offenders and generating evidence that may be forwarded to relevant authorities or law enforcement agencies. However, it may be argued that any steps in strengthening censorship or restricting access to Web-based hate content may only result in driving it deeper into the darker and less public realms of the Web, further removed from regulating mechanisms.

A Lawless Domain

Because of the inherent interconnectedness of the World Wide Web, material easily crosses state boundaries and bypasses traditional gatekeepers and censors. As discussed later in the Australian context, there are difficulties in enforcing orders to remove cyber racist material within a reasonable timeframe against third-party host platforms directly as well as in compelling an overseas author to comply with the laws of another jurisdiction. The cross-jurisdictional nature of online publication and the lack of harmonisation of regulation across states (Giannasi 2015) mean that it is possible to have a victim in one country, targeted by a cyber racist in another country, on a social media platform hosted in yet another jurisdiction. Dealing with any one instance of cyber racism effectively may therefore require coordination between law enforcement and government agencies from multiple countries as well as online host platforms, emphasising legal inconsistencies between jurisdictions (Mason and Czapski 2017; Facebook 2016b). Websites may even be intentionally set up in jurisdictions that have little law enforcement of hate speech, enabling perpetrators of cyber racism to evade censure (Corb 2015). However, the United Kingdom has witnessed some progression in cross-jurisdictional law enforcement measures. For instance, the case of *R v Sheppard and Whittle* [2010] EWCA Crim 65 examined the online publication of racially inflammatory material and established that, while the website was hosted by a server in the USA, everything in the case was related to England: the defendants operated in England, the offensive material had been generated and uploaded in England and the target audience was the English public. The jurisdiction, therefore, was English and the US-based website host was considered to be merely a stage in transmission (Rohlfing 2015; Giannasi 2015).

Lack of Consensus on Regulation and Criminalisation

One of the major challenges to regulation is the lack of consensus amongst ISPs, governments and policy makers as to what constitutes unacceptable and actionable online racism and when such behaviour should be regulated, much less criminalised. What is criminally offensive is not a fixed concept and must respond to and reflect community standards (Anderson 2011). Governments and online service providers grapple with setting standards of acceptable conduct on the Internet (Foxman and Wolf 2013). For instance, there are complexities surrounding material that is perceived as offensive to a target, but not actually criminal. Much racist online material occupies an ambiguous "grey area" between socially harmful and illegal activities and may often seek to be excused on the basis of free expression or even humour. The lack of consensus impedes the monitoring of, response to and prosecution of cyber racism, leaving victims vulnerable. In Australia, companies such as Facebook and Google have argued that it is not their role to set the standards—but rather the government should do that clearly, and then refer breaches to them for action. They argue they are most responsive to government-generated or government-supported complaints (Jakubowicz 2010).

Should cyber racism be a civil or criminal wrong? There are advantages to dealing with cyber racism through the criminal law. The victim does not carry the burden of enforcement and bystanders can play a role in bringing the matter to the police who can handle complaints quite quickly in some instances (McNamara 2002). Criminal sanctions and shaming individuals in court also sends a message that the state condemns racist behaviour. However, the criminal law contains no direct mechanisms for stopping the dissemination of racist material or halting its re-production again and again. Moreover, it tends to individualise the problem rather than drawing attention to the larger social problem within which racist speech occurs (McNamara 2002). Unlike civil vilification law, it does not generally identity the specific harm of racism to the community. Some have suggested that criminalisation has the potential to cement, rather than change, harmful behaviours by making martyrs of offenders (Bleich 2011; Louis and Hornsey 2015).

The effectiveness of criminal sanctions is also contingent on individuals being informed and empowered enough to report to police. This may pose particular problems for minority-group targets of racist conduct (Sivasubramaniam and Goodman-Delahunty 2008; Putt 2010). Some targets of cyber racism may therefore perceive their claims will not be dealt with seriously if taken to police. Problematically, there may also be a lack of public knowledge about applicable laws and forms of redress for cyber racism, whether criminal or civil ((AHRC 2013, p. 15). For instance, research conducted for the Law Commission in New Zealand found that only one in 10 people were aware of the laws or standards applying to harmful speech on the Internet, and 42% of people were not aware of what to do if confronted with a serious problem involving harmful speech online (The New Zealand Law Commission 2012).

Criminal law cannot, and should not, control all aberrant behaviour. Adopting a "minimalist" approach, we see the punitive and stigmatising effects of the criminal law as blunt instruments of "last resort" (Ashworth and Horder 2013, p. 40) for responding to cyber racism. Instead, other sources of control should first be considered that are more reliant on voluntary codes of conduct, civil preventative orders or administrative regulation, premised upon "morality, social convention and peer pressure" (Ashworth and Horder 2013, p. 33; see also Findlay et al. 2014).

In the online context, many champion the important regulatory role of content providers in offering more immediate and flexible responses to hate speech without the consequences associated with criminalisation (Citron and Norton 2011). But Rohlfing (2015) argues that challenges in regulating cyber racism lie in ISPs' and other content providers' resistance to monitoring offensive behaviour themselves. As we saw with our Facebook example earlier, online hosts and ISPs largely rely on victims reporting racist content, and it is only if that content is in breach of their Terms of Service and Codes of Conduct that it may be removed. Such a process inherently involves a time lag within which the abusive material may be further disseminated and go "viral" as it is shared across platforms and jurisdictions. The reliance on reporting is, according to Giannasi (2015), inefficient in addressing harm in a timely fashion. Empirical reports on cyber racism have been critical of the effectiveness of platforms in responding to the content reported to them (Oboler 2016; WJC 2017). Germany and then the European Union responded to a perceived failure of self regulation with agreements between themselves and platform providers to improve responsiveness (BBC 2015, Titcomb 2016). More recently Germany has introduced legislation imposing heavy financial penalties on platforms whose self regulation falls below standards acceptable to the state (McGoogan 2017). Our data nevertheless establish that "in platform" complaint and reporting is the first port of call for targets and witnesses of racism. We therefore argue that a degree of regulation over cyber racism is attainable through a network of legal and non-legal avenues. This involves recognising the inherent difficulty in striking a balance between enabling freedom of expression and protecting people from the harms associated with racist speech ((AHRC 2013; Bleich 2011). The dilemma, therefore, is how to regulate online behaviour that is harmful in itself and has the potential to incite offline/real-world violence, without undermining the principles of free speech. Is it possible to balance regulation with expressive freedom?

Free Speech Versus Freedom from Racism

Different jurisdictions adopt various approaches, yet the issue of free speech is central to debates regarding the regulation of racism (Bakalis 2016). At the "free speech" end of the spectrum is the USA that has the most entrenched, constitutional protections around the freedom to express offensive and provocative speech, underwritten by the First Amendment (Bleich 2011; Rohlfing 2015; Corb 2015). A concern regarding increased regulation is that any measures aimed at diminishing online racism could also be used as techniques of oppression and the denigration of free speech (Giannasi 2015). Nevertheless, any right to freedom of expression is not absolute (Bakalis 2016). European jurisdictions have slowly restricted freedom to express racist views over the past few decades in favour of "freedom from racism." This is exemplified by France, which takes a strong legislative approach and has had a number of recent high-profile prosecutions and debates regarding racist speech (Bleich 2011). Examples of this include French film star Brigitte Bardot, who has been convicted on charges of inciting racial hatred, and Marine Le Pen, leader of the French far-right National Front party, who was acquitted of racial hatred charges in December 2015. The killing of four cartoonists from French magazine *Charlie Hedbo* after publication of a satirical cartoon of the Prophet Mohammed, and subsequent arrests of persons accused of inciting racial hatred or condoning the attacks, has sparked further debate on these laws (Reuters 2008; Bisserbe 2015; Lawless 2015). Holocaust denial laws, in place in a number of European jurisdictions including Germany, France, Austria, Belgium, Luxembourg, the Czech Republic, Poland and Slovakia, represent one of the most controversial restrictions on speech, prohibiting people from contesting the occurrence of the Holocaust on the basis that this represents racism towards Jewish people and survivors (Bleich 2011; Mason and Czapski 2017).

These challenges suggest that tackling cyber racism requires a holistic approach that seeks to build resilience to racism amongst likely targets (resilience in the face of victimisation) and amongst those who are at risk of perpetrating racism (arming potential perpetrators and bystanders with the skills to reject racism) (Nissim 2014). While law - particularly law that imposes prohibitions - has a role to play in this resilience-building process, on its own it is not enough. Indeed, over-regulation can be counterproductive by undermining the resilience and ability of people to operate effectively within a community (Ebum 2008). In other words, over-reliance on legal intervention may undermine the capacity of both virtual and real-world communities to discern, resist and challenge racism on their own terms. Any enhancement to existing regulation should be accompanied by greater education for Internet users about respectful behaviours online, and the mechanisms by which harmful content can be dealt with, both through and beyond legal channels. On this basis, tackling racism online necessitates a multifaceted approach, which engages law enforcement authorities, online content providers, perpetrators, bystanders, victims and communities.

To understand how to achieve this it is critical to appreciate the current regulatory terrain as it operates in a given jurisdiction, thereby allowing an appreciation of where the gaps lie in that domain. In the following sections, we do this by reviewing the law in Australia as it pertains to cyber racism, before considering international protocols and online platforms such as Facebook more generally. We then bring these analyses together to identify the regulatory gaps, keeping the Australian example in mind.

The Current Legal and Regulatory Terrain: The Example of Australia

While Australia can be seen to regulate cyber racism at both Federal and State levels, it is important to note that Australia has not signed the *Additional Protocol to the Council of Europe Cybercrime Convention concerning acts of a racist and xenophobic nature committed through*

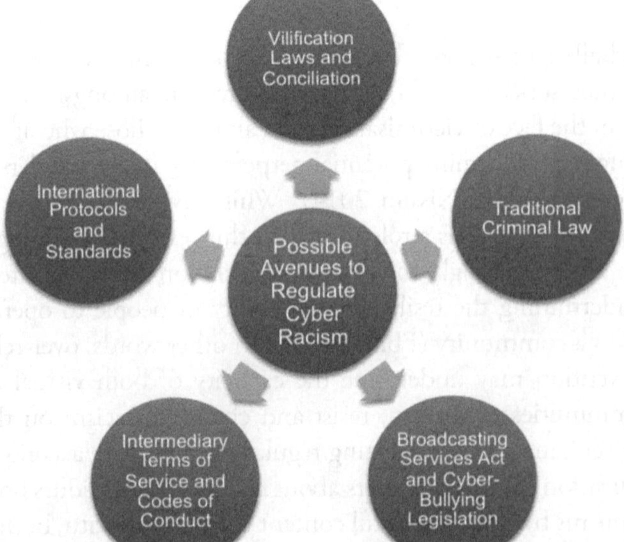

Fig. 8.1 An overview of the Australian legal and regulatory terrain

computer systems (Council of Europe 2003a, b). Instead, Australia relies on a mix of regulation including traditional criminal laws, hate speech—what Australia refers to as "vilification"—laws, industry Terms of Service and Codes of Conduct, and civil regimes. The following diagram represents the possible avenues to regulate cyber racism in Australia (Fig. 8.1).

Vilification Laws and Conciliation

At the Federal level, Section 18C of the Racial Discrimination Act 1975 (C'th), was implemented in order to give effect to Australia's obligations under the ICERD (1965), makes it a civil wrong for a person to do an act which is reasonably likely to *offend, insult, humiliate* or *intimidate* a person or group on the basis of their race, colour, national or ethnic origin. The act must be done "other than in private," which has been interpreted to include conduct occurring online, such as material published on a website that is not password protected (*Jones v Toben* [2002] FCA 1150, [74]; *Jones v Toben* [2003] FCAFC 137). The construction of the Internet as public is important in extending the reach of regulatory measures. Under Federal legislation, the impact of the act is measured objectively from the perspective of a hypothetical reasonable person in the position of the applicant or the applicant's victim group, thereby applying community standards rather than the subjective views of the complainant (AHRC 2014b). The conduct in question must cause "profound and serious effects, not to be likened to mere slights" (*Creek v Cairns Post Pty Ltd* (2001) 112 FCR 352, 356 (Kiefel J)). This Act does not operate extraterritorially; however, it seems that material, which is uploaded or hosted overseas but can be viewed by users in Australia, would fall within the bounds of the legislation (AHRC 2015; *Dow Jones and Company Inc v Gutnick* (2002) 210 CLR 575), and a similar argument is likely to apply to state and territory vilification legislation (e.g., Racial and Religious Tolerance Act 2001 (Vic) s 7(1)(b); s 24(3)(b)).

There are also racial vilification statutes in every Australian state and territory (with the exception of the Northern Territory), intended to operate concurrently with Federal laws. Most jurisdictions have both civil and criminal provisions that seek to render it unlawful for a person by a public act, to incite hatred towards, serious contempt for, or severe ridicule of a person or group on the grounds of race (Rees et al. 2014). As with the Federal laws, a "public act" is broadly defined and has been interpreted to encompass the publication of material online (e.g., Anti-Discrimination Act 1977 (NSW) s 20B; *Collier v Sunol* [2005] NSWADT 261). It is worth noting that the conduct covered in the Victorian legislation expressly includes the use of the Internet or e-mail to publish or transmit statements or material, and renders all racial vilification unlawful, whether in or outside Victoria, except where the person intended the acts be done in private (Racial and Religious Tolerance Act 2001 (Vic) ss 7, 7(2)(b), 12). Generally, there is a high harm threshold for state/territory legislation (higher than the threshold under Federal laws), as the complainant must show that a third party could have been incited to feel hatred towards the victim group as a result of the accused's conduct. This is difficult to prove and unsatisfactory for the victim, being divorced from any assessment of the offender's intention in performing the act, so it is highly questionable whether state/territory vilification laws would provide effective redress for victims of cyber racism (Rees et al. 2014; AHRC 2014b).

These human rights laws carry the "practical regulatory burden" for dealing with racist speech in Australia (Gelber and McNamara 2015, p. 636). Significantly, they place an emphasis on the conciliation of complaints wherever possible (Rees et al. 2014). Alternate forms of dispute resolution including arbitration and mediation are commonplace in Australia as precursors or replacements to litigation (Sourdin 2016). For example, at the Federal level, the AHRC is responsible for investigating racial hatred complaints. Globally, online dispute resolution systems such as modria.com used by eBay and PayPal are increasingly utilised. Recourse to conciliation can be undoubtedly advantageous, allowing for harmful conduct to be dealt with quickly and informally without resort to the court system. For example, the AHRC (2014a, b) points to the efficacy of this model for cyber racism:

- A complainant of Asian background reported a website that advocated violence against Asian people. The AHRC contacted the ISP to establish the identity of the website owner. Within a few days the website had been disabled by the ISP on account of it breaching their "Acceptance Use Policy" and
- A complainant reported a user posting racially derogatory comments on a video posted on a file-sharing website. When the website was contacted, the comments were removed and the offending user was suspended.

Conciliation enables a dialogue between victim and respondent, such that the respondent can come to terms with the consequences of their actions and make amends that are agreeable to the victim. It is important to understand that victims of racist conduct often are not looking for the perpetrator to face heavy penalties, but simply a genuine apology acknowledging the harm and the removal of harmful material (Gelber and McNamara 2015, p. 637).

Traditional Criminal Law

Australia has telecommunications offences that address the use of carriage services to upload racist material onto the Internet. Jurisdiction may be extended in respect of certain offences so that if the offender is an Australian citizen, he or she can be prosecuted even if the conduct occurs wholly *outside* Australia (Criminal Code Act 1995 (C'th) s 474.17(1); see also ss 15.1, 474–475). Although there is no reported case law specifically concerning racially motivated conduct, this section has been employed extensively to deal with harmful conduct online, with 308 successful prosecutions between its introduction in 2005 and 2014 ((Parliament of Australia 2014, p. 52) (*Agostino v Cleaves* [2010] ACTSC 19)). Significantly, the offence has been directly and successfully applied in recent years to online conduct of a racially or religiously vilifying nature. For instance, in 2014, a Western Australian (WA) man was charged under the section for a series of abusive tweets on Twitter directed at a sportsman of Fijian heritage, in which he referred to the player as a "black nigger" (AAP 2014). For less serious cases, offensive language and offensive conduct provisions can theoretically be used to prosecute racially offensive behaviour online, but are rarely used to do so (Brown et al. 2015).

The traditional criminal laws of Australian state and territory jurisdictions provide other possible avenues for prosecuting racial online threats, and various offences exist relating to threats of violence. For instance, some offences embrace the concept of "psychic" assault whereby a person is threatened with unwanted physical contact, although there must typically be an apprehension of imminent violence (Citron 2014) (*Knight v R* (1988) 35 A Crim R 314). In Western Australia, a jurisdiction which adopts an exclusively criminal approach in this area, there have been successful prosecutions for racially vilifying behaviour including a man who posted an antisemitic video on YouTube (*O'Connell v The State of Western Australia* [2012] WASCA 96). Additionally, legislation provides offences that target stalking and intimidating behaviours using technologies (see also Crimes (Domestic and Personal Violence) Act 2007 (NSW) ss 7, 13). In what was reportedly Australia's first prosecution of cyberbullying, a man was convicted under the Victorian stalking legislation in April 2010 over threatening text messages sent to a young person who eventually committed suicide (Milovanovic 2010).

Broadcasting Services Act and Cyberbullying Legislation

Another potential avenue of recourse for victims of cyber racism is the online content scheme within the Broadcasting Services Act 1992 (C'th), regulated by the Office of the Children's E-Safety Commissioner. This scheme imposes obligations upon ISPs operating in Australia and Internet content hosts in relation to harmful Internet material. While the primary purpose of the scheme is to protect children from Internet pornography, it would seem that the scheme may have incidental application to cyber racism. The scheme deals with offensive online material by issuing "take down" or filtering notices, and it seems conceivable that some cyber racist publications and films could fit within the scheme (Mason and Czapski 2017). Also administered by the Children's E-Safety Commissioner are cyberbullying provisions that enable the commissioner to request the removal of material from social media sites (Enhancing Online Safety for Children Act 2015 (C'th)). Whilst the scheme only applies to seriously harassing, threatening, intimidating or humiliating content targeted at Australian minors, it could encompass content that is racist in nature, though it appears not to have been. However, the legislation deals with material that is directed at a *particular* Australian child and, therefore, cannot assist with cyber racist material directed towards racial groups (Mason and Czapski 2017).

In summary, there are several limbs to the legislative and regulatory framework in Australia that can be harnessed to deal with cyber racism. As we will examine in subsequent sections, despite the quantity of that extant legislation and regulation, the most efficient avenues for victim redress and to trigger law enforcement attention are often simply via social media services.

International Protocols and Standards

European governments have been proactive in confronting hate speech and racism through judicial and non-judicial responses and through the lens of human rights (Stephanos 2014). There are a number of international protocols and standards in place that aim to deal with harmful content online, including racist content. The core example is the *Additional Protocol to the Council of Europe Cybercrime Convention concerning acts of a racist and xenophobic nature committed through computer systems* (Council of Europe 2003a, b). This instrument requires state parties to criminalise within their domestic laws the dissemination of racist or xenophobic material through a computer system. The *Additional Protocol* recognises the tensions between computer systems that "offer an unprecedented means of facilitating freedom of expression and communication around the globe" while also acknowledging the risks of abuse of such computer systems and the ease of disseminating racist and xenophobic propaganda. In this regard, it urges states to establish criminal offences to counteract such behaviours and creates a legal framework for the cross-jurisdictional prosecution of cyber racism (AHRC 2013, p. 22). The European Commission against Racism and Intolerance (ECRI), a specialist human rights monitoring body, recommends that states ratify the *Additional Protocol,* yet in the ECRI *General Policy Recommendation No. 15 on combating Hate Speech* (2015), it includes recommendations to raise public awareness of respect for pluralism in society and supports self-regulation of private and public institutions in combating hate speech and establishing ethical online practices (ECRI 2015, p. 9, 12; see also Council of Europe 2012-14). Nevertheless, referencing the European Union Council Framework Decision (2008), ECRI supports intervention through criminal laws. As previously mentioned, although Australia is party to the *Cybercrime Convention* that seeks to combat cybercrime, it has declined to sign or ratify the *Additional Protocol* (AHRC 2013).

Internet Service Providers: Terms of Service and Industry Codes of Conduct

Despite the availability of various legal avenues, one of the most important paths of regulation for harmful content online are the Terms of Service and Codes of Conduct provided by ISPs, browsers and search engines, such as Google, and online content providers, such as Facebook. Online platforms typically have a set of terms that govern the behaviour of users that subscribe to their service, with stipulated mechanisms for reporting or dealing with harmful content. In many instances, these Terms of Service are the first port of call for complainants rather than legal redress. As Klein puts it, Terms of Service often constitute the "only real law of the land in cyberspace" (Klein 2012, p. 442).

There are numerous examples of Terms of Service that could address cyber racist content. In the Australian context, individual ISPs have Terms of Service that implicitly, if not explicitly, encompass racist speech or the posting of racist content. For example, TPG's Standard Terms and Conditions prohibit customers from using their service to transmit "any material which interferes with other users or defames, harasses, threatens, menaces, offends or restricts any person," and indicates that they may suspend or discontinue an account if it has been used for "offensive and/or illegal activities under State and/or Commonwealth laws" (TPG 2016, clauses 7.9, 7.10). It is positive that ISPs have cooperated with the Australian AHRC's conciliation scheme to remove racist material, in the process deeming such material to be contrary to their acceptable use policies (AHRC 2014b).

Of course, ISPs and content hosts are governed by relevant domestic laws. The picture is more complex when we look at major content hosting platforms that are based overseas, and the way in which their Terms of Service operate and interact with any domestic legal system. After all, any Code of Conduct is voluntary in nature, and it would be difficult to enforce a judgement against a platform operating entirely from outside the jurisdiction. Most of the world's largest social media platforms, such as Facebook, Instagram and YouTube, are based in the USA, and commentators argue that any reticence by these platforms to deal with harmful content can be read in light of the high value accorded to free speech in the USA, however repugnant or offensive (Citron and Norton 2011; Oboler and Connelly 2014).

Let us take Facebook as an example to examine the efficacy of its Terms of Service for dealing with cyber racist content. In addition to having over 15 million active users in Australia, Facebook has also been identified as a major site for the proliferation of cyber racism (Dunn et al. 2014). Purtill (2016) writes "in the echo chamber of Facebook" racist perceptions become consensus. Yet all individuals and entities that make use of Facebook agree to the Terms of Service contained within its *Statement of Rights and Responsibilities* (Facebook 2015), committing not to use the service to:

- bully, intimidate or harass any user;
- post content that is hate speech, threatening, or pornographic; incites violence; or contains nudity or graphic or gratuitous violence and
- use Facebook to do anything unlawful, misleading, malicious or discriminatory.

The company's *Community Standards* (Facebook 2016a) were updated in 2015 to clarify what was and was not permitted on their platform. Relevantly for instances of cyber racism, Facebook states that it will remove hate speech content which *directly attacks* people based on, amongst other things, their race, ethnicity, national origin or religion (see Fig. 8.2).

The Community Standards further provide that:

> Organisations and people dedicated to promoting hatred against these protected groups are not allowed a presence on Facebook. As with all of our standards, we rely on our community to report this content to us……We allow humour, satire or social commentary related to these topics, and we believe that when people use their authentic identity, they are more responsible when they share this kind of commentary. (Facebook 2016a)

With an incredibly large number of users and a huge volume of content being posted each day, Facebook relies on communities to report content

Fig. 8.2 Facebook online statement on Hate Speech (2016)

that potentially violates its terms. Facebook allows its users to flag material they deem offensive in relation to various types of content, including posts, messages, photos and videos (Facebook *How to Report Things* 2016; Facebook *Tools for Addressing Abuse* 2016). Once an individual indicates that they do not wish to see a particular type of content, further dialogue boxes allow the user to indicate why they think the content should not be on Facebook and report the content to Facebook for review. Facebook reviews all reports of inappropriate content and removes content, typically within 72 hours, if they deem it to have violated the *Community Guidelines*.

Although Facebook's policies ostensibly prohibit racially abusive material and other kinds of harassing or offensive material, the social media platform has come under fire on occasions for its failure to remove certain material which most people would deem offensive or even hate speech (Oboler 2012, 2013, 2016). A pertinent Australian example involves Facebook's failure to deal effectively with "Aboriginal Memes," a page which denigrated Indigenous Australians (Sykes 2012). Facebook initially failed to remove the page on the basis that it did not breach its Terms of Service. The page was then briefly taken down, but re-emerged shortly afterwards with a "[Controversial Humor]" tag in the page title. The outcry surrounding the page saw an online petition for its removal gather over 20,000 signatures, and an investigation commenced by the Australian Communications and Media Authority (ACMA) before it was finally removed by Facebook in response to the backlash (Sykes 2012; Online Hate Prevention Institute 2012; Oboler 2012; Chester 2014).

The volume of users on Facebook makes it difficult to efficiently police content, and, at times, it appears that Facebook only removes harmful material when there is significant public backlash. In 2016, Facebook quite speedily took down two Australian pages, which had turned from "community noticeboards to uncensored soapbox" (Tomlin 2016) pages dominated by non-Indigenous authors. Both were removed after identification as "sources of vigilante threats and racial vilification" against Indigenous people in a WA goldfields town (Tomlin 2016). For instance, one person suggested a solution to criminal activity allegedly attributed to the Indigenous population: "What about we get some white cops from America .... lol .... Not really funny but might work;" they also wrote: "Things will never change unless there is annual cull" (Purtill 2016). Local law enforcement vented frustration over the Facebook pages, calling them "uncontrolled" and "inflammatory" (Clarke 2016). Then, once news of the unlawful killing of a 14-year-old Indigenous boy, Elijah Doughty, hit these pages, comments included: "Good job you thieving bastard. Don't think you'll be touching another bike anytime soon ahaha. About time someone took it into their own hands hope it happens again" (Purtill 2016). The administrator of one of the pages is reported to have denied any responsibility for moderating the offensive content, and commented, defiantly, on the difficulties in completely deleting such pages: "You can close this site down, but I guarantee another 10 will pop up" (Tomlin 2016). Nevertheless, a police commander commented: "You can't just put up these sites and turn a blind eye to the comments that get put on them—that doesn't cut it for me or anyone else in the community…Don't stick your head in a hole and pretend you're not responsible; at least at a moral level" (Tomlin 2016).

Even when platforms such as Facebook remove content, there are other issues. Social media platforms may block content within the jurisdiction in which it is unlawful, rather than removing the material altogether, meaning it can still be accessed overseas, or, for instance, through a Virtual Private Network (VPN) (Oboler and Connelly 2014, p. 119). As Oboler and Connelly (2014, p. 119) point out, even where content is not deemed to breach the Terms of Service, it may still need to be assessed by the laws of the reporting jurisdiction, leading to the same content possibly being assessed multiple times according to multiple different laws. Further issues may also arise where content draws upon area-specific material, such as derogatory slang or local tropes, which may not be apparent to persons employed in a global content moderation team (Oboler and Connelly 2014). This was certainly the case for "Aboriginal Memes."

These difficulties all point to the need for ongoing engagement between social media platforms, governments and civil society, so as to better demarcate and encourage socially acceptable grounds of behaviour online. Notably, major platforms *have* responded to protracted government pressure to improve their moderation practices. As discussed above, in late 2015, the German government announced a landmark agreement with Facebook, Twitter and Google under which these platforms agreed to remove hate speech from their platforms within 24 hours in Germany. This followed a surge in racist and xenophobic speech in Europe aimed at immigrants and Muslim populations (Lukerson 2015). In early 2016, Facebook launched its European Online Civil Courage Initiative and pledged over 1 million euros to organisations and researchers seeking to understand and disrupt online extremism (Chan 2016). These moves came in the face of extreme pressure from the German government including the threat of criminal proceedings against domestically based senior executives in light of a steep rise in hate crimes against immigrants linked to the online incitement (OHPI 2016). As mentioned, by mid 2017 Germany had moved its expectations of social media companies from a voluntary agreement into legislation backed by fines of up to €50m for non-compliance by large platforms (McGoogan 2017). Undoubtedly, any effort to better regulate cyber racism must include and recognise the important role played by online content providers, in conjunction with jurisdictional legal frameworks.

A Gap in Regulation

It is critical that we find ways to promote respect for human rights online. In Australia, for example, the above discussion reveals a gap in the present regulatory environment. Telecommunications offences, and, in rare instances, criminal vilification laws, have been applied to deal with serious cases of cyber racism, hitting individuals with fines, community service or custodial terms. Yet the vast majority of the regulatory burden for cyber racism currently falls on the conciliation/civil rights regulatory model.

But does this civil rights model place too heavy a burden on the victim to initiate and pursue proceedings (McNamara 2002)? Complaints cannot be brought by a third party or a bystander, but must be brought by the victim or the victim group. Nor does any state authority have the ability to initiate a complaint (Gelber and McNamara 2015, p. 637). Research into claims brought under civil vilification laws throughout Australia over a 20-year period found that successful complaints/litigation usually required an individual of extraordinary resolve to pursue the claim, backed by a well-resourced, respected community organisation (Gelber and McNamara 2015, p. 646). Although conciliation may be mooted as a quick and efficient mechanism of dispute resolution, many complaints were terminated on account of procedural barriers and the lengthy time it takes in some jurisdictions to reach conciliation (Gelber and McNamara 2015, p. 643–4). This is especially problematic in the online context, given the ease with which harmful material can proliferate. The system also relies on the willingness of perpetrators to comply with a voluntary undertaking, meaning it can be stalled by insincere or recalcitrant participants. Critically, because conciliation is necessarily a private and confidential process, it cannot achieve the educational and "standard setting" objectives that lie behind the enactment of racial vilification legislation (McNamara 2002). Only 2% of matters under civil vilification laws are resolved in a tribunal or court, such that their outcomes are made public (Gelber and McNamara 2015, p. 643). Yet, in spite of this low volume, those decisions have created important precedents and have been employed directly by advocates to change behaviour. Laws that expressly condemn racial vilification also have important symbolic value irrespective of how often they are used (Gelber and McNamara 2015, p. 656).

As our data evidence, for many targets and witnesses of cyber racism, the voluntary Terms of Service and Codes of Conduct of online content and social media services provide the first port of call for reporting a complaint. Although these kinds of informal regulatory arrangements may be far more responsive and flexible when compared with legal and quasi-legal schemes, reliance on private entity terms of use raises its own set of difficulties. As discussed above, these services are not automatically beholden to the legal standards of non-host jurisdictions. Some platforms may not have Terms of Service that adequately encompass cyber racist behaviour to the standard expected under the relevant domestic law, or else may not adequately enforce those standards. Where such a platform fails to remove racist content, there may be little recourse for a victim of cyber racism, especially where the apparent perpetrator hides behind a pseudonym or is located overseas.

In sum, cyber racism is inherently difficult to regulate; whilst various offences and laws deal with online and racist behaviour, there is no enforceable regime which deals expressly and specifically with *cyber racism*, and which makes both perpetrators and content providers accountable through a unified scheme. It is abundantly clear that racism plays a damaging role in contemporary society, an issue that has both been exacerbated and taken on new forms as a result of the Internet. This gap in regulation is one that must be addressed.

Addressing the Gap: Regulating for Resilience

In this section, we explore one option for closing the regulatory gap that we have exposed earlier. This proposal has the Australian legal context firmly in mind but is also adaptable to jurisdictions facing comparable regulatory dilemmas.

We support a regulatory framework for the Internet that promotes respect for human rights and the articulation of community standards. We seek to raise the bar for acceptable online conduct, foster community resilience and engage multiple stakeholders, recognising the inherent difficulties associated with monitoring and policing the Internet. Such a strategy is at variance with responses that see the criminal law as being the first or immediate answer. In a very practical sense, the problem with relying on criminalisation is that there is simply too much offensive and hostile material online for law enforcement and prosecutors to process (Giannasi 2015). We need a range of different remedies, depending on the situation—takedown notices for instance—and recognition of the whole spectrum of legal and non-legal regulatory channels presently being used to curb harmful speech.

We need strategies that encourage free use of the Internet while regulating, moderating and, moreover, setting standards. In addition, we need strategies that empower communities and individuals to, firstly, resist stigmatising and negative content, and secondly, to respond in a manner that does not further exacerbate the problem. This approach is very much in accord with the Council of Europe's "No Hate Speech Movement" (2012–14) that recognises the many benefits of the Internet in facilitating positive engagement and communication but also as an agent for disseminating racism and hatred. In supporting human rights and dignity, that movement seeks to raise awareness of hate speech, provide advocacy and intervention and moreover equip and educate young people to manage and overcome this burgeoning problem. This approach to cyber racism could be seen akin to public order offence regulations that seek to maintain public peace, social welfare and civic tolerance in public places (Anderson 2011), given that the notion of public place has been extended by cyberspace. Indeed, Bakalis (2016, p. 265), in discussing the regulation of cyber hate in England, draws a distinction between regulating against personal harm to individuals in a private domain versus harms inflicted on the public in the sense that hate speech pollutes public space and undermines social harmony.

Online spaces can probably never be fully regulated. Rather, we need models in place to monitor and condemn harmful behaviours online, and which empower victims and communities by helping them to recognise and challenge racism, on their own terms, without excessively curbing speech. As discussed earlier, our survey data show that Internet users are prepared to put up with a degree of harm in return for freedom of expression. At the same time, there is widespread community support for laws against racial hate, and a continued acknowledgement of harms associated with racist behaviours.

We should not downplay the key role of online content providers, such as Facebook, in dealing with harmful content. Major host platforms have demonstrated that they can adapt to concerns over racist and hateful content proliferating across their platforms, even if their response only occurs following significant public, commercial or government pressure.

Their content-reporting channels, while needing improvement, are able to respond to complaints and to changing trends and standards relatively quickly (Dunn et al. 2014).

At the same time, the law plays an important role in demarcating harmful behaviour and developing community standards, and provides a means of enforcement where voluntary mechanisms fail. The law must be used to build community resilience, rather than erode it. As we noted earlier, laws that are too paternalistic can undermine resilience by denying agency and the capacity to discern and resist racist harassment (Ebum 2008). The use of the law, including criminal law, is only one means of regulating behaviour. It should be a last resort given that many users of the Internet and of major social media platforms are young people, who we should ideally be seeking to educate on respectful online behaviours, rather than punishing them.

A Civil Penalty Regime for Cyber Racism?

One way to address the kind of regulatory gap that we have identified earlier is to introduce a civil penalties scheme that offers a multipronged approach. A possible model for such a scheme is the new Australian cyberbullying legislation Enhancing Online Safety for Children Act 2015 (C'th) and the New Zealand Harmful Digital Communications Act 2015. Importantly, while there are differences, the Australian and New Zealand schemes both include mechanisms that *engage* with those who post harmful content and the platforms that host that material. The Australian cyberbullying regime places an initial obligation on persons to report harmful material to the relevant social media service—it is only when that material is not removed within 48 hours that the administering agency can act to request or order the platform to remove the content. Failure to comply can result in a fine or, if necessary, a court injunction for social media platforms and end users. Crucially, this approach to cyberbullying integrates an educative function, and online content providers are incentivised to improve their online safety and reporting practices.

Such a scheme could comprise the following elements:

1. Establish a harm threshold that reflects community standards;
2. Ensure that the first port of redress for cyber racist content is "in platform" with online content providers;
3. Place pressure on online content providers to more effectively police online conduct, including liability for failure to respond;
4. Incorporate penalties for end users who fail to remove harmful content;
5. Allow for third parties to intervene and report cyber racist content;
6. Provide avenues to educate Internet users about their rights and responsibilities with respect to appropriate behaviour online and
7. Improve the ability to record and monitor harmful online behaviours.

A Harm Threshold That Reflects Community Standards

Any attempt to regulate cyber racism should begin with a well-defined and appropriate threshold of harm for prohibited conduct. In Australia, racial vilification laws offer the most logical starting point for setting an appropriate threshold of harm in a civil penalties scheme. In particular, s 18C of the *RDA* sets a national standard that has been in place for over 20 years (Soutphommasane 2015). As we have noted throughout this book, both Internet users and the general community have recently expressed support for the threshold of illegality set up by racial vilification laws. As Gelber and McNamara (2015, p. 167) conclude, a "very large majority of the public supports the idea that hate speech laws are an appropriate component of the framework within which public debate takes place in Australia."

To be captured by this model, comments would need to "offend, insult, humiliate or intimidate" a person or group—in a way that has profound and serious effects that are more than mere slights—on the basis of their race, colour, national or ethnic origin. All of this would be assessed by a "hypothetical" reasonable person in the position of the victim group. It would not be necessary to show incitement or subjective fault on the part of the person who made the comments (McNamara 2002).

No law on its own will be able to fully reconcile strong differences of opinion over an appropriate threshold for intervention, which is why it is important to incorporate a combination of legal and non-legal measures into the regulation of cyber racism.

Engaging Existing Reporting Mechanisms

Under a civil penalties scheme, people would be required to make reports about harmful content directly to online content hosts, relying upon their existing Terms of Service and reporting mechanisms. It is only where this proves ineffectual or inapplicable that the state can intervene as necessary. This recognises the key role online content providers such as Facebook already play in regulating the vast volume of content online.

Placing Pressure on Online Content Providers

Enacting firm legislation puts pressure on online content providers to improve their mechanisms for dealing with racist conduct and enforce existing codes more effectively, so as to avoid civil penalties. Under the Australian cyberbullying scheme, large social media platforms risk a fine if they fail to comply with a takedown notice. In its first six months of operation, the Australian cyberbullying legislation shows promise in this regard, with most platforms complying with a takedown notice within 12 hours of requests, making it unnecessary to fall back on civil penalties (Francis 2015). The recent response of Google, Facebook and Twitter to pressure from Germany to expedite their processes for removing racist hate speech on their European platforms is also testament to government efforts in this space, even if the response from the platforms has not fully met the aspirations of those governments (Lukerson 2015). The new penalties under German law will likely see increased efforts towards compliance by the platforms.

Penalties for Perpetrators of Cyber Racism

Where the perpetrator is identifiable and fails to engage with the victim to repair the harm that is caused, the state can also intervene with the threat of civil penalties to enforce any orders. This improves upon a conciliation model, for which enforcement against perpetrators remains a barrier to efficacy. In some cases, having a scheme which acts as an alternative to conciliation may be preferable for instances of racism, particularly where the victim might not wish to confront the perpetrator, or where the perpetrator may not be willing to engage in that process (or is unable to be found). Perpetrators who recognise their mistake may also prefer removing the content and paying a small fine than engaging in protracted discussion. Having a legal threat overhead is likely to force compliance with orders to take down material without having to resort to a court settlement. It also avoids unnecessary criminal sanctions.

Allowing for Third-Party Intervention

A civil penalty regime can also be drafted in such a way that complaints can be brought by third parties or by the state. This distinguishes the regime from most existing conciliation procedures and civil vilification laws, which require an affected victim or victim group to initiate the claim, and where regulatory agencies provide no assistance in navigating the court process. Involving bystanders and non-victim parties in the process is a critical part of building community capacity to identify and respond to racist behaviour (Nissim 2014), taking the pressure off those who are its targets. A civil-based penalty regime to deal with cyber racism could provide an effective form of redress for harmful racist content without resorting to overly punitive or censoring measures.

Mechanisms to Educate Internet Users

Under a civil penalty scheme, any order may be made available publicly and would therefore be able to play an educative role regarding respectful online behaviour. In this way, the scheme can be used to foster community standards and feed into broader educative initiatives and resilience-building strategies conducted by the overseeing body. The advantages can be seen in the voluntary reporting system FightAgainstHate.com and in the publications of the Online Hate Prevention Institute, which explain racist content and encourage its reporting as discussed in Chap. 7. It should be noted that these publications usually redact the names of perpetrators as well as victims. Orders made under the proposed scheme would play a similar role, being available in the public domain to play an educative role about inappropriate online behaviour. The ability to publically name a serial offender may provided a form of escalated response under a civil penalty scheme.

Education is critical in empowering and mobilising individuals and communities to proactively report and self-regulate online behaviours (Council of Europe 2012–14, Bakalis 2016, European Commission against Racism and Intolerance (ECRI) 2015, Stephanos 2014). Often the best response to cyber racism is for people to cease reading or contributing to pages: "If everybody walks away and one man is left shouting in the street, then he has no audience" (Dr Helen Cripps reported in Tomlin 2016). In this regard, we note Facebook's mission to encourage respectful behaviour, discussed earlier, as well as ECRI's (European Commission against Racism and Intolerance (ECRI) 2015) goal of inspiring respect for pluralism in society and ethical self-regulation. These all represent strategies to raise the bar for acceptable online conduct, set community standards and ultimately build resilience.

Better Recording and Monitoring of Online Behaviour

Finally, this scheme provides a channel to record and monitor cyber racist activity as it is reported (Online Hate Prevention Institute 2015). Better data enhance our efforts to understand and combat such harmful material.

Conclusion

A coherent, targeted and globalised response to cyber racism is overdue (Bakalis 2016). Cyber racism poses its own distinct challenges for regulators—not only through the ease in which hateful content can spread online, but through the challenges associated with anonymity and disinhibition, the volume of Internet material and the difficulties associated with different jurisdictional attitudes towards the regulation of free speech.

This chapter has surveyed the framework of legal and regulatory channels in place to deal with cyber racism with specific reference to Australia, to identify how this issue can be tackled more effectively. We recommend a multilayered approach to regulation, which recognises the importance of developing community standards and Codes of Conduct, buttressed by civil and criminal intervention. We argue for the strengthening of administrative remedies over more criminalisation, combined with strategies for promoting ethical behaviour online and an awareness of human rights. Ultimately, our model favours prevention over prosecution. Implementing a civil penalties type model for dealing with cyber racism has the potential to bridge a regulatory gap by providing an enforceable scheme that deals expressly with cyber racism and targets both perpetrators and third-party hosts of racist content (of course, such a scheme could be expanded to address online expressions of other forms of prejudice). This provides options for enforcement once other conciliatory or compliance measures are exhausted.

Together, these approaches help build resilience by developing safe and effective ways for the targets of racism to challenge it (Nissim 2014, p. 3), and by sending the right message to those who engage in racism on the Internet or are bystanders. Returning to Facebook's preparedness to take down two racist pages based in Western Australia in 2016, the relative speed of this action served community expectations and acted as a warning: "don't think you can get away with this" (Tomlin 2016). That is the message which must be sent by an effective scheme of regulation that promotes resilience to cyber racism.

Notes

1. In 2014, during debates regarding proposed amendments to racial discrimination legislation, the Australian Attorney-General George Brandis controversially asserted that: "People do have a right to be bigots" (Griffiths 2014).

References

Case Law

Agostino v Cleaves. (2010). ACTSC 19
Collier v Sunol. (2005) NSWADT 261
Creek v Cairns Post Pty Ltd. (2001). 112 FCR 352
Dow Jones & Company Inc v Gutnick. (2002). 210 CLR 575
Jones v Toben. (2002). FCA 1150, [74]
Jones v Toben.(2003). FCAFC 137
Knight v R. (1988). 35 A Crim R 314
O'Connell v The State of Western Australia. (2012). WASCA 96
R v Sheppard and Whittle. (2010). EWCA Crim 65

AAP. (2014). Man Charged Over Racist Tweets to Nic Naitanui. *Perth Now (online)* [GM]. http://www.perthnow.com.au/news/western-australia/man-charged-over-racist-tweets-to-nic-naitanui/story-fnhocxo3-1226861339231. Accessed 17 May 2017.

AAP. (2016). Man Who Abused Nova Peris in Facebook Rant Pleads Guilty. *The Australian.* http://www.theaustralian.com.au/national-affairs/man-who-abused-nova-peris-in-facebook-rant-pleads-guilty/news-story/511c66ab20f02d38ad2722381d1244e9. Accessed 16 May 2017.

AHRC. (2015). *Cyber Racism and Community Resilience Project, Working Paper: Civil and Criminal Racial Vilification Provisions* (Unpublished Working Paper, AHRC, July 2015) 8.

Anderson, J. (2011). *Criminal Law Guidebook.* Oxford: Oxford University Press.

Ashworth, A., & Horder, J. (2013). *Principles of Criminal Law.* Oxford: Oxford University Press.

Australian Human Rights Commission (AHRC). (2013). Human Rights in Cyberspace. *Background Paper*, AHRC. https://www.humanrights.gov.au/our-work/rights-and-freedoms/publications/background-paper-human-rights-cyberspace. Accessed 1 Mar 2017.

Australian Human Rights Commission (AHRC). (2014a). *Overwhelming Majority Reject Change to Racial Vilification Law. Humanrights.gov.* https://www.humanrights.gov.au/news/stories/overwhelming-majority-reject-change-racial-vilification-law. Accessed 17 May 2017.

Australian Human Rights Commission (AHRC). (2014b). *Racial Discrimination: Know Your Rights*. Sydney: Australian Human Rights Commission. http://www.humanrights.gov.au/sites/default/files/Racial. Discrimination_2014_Web.pdf. Accessed 17 May 2017.

Bakalis, C. (2016). Regulating Hate Crime in the Digital Age. In *The Globalization of Hate: Internationalizing Hate Crime?* [Online]. Version. http://www.oxfordscholarship.com/view/10.1093/acprof:oso/9780198785668.001.0001/acprof-9780198785668-chapter-17

BBC. (2015). Facebook, Google and Twitter Agree German Hate Speech Deal. *BBC News* (online). http://www.bbc.com/news/world-europe-35105003. Accessed 20 Aug 2017.

Bisserbe, N. (2015). Marine Le Pen Acquitted of Inciting Racial Hatred. *The Wall Street Journal* [GM]. http://www.wsj.com/articles/marine-le-pen-acquitted-of-inciting-racial-hatred-1450193282. Accessed 17 May 2017.

Bleich, E. (2011). *The Freedom to be Racist? How the United States and Europe Struggle to Preserve Freedom and Combat Racism*. Oxford: Oxford University Press.

Bocij, P., & McFarlane, L. (2003). Cyberstalking: The Technology of Hate. *Police Journal, 76*(3), 204.

Bodkin-Andrews, G., Newey, K., O'Rourke, V., & Craven, R. (2013). Promoting Resiliency to Counter Racism: The Lived Wisdom Within Aboriginal Voices. *InPsych: The Bulletin of the Australian Psychological Society Ltd., 34*(4), 14–15.

Brown, D., Farrier, D., NcNamara, L., Steel, A., Grewcock, M., Quilter, J., & Schwartz, M. (2015). *Criminal Laws: Materials and Commentary on Criminal Law and Process of New South Wales*. Annandale: Federation Press.

Chan, M. (2016). Facebook Launches Initiative to Stop Extremist Posts in Europe. *Time (onlIne)*(19 January 2016). http://time.com/4184559/facebook-initiative-extremism/. Accessed 4 Oct 2016.

Chester, R. (2014). Facebook Shuts Vile Aboriginal Memes Page, Despite Earlier Claiming It Didn't Constitute "Hate Speech". *News.com.au*. http://www.news.com.au/technology/facebook-shuts-vile-aboriginal-memes-page-despite-earlier-claiming-it-didnt-constitute-hate-speech/story-e6frfrnr-1226811373505. Accessed 15 June 2016.

Citron, D. (2014). *Hate Crimes in Cyberspace*. Cambridge: Harvard University Press.

Citron, D., & Norton, H. (2011). Intermediaries and Hate Speech: Fostering Digital Citizenship for Our Information Age. *Boston University Law Review, 91*(4), 1435–1439.

Clarke, T. (2016). Facebook Racists Blamed for Inciting Kalgoorlie Riot May Be Charged. *The West Australian*. https://au.news.yahoo.com/thewest/wa/a/32498289/facebook-racists-blamed-for-kalgoorlie-riot-may-be-charged/–page1. Accessed 13 Dec 2016.

Corb, A. (2015). Online Hate and Cyber-Bigotry: A Glance at our Radicalized Online World. In N. Hall, A. Corb, P. Giannasi, & J. Grieve (Eds.), *The Routledge International Handbook of Hate Crime*. Abingdon: Routledge.

Council of Europe. (2003a). Additional Protocol to the Convention on Cybercrime, Concerning the Criminalisation of Acts of a Racist and Xenophobic Nature Committed Through Computer Systems. *ETS No.189*. http://www.coe.int/en/web/conventions/full-list/-/conventions/treaty/189. Accessed 23 July 2015.

Council of Europe. (2012–14). No Hate Speech Movement. *Nohatespeech movementcampaign*. http://www.nohatespeechmovement.org/campaign. Accessed 9 Oct 2016.

Dooley, J., Pyzalski, J., & Cross, D. (2009). Cyberbullying Versus Face-to-Face Bullying: A Theoretical and Conceptual Review. *Zeitschrift für Psychologie/Journal of Psychology, 217*(4), 182–188.

Dunn, K. M., Paradies, Y., & Atie, R. (2014). *Preliminary Results: Cyber Racism and Community Resilience: The Survey*. Results Presented at Research Reporting Workshop. University of Technology Sydney.

Ebum, M. (2008). Litigation for Failure to Warn of Natural Hazards and Community Resilience. *The Australian Journal of Emergency Management, 23*(2). https://ajem.infoservices.com.au/items/AJEM-23-02-04. Accessed 4 July 2016.

European Commission Against Racism and Intolerance (ECRI). (2015). *Combating Hate Speech* (ECRI General Policy Recommendation No.15). ECRI. https://www.coe.int/t/dghl/monitoring/ecri/activities/GPR/EN/Recommendation_N15/REC-15-2016-015-ENG.pdf. Accessed 19 Nov 2016.

Facebook. (2015). Statement of Rights and Responsibilities. *Facebook*. https://www.facebook.com/legal/terms. Accessed 3 June 2016.

Facebook. (2016a). Community Standards. *Facebook*. https://www.facebook.com/communitystandards. Accessed 7 Sept 2016.

Facebook. (2016b). Data Policy. *Facebook*. https://www.facebook.com/policy.php. Accessed 22 July 2016.

Facebook. (2016c). Newsroom—Company Information. *Facebook*. http://newsroom.fb.com/company-info/. Accessed 4 Mar 2017.

Findlay, M., Odgers, S., & Yeo, S. (2014). *Australian Criminal Justice*. Victoria: Oxford University Press.

Fitzpatrick, S. (2016). Federal Election 2016: Man Charged Over Racist Nova Peris Tirade. *The Australian*. http://www.theaustralian.com.au/federal-election-2016/federal-election-2016-one-in-three-indigenous-suffer-abuse/news-story/c8fc8a92386c35bc47326e71e6f2d44b. Accessed 17 May 2017.

Foxman, A., & Wolf, C. (2013). *Viral Hate: Containing Its Spread on the Internet*. London: Palgrave Macmillan.

Francis, H. (2015, October 8). Facebook, Google Sign up to Anti-cyberbullying Scheme, May Face Penalties. *Sydney Morning Herald*. http://www.smh.com.au/technology/technology-news/facebook-google-sign-up-to-anticyberbullying-scheme-may-face-penalties-20151008-gk40xr.html. Accessed 16 Nov 2016.

Gelber, K., & McNamara, L. (2015). The Effects of Civil Hate Speech Laws: Lessons from Australia, *49*(3), 631–664. http://onlinelibrary.wiley.com/doi/10.1111/lasr.12152/abstract. Accessed 12 Feb 2017.

Giannasi, P. (2015). Policing Hate Crime. In N. Hall, A. Corb, P. Giannasi, & J. Grieve (Eds.), *The Routledge International Handbook of Hate Crime* (pp. 331–342). Abingdon: Routledge.

Griffiths, E. (2014, March 24). George Brandis Defends 'Right to Be a Bigot' Amid Government Plan to Amend Racial Discrimination Act. *ABC News*, http://www.abc.net.au/news/2014-03-24/brandis-defends-right-to-be-a-bigot/5341552. Accessed 21 July 2016.

Jakubowicz, A. (2010). Cyber Racism, Cyber Bullying, and Cyber Safety. *Conversation at the AHRC Cyber-Racism Summit 2010*, andrewjakubowicz.com. https://andrewjakubowicz.com/publications/cyber-racism-cyber-bullying-and-cyber-safety/. Accessed 5 June 2010.

Jakubowicz, A., Dunn, K., & Sharples, R. (2017). Australians Believe 18C Protections Should Stay. *The Conversation*, *The Conversation*. https://theconversation.com/australians-believe-18c-protections-should-stay-73049. Accessed 17 Feb.

Jakubowicz, A., Dunn, K. M., Mason, G., Paradies, Y., Bliuc, A., Bahfen, N., & Connelly, K. (2014). Submission to the Commonwealth Attorney General's Department, Amendments to the Racial Discrimination Act 1975. In Department, A.G.S. (Ed.).

Jakubowicz, A., Dunn, K. M., Mason, G., Paradies, Y., Bliuc, A., Bahfen, N., & Connelly, K. (2016). Cyber Racism and Community Resilience Research Project (CRaCR) Submission. *Freedom Speech in Australia* Submission 54. http://www.aph.gov.au/Parliamentary_Business/Committees/Joint/Human_Rights_inquiries/FreedomspeechAustralia/Submissions

Kerr, O. (2008). Criminal Law in Virtual Worlds. *University of Chicago Legal Forum*(1), pp. Article 11. http://chicagounbound.uchicago.edu/uclf/vol2008/iss1/11. Accessed 15 July 2015.

Klein, A. (2012). Slipping Racism into the Mainstream: A Theory of Information Laundering. *Communication Theory, 22*(4), 427–448. http://onlinelibrary.wiley.com/doi/10.1111/j.1468-2885.2012.01415.x. Accessed 24 Nov 2016.

Lawless, J. (2015). Charlie Hebdo Raises the Question: Is Hate Speech Protected in France?. *Huffington Post (online).* http://www.huffingtonpost.com/2015/01/27/free-speech-france-charlie-hebdo_n_6550240.html. Accessed 2 Mar 2017.

Louis, W., & Hornsey, M. (2015) Psychological Dimensions of Racial Vilification and Harassment. *Paper Presented at 40 Years of the Racial Discrimination Act 1975 (C'th) Conference. Sydney.* https://www.humanrights.gov.au/our-work/race-discrimination/publications/perspectives-racial-discrimination-act-papers-40-years. Accessed 27 Oct 2016.

Lukerson, V. (2015). Facebook, Google Agree to Curb Hate Speech in Germany. *Time (online).* https://www.humanrights.gov.au/our-work/race-discrimination/publications/perspectives-racial-discrimination-act-papers-40-years. Accessed 3 Apr 2016.

Mason, G., & Czapski, N. (2017) Regulating Cyber Racism. *Melbourne University Law Review (Advance), 41*(1). http://law.unimelb.edu.au/mulr. Accessed 17 May 2017.

McGoogan C. (2017). Germany to Fine Facebook and YouTube €50m if They Fail to Delete Hate Speech. *The Telegraph* (online). http://www.telegraph.co.uk/technology/2017/06/30/germanyfine-facebook-youtube-50m-fail-delete-hate-speech/. Accessed 20 Aug 2017.

McNamara, L. (2002). *Regulating Racism: Racial Vilification Laws in Australia.* Sydney: Institute of Criminology, University of Sydney.

Milovanovic, S. (2010). Man Avoids Jail in First Cyber Bullying Case. *The Age.* http://www.theage.com.au/victoria/man-avoids-jail-in-first-cyber-bullying-case-20100408-rv3v.html. Accessed 11 Aug 2016.

Nissim, R. (2014). Building Resilience in the Face of Racism: Options for Anti-racism Strategies. *Australian Policy Online and Australian Human Rights Commission.* Australian Human Rights Commission. http://apo.org.au/files/Resource/rnissim_buildingresilienceinthefaceofracism_oct_2014.pdf. Accessed 23 July 2015.

Nova Peris Accepts Apology from Disgraced Chiropractor Chris Nelson', *ABC News* (Sydney, 13 July 2016), archived at http://www.abc.net.au/news/2016-07-13/nova-peris-accepts-chiropractors-apology/7627240.

Oboler, A. (2012). *Aboriginal Memes and Online Hate.* Caulfield South: The Online Hate Prevention Institute. http://ohpi.org.au/aboriginal-memes-and-online-hate/.

Oboler, A. (2013). *Islamophobia on the Internet: The Growth of Online Hate Targeting Muslims* (Report IR13-7). Online Hate Prevention Institute. http://ohpi.org.au/islamophobia-on-the-internet-the-growth-of-online-hate-targeting-muslims/. Accessed 23 Mar 2017.

Oboler A. (2016). *Measuring the Hate: The State of Antisemitism in Social Media.* Online Hate Prevention Institute. http://ohpi.org.au/measuring-antisemitism/. Accessed 20 Aug 2017.

Oboler, A., & Connelly, K. (2014, 10–12 December). Hate Speech: A Quality of Service Challenge. In *IEEE Conference on e-Learning, e-Management and e-Services* (pp. 117–118). Melbourne.

Online Hate Prevention Institute. (2012). PRESS RELEASE: Discussions with Facebook over Aboriginal Memes. *OHPI.* http://ohpi.org.au/press-release-discussions-with-facebook-over-aboriginal-memes/. Accessed 9 May 2016.

Online Hate Prevention Institute. (2015).Fight Against Hate. *OHPI.* https://fightagainsthate.com/. Accessed 13 Mar 2016.

Online Hate Prevention Institute. (2016). Facebook Launches a Campaign Against Hate Speech in Europe. *OHPI.* http://ohpi.org.au/facebook-launches-a-campaign-against-hate-speech-ineurope/. Accessed 20 Aug 2017.

Paradies, Y. (2006). Defining, Conceptualizing and Characterizing Racism in Health Research. *Critical Public Health, 16*(2). http://www.tandfonline.com/doi/abs/10.1080/09581590600828881.) Accessed 27 Feb 2014.

Paradies, Y., Chandrakumar, L., Klocker, N., Frere, M., Webster, K., Berman, G., & McLean, P. (2009). *Building on our Strengths: A Framework to Reduce Racial Discrimination and Promote Diversity in Victoria Melbourne.* Victoria: Victorian Health Promotion Foundation. http://ro.uow.edu.au/scipapers/4674/.

Parliament of Australia. (2014). *Enhancing Online Safety for Children Bill 2014, Explanatory Memorandum* Canberra: House of Representatives. https://tinyurl.com/ExpMemOnlineSafChild. Accessed 15 Sept 2016.

Purtill, J. (2016). Racist. Violent. Deleted: The Facebook Posts Dividing Kalgoorlie. *Triple J Hack.* http://www.abc.net.au/triplej/programs/hack/the-facebook-posts-dividing-kalgoorlie/7805346. Accessed 13 Feb 2017.

Putt, J. (Ed.). (2010). *Community Policing in Australia.* Canberra: Australian Institute of Criminology.

Rees, S., Rice, S., & Allen, D. (2014). *Australian Anti-discrimination Law* (2nd ed.). Sydney: The Federation Press.

Reuters. (2008). Ex-film Star Bardot Gets Fifth Racism Conviction. *Reuters (online).* http://in.reuters.com/article/idINIndia-33883520080603. Accessed 24 July 2015.

Rohlfing, S. (2015). Hate on the Internet. In N. Hall et al. (Eds.), *The Routledge International Handbook on Hate Crime* (p. 293). New York: Routledge.

Sivasubramaniam, D., & Goodman-Delahunty, J. (2008). Ethnicity and Trust: Perceptions of Police Bias. *International Journal of Police Science & Management, 10*(4). http://journals.sagepub.com/doi/abs/10.1350/ijps.2008.10.4.094?journalCode=psma. Accessed 3 Aug 2015.

Sourdin, T. (2016). New Professor's Talk. *University of Newcastle*. https://www.newcastle.edu.au/community-and-alumni/community-engagement/events-and-public-lectures/new-professor-talks. Accessed 22 Nov 2016.

Soutphommasane T. (2015). A Brave Act. In *Conference Paper, 40 Years of the Racial Discrimination Act 1975 (Cth) Conference*, 19–20 February 2015, 7, 9, archived at https://www.humanrights.gov.au/sites/default/files/document/publication/Conference%20Papers_Publication%20layout%20Final_0.pdf.

Stephanos, S. (2014). Combating Religious Hate Speech: Lessons Learned from Five Years of Country Monitoring by the European Commission Against Racism and Intolerance (ECRI). *Religion and Human Rights, 9*(2-3), 139–150.

Stephens, K. (2015). Mariam Veiszadeh Now the Target of US Anti-Islam Site. *The Queensland Times*. http://www.qt.com.au/news/mariam-veiszadeh-now-target-us-anti-islam-site/2555598/. Accessed 23 Sept 2015.

Sykes, E. (2012). Racist Facebook Page Deactivated After Outcry. *ABC News (Online)*. http://www.abc.net.au/local/stories/2012/08/08/3563446.htm. Accessed 17 Apr 2017.

The New Zealand Law Commission. (2012). Harmful Digital Communication: The Adequacy of the Current Sanction and Remedies. *Ministerial Briefing Paper*.

Titcomb J. (2016). Facebook and Twitter Promise to Crack Down on Internet Hate Speech. *The Telegraph* (online). http://www.telegraph.co.uk/technology/2016/05/31/facebook-and-twitter-promiseto-crack-down-on-internet-hate-spee/. Accessed 20 Aug 2017.

Tomlin, S. (2016). Facebook Pages Shut Down After Violence in Kalgoorlie. *ABC News*. http://www.abc.net.au/news/2016-09-02/toxic-social-media-culture-in-the-spotlight-kalgoorlie-riot/7805050. Accessed 24 Sept 2016.

TPG. (2016). Standard Terms and Conditions. *TPG Internet*. https://www.tpg.com.au/terms_conditions/standard

Tynes, B., Umana-Taylor, A., Rose, C., Lin, J., & Anderson, C. (2012). Online Racial Discrimination and the Protective Function of Ethnic Identity and Self-esteem for African American Adolescents. *Developmental Psychology, 48*(2), 343–355. https://www.ncbi.nlm.nih.gov/pubmed/22369340. Accessed 14 June 2016.

Weaver, S. (2011). Jokes, Rhetoric and Embodied Racism: A Rhetorical Discourse Analysis of the Logics of Racist Jokes on the Internet. *Ethnicities, 11*(4), 413–435.

Wilson, L. (2013). Muslim Cleric Man Haron Monis Loses Bid to Have Charges Over Offensive Letters Quashed. *The Australian*. http://www.theaustralian.com.au/news/nation/muslim-cleric-man-haron-monis-loses-bid-to-have-charges-over-offensive-letters-quashed/news-story/bc1d487b-75339d526ed92105f37193d0. Accessed 17 May 2017.

World Jewish Congress. (2017). Tehe Rise of Antisemitism on Social Media. *WJC. New York*. Online at http://www.crif.org/sites/default/fichiers/images/documents/antisemitismreport.pdf. Accessed 20 Aug 2017.

Conventions and Legislation

Council of Europe. (2004). Convention on Cybercrime. *ETS No. 185*. https://www.coe.int/en/web/conventions/full-list/-/conventions/treaty/185. Accessed 18 May 2017.

Council of Europe. (2003b). Additional Protocol to the Convention on Cybercrime, Concerning the Criminalisation of Acts of a Racist and Xenophobic Nature Committed Through Computer Systems. *ETS No.189*. http://www.coe.int/en/web/conventions/full-list/-/conventions/treaty/189. Accessed 23 July 2015.

Council of Europe. (1950). Convention for the Protection of Human Rights and Fundamental Freedoms. http://www.echr.coe.int/Documents/Convention_ENG.pdf. Accessed 18 May 2017.

United Nations Human Rights Office of the High Commissioner. (1965). International Convention on the Elimination of All Forms of Racial Discrimination.http://www.ohchr.org/EN/ProfessionalInterest/Pages/CERD.aspx. Accessed 18 May 2017.

Anti-Discrimination Act 1977 (NSW)

Broadcasting Services Act 1992 (C'th)

Crimes Act 1900a (ACT)

Crimes Act 1900b (NSW)

Crimes Act 1958 (Vic)

Crimes (Domestic and Personal Violence) Act 2007 (NSW)

Criminal Code Act 1899 (Qld)

Criminal Code Act 1983 (NT)

Criminal Code Act 1995 (C'th)
Criminal Code Act 1924 (Tas)
Criminal Code Act 1913 (WA)
Criminal Law Consolidation Act 1935 (SA)
Discrimination Act 1991 (ACT)
Enhancing Online Safety for Children Act 2015 (C'th)
Harmful Digital Communications Act 2015 (NZ)
Racial Discrimination Act 1975 (C'th)
Racial and Religious Tolerance Act 2001 (Vic)

9

Conclusion: Future Directions in Building Community Resilience

Framing Cyber Racism as a Problem That Can Be Addressed

Cyber racism has kept the United Nations (UN) in a balance of indecision, infuriated governments, terrified victims, enabled perpetrators to grow their follower-swarms and accelerated the spread of radical nationalist and transnational social movements. Cyber racism has normalised hate, resurrected ancient antipathies, reignited dormant political aggression and contributed to disruption of the political order wherever the Internet has spread. By emboldening its propagandists, hardening its bystanders and fragmenting its targets, the spread of race hate online has changed the shape of expected civility, and exposed the limited capacity of key stakeholders to reverse the tides of separatism and intolerance. Like pouring icy water on heated rock, cracks thought sealed over a generation ago have reappeared, the sound of their emergence (((echoing))) everywhere.

Our research has demonstrated that the Internet has opened up opportunities for hate speech that are increasingly confronting. If over one-third of regular users of the Internet in Australia (with similar proportions

A. Jakubowicz et al., *Cyber Racism and Community Resilience*, Palgrave Hate Studies,
DOI 10.1007/978-3-319-64388-5_9

reported in the USA and Germany) are likely to have encountered, what they recognise as racism or race hate speech, at a sufficient level of intensity for them to recollect it and reflect on it, then we would argue the extent of unrecognised or unacknowledged experiences is potentially much wider. Our research shows that people who hold racist views are less likely to recognise online encounters as racism, especially where they define racism solely as advocacy of violence. People with racist views are less likely to empathise with the targets of racism, thereby not recognising or acknowledging it even when encountered.

A majority of people who report encountering racism laugh it off, let it slide or perhaps block the originator. A small minority report the originator to onsite standards monitors, usually without receiving satisfaction. A tiny proportion takes further action, usually by approaching offsite regulators or other officials. This pattern of responses suggests that a certain level of racism has become normalised, an acceptable tide-mark of intolerance and anger. As discussed in Chap. 4, racist groups seek this outcome, continuing to push the limits to ensure the tideline of "normal" racism surges upward. Our focus groups in Chap. 5 demonstrated that many people who felt they were the targets of racism withdrew from situations where they might encounter it in the future. Over time, the Internet has become dotted with myriads of self-selecting like-minded groups, some formed by active racists, many formed by refugees from hate, seeking their own-made sanctuaries.

Racism does not only encompass processes of exclusion or marginalisation by majorities of minorities. Minorities (the term is, of course, dependent on context) can also develop hate-filled discourses directed at those they see as their enemies, and those who they feel may be oppressing them. Such systematic "Othering" can occur in many situations where the common humanity of people becomes lost in a sea of stereotypes and offensive epithets.

The internalisation of racial discrimination through exposure to targeted racism online has many serious impacts on its victims. A sense of self-worth, or capacity to express oneself and achieve one's goals, can all be eroded, as racist language and world views permeate society—in schools, in workplaces and in public. Some racism may be acceptable enough where little damage is done and there are no apparent serious consequences, yet the repeated acquiescence to hate speech as part of the background noise of the Internet has a debilitating effect on civility, empathy and social cohesion. Moreover, by isolating individuals or groups and forcing them to withdraw from the new digital commons offered by the Internet, race hate erodes the social capital and trust necessary for effective reciprocal recognition of all those participating in the social realm. Ultimately, it corrodes the underpinnings of democracy, abandoning the political debate to the last person standing able to operate in a hate-filled world.

Where racism occurs as part of the planned implementation of a political program, the consequences are likely to be rather more serious. Building resilience requires a recognition of how racism grows, the way it draws into its shadow those who find it attractive and how it "twists the knife," while understanding the difficulties in developing effective responses and strategies of empowerment. Importantly, as racism online intensifies, the probability of it "leaping" the gap into the real world grows, a process that has already occurred many times.

In this final chapter, we recap and extend our examination of the pathways to resilience, bringing together theoretical models of community development with resilience building and exemplars from a range of situations and countries around the world in relation to online racism. We explore strategies and responses from within communities experiencing racist harassment, as well as collaborating social activists who provide technical capacities and organisational support. We also identify how governments and corporates have responded, and their roles in contributing to the enhancement of community resilience.

The biggest challenge appears to lie with the role and responsibility of the major Internet sites such as Facebook, Google (Alphabet), YouTube, Twitter, Instagram and similar providers. The global Internet industry depends on as free an access to the Web as is feasible; indeed such freedom forms the mantra underpinning the values of the Internet. Yet so committed have the developers of the Internet been to these freedoms (and the value created) that they have produced a cyberspace where it has become almost impossible to stop the spread of hate. Freedom of speech has too often transmogrified into a freedom to hate without responsibility. We further explore the way in which community standards operate on a range of sites and test whether the claims that are made by providers are indeed a good fit with what occurs in practice.

The legal environment within states, and across their borders in unions such as the Council of Europe or the UN, has for over two decades sought to regulate race hate speech online. We have shown that this has not proved to be an easily managed exercise, given differing national histories of racism, differing levels of tolerance towards prejudice and differing priorities for intervention. However, each of the jurisdictions has devised some collection of regulatory principles and practices, backed by a range of penalties, and driven by a series of proactive programs. We ask then, what might an ideal regulatory environment need to cover if it were both to protect freedom of speech, while minimising exposure to extreme race hate online? Only in totalitarian regimes where there is effective technological and security capacity has the freedom to speak one's mind about race been constrained (and directed towards regime-acceptable national racisms).

In addressing race hate online, it is important to note that two other online spheres of hate have been at the centre of attention by governments over the past decade. Each of these dissolves into race hate speech at the borders, and may utilise racism as part of its repertoire of mobilisation and attack. Cyberbullying, especially among young people (Notar et al. 2013a, b), and the use of the Internet in the recruitment and reinforcement of radicalisation into religious fanaticism and racially focused power have, in general, generated far more robust and proactive interventions by governments and civil society, than has cyber racism. This has been an easier task because neither is mediated through a prism of freedom of speech (Stevens and Neuman 2009), and both are perceived as direct and immediate threats to safety and well-being, both individual and social.

An extensive audit of strategies available to respond to cyberbullying, especially in high schools, demonstrated that there were technical and social interventions that could reduce the incidence and modify the impact of bullying online (Notar et al. 2013b). The key elements (which recur in many studies) relate to a focus on:

- Building empathy and understanding among online users (who included perpetrators, witnesses and victims);
- Teaching and practicing online safety skills; and
- Equipping users with strategies to "reject digital abuse in their lives."

Meanwhile, responses to online radicalisation have shown how the Internet facilitates and accelerates radicalisation, but the radicalisation itself was only finalised through real-world social contact and integration into real-world networks and groups (Stevens and Neuman 2009). Nevertheless, the Internet played key roles, by:

- illustrating and reinforcing ideological messages and narratives;
- easing the process of joining and integrating into conspiratorial networks; and
- creating new social environments in which otherwise unacceptable views and behaviour come to be normalised and internalised into the world views of the participants.

This International Centre for the Study of Radicalisation and Political Violence (ICSR) study proposed that an anti-radicalisation strategy would need to deter producers of material by using legal and technical avenues to block, take down and prosecute perpetrators. Moreover, an Internet Users' Panel was proposed, resourced by the industry, to monitor the implementation of standards by platforms, and ensure the widespread awareness and ease of use of complaints procedures, thereby "empowering" users to have more influence on how the Internet worked for them. Part of the empowerment would be the deepening of media literacy, so that users would more readily identify and resist recruitment. There would also need to be a funding mechanism to build grass-roots sources for "positive messages" to counter extremism. The report concluded by emphasising that action required collaboration and partnerships between the industry, government and the community.

Many of these ideas have percolated into discussions about building resilience against racism, though, interestingly, the insights from studies of Internet racism have had less influence on cyberbullying and counter-radicalisation. In the Australian context, the E-Safety Commissioner does not mention racism, while the counter-terrorism narratives have been criticised for being racist, in that they target minority groups and thereby reinforce terrorist claims to be victims of ethno-religious hate which helps them recruit new members. Moreover much terrorist rhetoric is itself racist, focussed on ethnographic-religious groups that they target, but is usually not described in this way (Briskman 2015).

Civil society organisations and social activist movements have played critical roles in devising and implementing ways of identifying, monitoring and opposing race hate online. We have shown that such institutions have been willing to move beyond governments, which are sometimes constrained from acting due to domestic legal and political issues. Often driven by social movements frustrated with the lethargy evident in the responses of corporations or the timidity exposed in the priorities of governments, civil society has developed initiatives and advocated for interventions directed towards limiting the impact of racism and empowering its victims to push back.

Cyber racism has been produced by the intimate relations of power generated by the state, the economy and the civil society. It is in that embrace where we must now search for its containment.

Pathways Towards Resilience

If we consider our various "targeted" communities and individuals introduced in Chaps. 3 and 5, and their experiences and responses to their encounters with more extreme versions of race hate and the more repeated forms of everyday racism, then we can lay out both the processes of encounters and the dynamics of engagement, reaction and recovery. What has happened to them that they need to find a capacity for resilience? Resilience at the personal level can mean either a regenerative process of strengthening and affirmation (Luthar 2006) or:

> being … (a) disconnected from others because they could not trust, (b) isolated because they had inadequate or no support systems, and (c) insulated because the emotional pain was too much to bear. (Hunter and Chandler 1999)

The idea of resilience has been further refined, requiring both significant adversity and positive adaptation to be present (Luthar 2006). Resilience can also be thought of as "relative resistance to psychosocial risk experiences" (Rutter 2000), while community resilience would refer to the "the capacity of a distinct community or cultural system to absorb disturbance and reorganise while undergoing change so as to retain key elements of structure and identity that preserve its distinctness" (Healy 2006).

Indigenous Australians have the highest rate of experiences of racism of all our research subjects, a finding born out relentlessly by other studies of Australian racism (Griffiths and Pedersen 2009). Indigenous Australians have shorter lives, poorer health, higher rates of infant mortality, massively higher rates of incarceration and significantly lower incomes and capital assets compared with the rest of the Australian population. They tend to experience lives that are more violent, and their contacts with the legal system are often far more punitive (Williams 2009; Grimshaw and McGregor 2006, Paradies 2017). Online harassment and intimidation of Indigenous people occur far more widely and with far greater hostility than against other Australian minority groups (Oboler 2012).

Indigenous Australians' experiences of the Internet have included on the one hand, exposure to stereotype, demands that they violently self-harm, self-denigration, being the butt of jokes and put-downs and being on the look-out for attacks, and on the other, finding few safe places they have discovered, opportunities to acquire new skills, share affirming narratives and develop their cultures. Resilience building among Aboriginal Australians has become something of a survival strategy, as it has elsewhere. Writing of resilience among Aboriginal peoples in Canada, factors were identified that might illuminate the underpinnings for resilience at the community and cultural levels, as well as for the individual and family (Ledogar and Fleming 2008). For Indigenous communities, racism or perceived discrimination represents a significant risk for vulnerability, as does:

> historical loss or trauma and unresolved historical grief.... Resilient reintegration occurs when one experiences some insight or growth as a result of disruption. It results in the identification or strengthening of resilient qualities. (Ledogar and Fleming 2008)

Thus, a resilience strategy for Indigenous groups experiencing a traumatic confrontation with online racism, compounded by structural discrimination in the offline world, would need to address both the direct experience of racism online and the underlying grief over historical loss that remains pervasive. As we reflect on the contemporary experience of Indigenous people with racism, the example of IndigenousX, examined in Chap. 7, may well be instructive.

An OHPI examination of Aboriginal Memes and Facebook responses that began circulating in mid-2012 revealed a wide range of online hate. The analysis demonstrated that non-Aboriginal people, especially the Facebook community standards guardians, initially perceived racism against Aborigines when it came through visual representations superimposed with tag lines ("memes") as humorous. Discussing the lack of reaction by Facebook to complaints about the memes, Oboler (2012) argued that:

- Social media magnified and distributed hate;
- Facebook gained an income from the persistence of hate pages that drew visitors and
- The small groups that posted hate could be identified and banned, but rarely were.

Random attacks on Indigenous people online, the circulation of racist memes and the creation of a YouTube "genre" of "Abo fighting" by a poster named "Cyberboong" demonstrate that Indigenous people are constantly subject to racist stereotyping, sometimes self-destructively, and, at times, from their own group.

IndigenousX has survived and prospered, we would suggest, because it has discovered a strategy that combines the criteria laid out earlier for resilient regeneration:

1. The Twitter feed has been protected from trolling;
2. The curator has found and involved around 50 different lead posters each year, sustaining interest and novelty (keeping the swarm engaged);
3. The focus has been on their daily lives and passions, personalising and building engagement further;
4. The overarching narrative has been one of cultural affirmation, rejection of stereotypes and strengthening of links between Indigenous people; and

5. The brand has gained in stature and influence, for example, providing regular material to *The Guardian* newspaper online.

In addition, the site has not succumbed to the other danger of less optimal resilience, where "survival tactics of violence, high risk behaviors, and social and emotional withdrawal" (Hunter and Chandler 1999) could otherwise have created and intensified a discursive environment of racialised hostilities to non-Indigenous people.

In their analysis of resilience as a system of communication, Houston and colleagues identified four critical elements—two were about communication and two about community (Houston et al. 2015). The two communication elements encompassed systems and resources, and strategic processes, while community referred to both the relationships existing in the community and its attributes. Some of these may prove to be malleable, while others may turn out to be resistant to change.

Communication systems and resources identified all the media present for a community, what official sources of information existed and how were they communicated, and the organisations that citizens had created that could participate in the communication process. These organisations could exist "in the real world," online or operate in both spheres. However, they would need to be aware they were part of a communication system, and that they were embedded in the dynamics of building and supporting regenerative resilience.

The strategic communication processes would need to include community planning for the risks ahead, and how communities would respond. They would need to understand how to act in a communal manner. Furthermore, there would need to be planned time for critical reflection on challenges and potential responses, shared problem solving and identified avenues for community empowerment. Preparedness planning, that accessed and disseminated accurate information supported by ongoing education about risk identification, would contribute to resilience awareness.

At the heart of the communication model sits the complex of community relationships, and their roles in supporting resilience and capacities to respond. While some communities actually do exist in a place and people do interact with each other in many offline contexts, rather more online communities form only for the purposes of a single interaction, or are sustained through groups on Facebook or through follower lists on Twitter or Reddit or YouTube. The transience of such communal moments contributes to the difficulties associated with community development on the Web. However, the findings of the Houston group's research demonstrate that an investment in social support will realise significant improvements in social capital, both in the type that builds intracommunal bonds, and also in the forming and extension of weaker links that nevertheless connect people to other potential allies and collaborators through bridging social capital (Aldrich and Meyer 2015). Such weaker links would include partnerships with key political actors, good relations with other media organisations and the value that would accrue from partnerships with public agencies and private corporations.

We now explore how different elements of the state/economy/civil society nexus might facilitate the building of resilience among targeted communities, and in collaboration with the partners they might secure. At the heart of the system lie the corporate providers, whose focus remains on building the returning user bases they require for profitability, while trying to avoid reputational damage. Increasingly, they are turning their attention towards how they will deal with issues of race hate online.

Industry Platform Codes and Practices

The primary defence offered by Internet service providers to criticisms of their failure to protect users from racist harassment lies in their terms of service and the community standards attached to them. People using the service who breach those standards can be temporarily blocked, have posts removed or ultimately be banned. Most community standards reflect the balance that the provider determines between freedom to post and threat to other users. The preferred user response for providers is for the user to block their exposure to the offensive post, leaving it available for other users to access. The responsiveness of providers to complaints continues to be a matter of debate, with OHPI identifying all three of the main providers (Facebook, Twitter and YouTube) as problematic when they were provided with evidence of antisemitic material (Oboler 2008). Moreover, in 2017, the Simon Wiesenthal Institute in launching its annual Digital Terrorism and Hate Report, noted that "The Social Media giants have thus far failed to curb the online marketing campaigns of hate and terror, leaving us all at greater risk. Moving forward we will need to build a strong public-private consortium to stop the hate" (Simon Wiesenthal Center 2017). In its Digital Terrorism and Hate Report card, the Center awarded Facebook a B-, Twitter a C+, and Google/YouTube a D as grades for their action on Hate online.

Facebook has refined both its standards and its protections for users, making the default for groups "private," whereas they used to be "public", while more clearly detailing the standards to be applied. While the standards are global, their application seems to vary by jurisdiction, depending on local laws (Simon Wiesenthal Institute 2009). Germany has the most stringent legal constraints, with threats of legal action against Facebook including criminal prosecutions of senior executives (Maier 2016; Oboler 2016) focusing the provider's attention on how it might reduce the spread of anti-immigrant racism in Europe.

In a Note on Facebook in May 2013, the company addressed its policies on "Controversial, harmful and hateful speech." It noted that:

> Many different groups which have historically faced discrimination in society, including representatives from the Jewish, Muslim, and LGBT communities, have reached out to us in the past to help us understand the threatening nature of content, and we are grateful for the thoughtful and constructive feedback we have received… We prohibit content deemed to be directly harmful, but allow content that is offensive or controversial.

Facebook continued that it prohibited hate speech, which it defined as:

> direct and serious attacks on any protected category of people based on their race, ethnicity, national origin, religion, sex, gender, sexual orientation, disability or disease. We work hard to remove hate speech quickly, however there are instances of offensive content, including distasteful humor, that are not hate speech according to our definition.

Noting that the posting of insensitive or cruel content:

> often results in many more people denouncing it than supporting it on Facebook. That being said, we realize that our defense of freedom of expression should never be interpreted as license to bully, harass, abuse or threaten violence.

Facebook also stated that:

> Organisations and people dedicated to promoting hatred against these protected groups are not allowed a presence on Facebook.

Despite the investment in industry leading systems to identify and remove hate speech, Facebook recognised that:

> it has become clear that our systems to identify and remove hate speech have failed to work as effectively as we would like …. [and] content is not being removed as quickly as we want. (Levine 2013)

In the years that followed that post, over 2000 comments were left on the statement, including one in December 2016. This referenced a story in DW News (DW 2016), which reported that the International Auschwitz Committee had accused Facebook of having an "arrogant approach to hate speech," after reports that nearly half the posts flagged as "hate speech" had not been deleted within the 24 hours agreed between the German Government and online firms late in 2015 (DW 2015). Furthermore, YouTube was much more lethargic, having removed only 10% of reported posts.

As previously noted, YouTube has a much more dire reputation than Facebook for failing to remove offensive and racist material. YouTube also makes it clear that it will allow material that is offensive, although it does not:

> support content that promotes or condones violence against individuals or groups based on race or ethnic origin, religion, disability, gender, age, nationality, veteran status, or sexual orientation/gender identity, or whose primary purpose is inciting hatred on the basis of these core characteristics.

This can be a delicate balancing act, but if the primary purpose is to attack a protected group, the content crosses the line. (The YouTube Team 2017)

Again, YouTube, owned by Google Alphabet, prefers its users to block offensive material, thereby protecting the user from exposure but allowing the offensive material to remain for others. Only if that is not satisfactory, should users then move to report it. Nevertheless, allegedly antisemitic and racist vloggers have been permitted to remain online and develop major profiles. Take the vlogger Pewdiepie, for example. With 15 million YouTube followers, he has fought back against critical attacks on him by posting a vlog incorporating "Sniper Elite 4:Fuhrer" under the heading "I kill Hitler*not clickbait." He claimed that telling jokes about supposedly hunting Hitler was designed to show he was not himself a Nazi sympathiser. He had recognised the vlogs were offensive, but he denied they were "hateful." He had previously been criticised for pro-Nazi jokes, and had been described by *The Daily Stormer* as "our guy" who "normalizes Nazism, and marginalizes our enemies" (England 2017).

Twitter represents another dimension of online communication altogether. While once mainly used by liberals and minorities to communicate (and by ISIS in Iraq and Syria notably at the battle for Mosul in 2014) (Brookings and Singer 2016), it has seen the exit of many previously tweeting journalists, and others, because it has allowed trolls fairly free reign (Stein 2016). It has, however, sought to push back against the alt-right infiltration of the platform, suspending key alt-right accounts in November 2016 (BBC News 2016). In January 2017, an alt-right coalition announced it would crowdfund a legal action against Twitter, despite the popularity of the platform among alt-right supporters, one of whom had said: "The benefits of Twitter are interacting with normies, influencing discussion and getting alt-right memes trending."

States and Legal Environments

Governments either alone or in tandem have tried many different approaches to limit the expansion and impact of race hate speech on the Internet. The global awareness of the issues emerged at the dawn of the Internet; various layers of action by the UN, the Council of Europe and individual governments, however, have only moved a small distance in seeking to produce a more civil and safe space. It is worth asking therefore what laws have emerged that are successful in reducing the extent and impact of race hate speech, and what rationales have been used to ensure that such limitations can be applied? We developed this inquiry in detail in Chap. 8 where we specify a strategy for the Australian political and legal context as an example of the thinking that might be applied more widely.

Racist speech, whether through language, images or symbols, is a tangible manifestation of racism. Calls for its regulation rest on claims that it inflicts three levels of harm:

1. Psychological harm to an individual's sense of dignity and safety (Paradies 2017; Berman and Paradies 2010);
2. Group harm to other members of the target community who may interpret such commentary as a sign of intolerance and disrespect (Sentencing Advisory Council 2009); and
3. Social harm that undermines the "public good of an inclusive society," which is implicit to liberal democracies and expressed through values of multiculturalism and equality. This violation of shared values makes racist speech a "public wrong" (Oboler 2014).

This understanding of racist speech belies two further complexities. The first is the application of the term "racism" to bias towards characteristics that fall outside claims related to propositions about biological race, such as ethnicity, language, nationality, tribal linkages, immigration status or religion. Without downplaying the distinct aetiologies of these different forms of prejudice, there is also considerable permeability between many of them—for instance, around the popular belief that characteristics are chosen or inherited. With some exceptions, the weight of Australian and international authority gives a fairly broad interpretation to the concept of race as one that should be "understood in the popular sense" rather than restricted to effectively meaningless genetic tests or assumptions derived from physiognomic diversity.

Second, difficulty remains in drawing a distinction between racial speech that warrants prohibition and that which does not, or, to put this another way, in the question of how to strike a balance between freedom of expression and protection from racist speech (Australian Human

Rights Commission (AHRC) 2013; Bleich 2011). Liberal democracies have taken a variety of nuanced approaches, all of which are grounded in different historical, political and institutional contexts. At one end of the spectrum is the USA, which has bucked the post-war trend towards limiting freedom of expression to protect minority groups by penalising racist speech, while at the same time having significant legislation to punish discrimination and racist action. Although not without limits, the USA has the most extensively entrenched protections around the freedom to express offensive and provocative speech, underwritten by the interpretation given to the First Amendment. In contrast, European jurisdictions have slowly but consistently expanded human rights protection with restrictions against the expression of racist views over the past few decades. This is exemplified by France, which takes a strong legislative approach and has had a number of high-profile prosecutions for racist speech in recent years.

As we discussed in Chap. 8, Australia has racial vilification laws in place in almost every jurisdiction, but there continues to be contestation about where to draw the legal line between acceptable and unacceptable speech. This was exemplified by a minority of submissions to the 2014 Federal Attorney-General's consultation on proposed amendments to section 18C of the *Racial Discrimination Act 1975* (Cth), which proposed its removal, compared with the majority arguing in favour of retention. These debates were reignited with the establishment of a Parliamentary Inquiry into Freedom of Speech in Australia in 2016, which examined, amongst other things, the operation of section 18C. The Committee was unable to reach agreement on its conclusions; the government nevertheless took legislation to Parliament in March 2017 where it failed to gain the support of a majority of senators. It is unlikely that any legal standard will completely resolve this contestation, which calls for a combination of legal and non-legal responses. It is notable, however, that a 2014 Nielsen survey shows that 88% of Australians believe it should be unlawful to offend, insult or humiliate others on the basis of race, though in submissions to the 2016/17 Inquiry by Parliament, competing surveys claimed rather different findings about contemporary public attitudes (Jakubowicz 2017; Jakubowicz et al. 2017). Those opposing the changes to the Act nevertheless reflected a majority view from which it can be deduced that the Australian public has a continuing appetite for formal regulation, a testament to law's importance in adjudicating a path between the differences of opinion that surround state-based sanctions of racial speech.

Due to its nature and the technology involved, cyber racism presents its own specific regulatory challenges. Material posted online is ubiquitous and relatively permanent. Information can be disseminated instantaneously, continuously and globally, reaching far greater audiences than practicable in the offline world (The New Zealand Law Commission 2012). Moreover, when material is published online, it can remain "cached" or stored, can potentially be accessed via search engines and can easily be duplicated. The removal of harmful content from a platform does not guarantee its erasure from cyberspace. This has led some to describe cyber hate as a "permanent disfigurement" on members of the targeted group (Waldron 2010).

The online world connects people with real or imagined communities of others who share their viewpoints, adding credibility and encouragement to those who already harbour discriminatory views (Keller 2009) and can fuel a depersonalised mob-mentality of harassment (Citron 2014). Those who harbour racist views are more emboldened to express those views and less willing to compromise where they perceive their view as a majority one (Pedersen 2010). As a vast information-gathering machine, the Internet can help enforcement by making it easier to identify some perpetrators. However, the use of "anonymous" profiles and pseudonyms, quite apart from having a disinhibiting effect (McFarlane and Bocji 2003), makes it difficult for victims of hateful conduct to seek informal redress, and has, at times, necessitated legal compulsion of Internet providers (sometimes successfully) to reveal the identities of perpetrators.

Further, the Internet often bypasses the traditional media gatekeepers that act as a check on the dissemination of unpalatable viewpoints, and content easily spreads without regard for state boundaries. The cross-jurisdictional nature of online "publication" makes it possible to have an Australian victim, targeted by a cyber racist in another country, on a social media platform hosted in a third country by a company incorporated in a fourth country. Dealing with any one instance of cyber racism may require coordination between law enforcement and government agencies from multiple countries as well as intermediaries such as online host platforms and connectivity providers, throwing to light legal inconsistencies between jurisdictions.

When the Council of Europe addressed the question of rising hate in Europe in 2010, they referred to the issues with social media within the context of the rising risks confronting the European project. They identified rising intolerance, rising support for xenophobic and populist parties, discrimination, minorities without rights, parallel societies, Islamist extremism and tension between freedom of speech and religious solidarity. They saw insecurity, immigration, distorted images of minorities in the media (now termed "fake news") and a crisis of leadership as the drivers. Reflecting on the changing dynamics of Europe, they reported:

> unfortunately, a growing number of Internet users are abusing the Web to spread racist or xenophobic propaganda, and to incite others to hatred—to such an extent that the Internet has now become the number one dissemination tool for hate speech and creating fear. To make matters worse, new interactive web 2.0 services allow extremists to leverage technologies such as blogs and video-sharing to promote their agenda on popular "social media" sites such as Facebook, MySpace, Twitter and YouTube. According to the 2010 "Digital Hate Report", the Internet grew 20% uglier in 2009….

In response to these insights, the Council urged its member states to ensure that:

> media literacy programmes are included as a core element in school curricula, and that children and young people are alerted to expressions motivated by racist, xenophobic and anti-Semitic or other related bias which they may encounter on the Internet. Member states should also make sure that law enforcement officers and prosecutors are trained to deal with similarly motivated hate crimes on the Internet, and work with the Internet industry to encourage it to take a more active role in addressing this issue. (Group of Eminent Persons of the Council of Europe 2010)

We are left then with the critical role to be played by civil society organisations, as they research, monitor, reflect and plan ways forward for liberal societies. The identification of such groups as critical players in enhancing the online community capacity to resist and push back against racism has now become the focus for many recent initiatives.

Civil Society and Social Activism

The first major civil pushback against cyber race hate emerged from within the global Jewish communities, prompted by the US Anti-Defamation League recognition of the first Internet problems (Capitanchik and Whine 1996; Hoffman 1996, 1997), and the Internet hate reports generated annually by the Simon Wiesenthal Institute (Simon Wiesenthal Center Snider Social Action Institute 2016). Although the Sharpeville massacre in South Africa had been the turning point for the international community in terms of racism, antisemitism had been one of the other stimuli for the global community accepting the International Convention to Combat All Forms of Racial Discrimination during the 1960s. The Internet has allowed antisemitism to re-emerge on a global basis, reigniting hostility and defamation of Jews through out the world. Antisemitism was also improperly justified by hostility to Israeli policies, though the hatred and antipathy against Jews drew on centuries of violence and marginalisation in Europe and elsewhere in the Middle East and North Africa (Back 2002; Simon and Schaler 2007; Oboler 2016; Fleishman and Smith 2016). By 2015, the international Jewish community through the Global Forum for Combating Antisemitism (GFCA), had identified an integrated package of interventions that should be implemented by the Internet industry, governments and civil society. Their proposals included the following:

- The global adoption of a common definition for hate speech and antisemitism;
- Criteria for blocking;
- Information guidelines;
- Industry policing of hate speech;
- Timely and effective hate processes;
- Preventing search engines finding hate sites and
- Pursuing, preventing and prosecuting terrorist hate speech.

Government Interventions

- Establish units aimed at combating cyber racism;
- Prosecute abuses more actively and
- Require corporates to adopt global standards.

Civil Society Interventions

Civil society would need to focus its attention on pressuring governments to pressure platforms (Global Forum for Combating Antisemitism 2015).

In many ways, the GFCA took on the ADL priorities, without the overlay of the US First Amendment.

Twenty years into the Internet and awash with the rising frustration at the failure of governments to hold back the tide of race hate speech, especially in Europe, the international community through UNESCO brought together the shared understanding of how online hate speech might be countered (Gagliardone et al. 2015). This UNESCO report argued that hate speech was most likely to flourish during periods of acute political tension, such as during elections where racial or ethnic groups might be in competition, or when social crises erupted—such as refugee arrivals, terrorist attacks and domestic incidences linked to ethnic group differences. The UNESCO project focused on dangerous speech, that is, hate speech, most likely to be a catalyst for intergroup violence.

Three forms of race hate speech gave rise to the concerns covered in the report:

a) Expression constituting an offence under international law that can be prosecuted criminally;
b) Expression not criminally punishable but that may justify a restriction and a civil suit and
c) Expression that does not give rise to criminal or civil sanctions but still raises concerns in terms of tolerance, civility and respect for others.

The report places particular emphasis on civil society and social steps rather than state-initiated legal measures. It identified four initiatives:

1. Research efforts to monitor how hate speech online emerges and spreads, developing early warning systems and methods to distinguish among different typologies of speech acts;
2. Coordinated actions by members of the civil society seeking to create national and international coalitions to address emergent threats connecting online hatred and violence offline;
3. Initiatives to encourage social networking platforms and ISPs to play a more robust role in actively responding to hate speech online; and
4. Media literacy campaigns and initiatives aimed at preparing users to interpret and react to hateful messages (Gagliardone et al. 2015, p. 6).

One of the most critical dimensions identified involved building media literacy through education. Such an approach bears out our findings in Chap. 3 that people who hold racist views are less likely to identify what they encounter as racism, partly because of the high bar they use to define racism (incitement to violence) and partly because they are less likely to empathise with the victims and thus form an internal narrative that links them to the victims in any positive way.

Components of the proposed initiatives include raising awareness among users of the legal issues involved, and of the range of social issues that users need to think about when they are online. In effect, the initiatives seek to minimise the disinhibition effect of being online, by triggering and supporting mindful reflection, active listening and thoughtful engagement. Such an approach builds empathy, increases recognition of the dangers of hate speech and thereby reduces active production of hateful speech.

Three broad educational goals involve identification, analysing and acting on race hate speech:

1. **Identification** helps engaged users recognise what is going on around them, helps them to frame it and move it on to their networks;
2. **Analysis** supports users to think about what they encountered, in order to understand how the hate speech was structured and what

impact it was trying to achieve, especially through the normalisation of hate in the everyday online world; and

3. **Action** motivates users to respond to hate speech by undermining it, writing against hate speech, countering the normalisation of hate speech and monitoring the wider media for hate speech and ensuring its presence is contested.

While many organisations were identified by the UNESCO report and their strategies celebrated, the range of such interventions alone may have had little impact on the extent of cyber racism. The platforms pushed towards resisting demands to censor material, while civil society tried to restrict access to platforms that enable dangerous speech. The tension between the different interests of stakeholders was not resolved, leaving more systematic responses still lacking implementation.

The First Amendment protection of freedom of speech has driven US civil society organisations towards considering how to combat race hate online without using the law. Ever since the pessimistic projection about the future of race hate on the Internet voiced by the SPLC (Richardson 1998), the SPLC and the ADL have sought to build coalitions armed with more sophisticated strategies of intervention.

The ADL has identified the platforms as the major problem, calling on them to define what they accept as hate speech, establish clear reporting mechanisms, act quickly when they find something in breach, make public the grounds on which material has been removed and notify complainants on what grounds they had rejected complaints (other than "doesn't contravene the standard"). In the USA where free speech remains a major legislative protection, the ADL argued for legislation on cyberbullying to address the hurt as a result of hate speech from a different perspective. Most importantly, the government needed to speak out against bigotry, a demand reiterated of the Trump presidency early in its term (Jewish Telegraphic Agency 2017). The debate over the responsibility of the presidency in condemning hate speech and racism intensified in August 2017, when President Trump vacillated on his condemnation of right wing racist demonstrators in Charlottesville, Virginia following the murder of an anti-racist demonstrator (Calmes 2017). Children also had to be alerted to the three dangers of cyber racism—they should be able to identify it and discard it, they should not be passive victims and they should not become inadvertent perpetrators. The ADL also developed a model of an active online user, geared up to resist cyber racism. That person (or group) should always flag offensive comment for review, and provide a clear and coherent reason for their choice. They should speak out at a level with which they feel comfortable, countering offensive viewpoints when they can, demonstrating an alternative and non-hateful perspective. This, the ADL argues, is always worthwhile, as sites with no positive material can recruit followers through the power of acquiescence. Applaud those who stand up against racism: talk, communicate, identify the rules and use them.

We have shown how online racism emerges from a process of community development, of lurkers swarming around sticky spots where targets are trashed and hate speech normalised and reinforced. However uncomfortable it may be for those who engage in it, counter-speech cuts into the assumptions of the racists and dilutes the taken-for-granted certainties promoted by the race hate propagandists. The ADL operates as a monitor of race hate in the USA, as do other organisations elsewhere. Their roles in sustaining awareness, advocating resistance and supporting resilience play a critical part in what has increasingly become a widespread and probably unending guerilla war. The SPLC has tracked hate groups in the USA since 1999. It argues that the resurgence in groups since 2015 has occurred as the earlier retreat to the Internet has now triggered an increase in on-the-ground activities (Southern Poverty Law Center 2017).

Close collaboration to combat hate speech, between the two main "pivots" of cyberspace, Google and Facebook, and civil society, began to emerge in 2015, and accelerated in 2016. In the UK, Facebook joined with Demos (a political issues foundation) to explore how they might together "counter that type of disagreeable or extremist content." Facebook had committed itself to supporting its choice of a "more effective way" to tackle this problem, namely, counter-speech, described as a "common, crowd-sourced response to extremism or hateful content" (Bartlett and Krasodomski-Jones 2015). In the research for their project, Demos identified 124 "populist right wing pages" across Europe, with over 25,000 posts. The researchers could only identify 26 counter-speech pages, with 2300 posts. The right-wing pages generated nearly 8 million interactions, the counter-speech just over half a million. Reflecting on the insights that they had drawn from the data, the authors proposed that counter-speech should use a lot more videos and pictures, develop a lot more constructive counter-speech and modify its strategies to "dramatically increase the reach of their messages." Such findings confirm our concerns about the challenges of building resilient online communities, which have not proven to "bite" effectively into the dialogues monopolised by hate speech propagandists.

Also in the UK, a partnership between Demos and Web developer Bold Creative approached the issue of digital citizenship within the framework of countering radicalisation. The concept of "digital citizenship" addressed to young people was designed:

> to build the resilience of participants to extremism and radicalisation online.... The intervention sought to teach young people how to recognise online propaganda and manipulation, understand how social media change how we communicate, and develop a sense of responsibility over their online social networks. (Reynolds and Scott 2016)

Across continental Europe, Facebook has been coming under considerable pressure from anxious governments, motivated by rapidly spreading ultra-nationalist and White power social media networks focused on ending the European project, expelling immigrants and suppressing Muslims and other minorities. In early 2016, Facebook launched another "counter-speech" project, the "Online Civil Courage initiative," which was designed to challenge hate speech and extremism online. A year later, it had 40,000 "likes" and a similar number of followers. The project included a Facebook page, which was often hit by trolls, but sustained through hashtags such as #DontHateCommunicate, and a training program for 30 activists from across Europe "to do speech effective against hatred." The first evidence suggested that such initiatives were not driven by the same social movement energy that inflames the hate speakers.

However, a joint project with Facebook, Alphabet and Twitter focusing on support for three NGOs had rather more positive outcomes (Toor 2016). The aim was to "assist small and medium size NGOs in producing, disseminating and evaluating counter-narratives and counter-speech." One project, ExitUSA, which has some similarities with the Australian All Together Now Exit White Power project, focuses on helping White power members to leave the movement, which proved very effective. The project was operated by the UK-based Institute for Strategic Discourse and recommended:

- The allocation of modest funding;
- The use of in-depth research with target audience and behaviours;
- Training staff in using social media marketing tools and
- Coordinating online and offline events, but being aware of bots that might distort results and mislead about the impact.

Each of the social media platforms had its own strengths and weaknesses, which was why the integrated strategies proved more effective (Silverman et al. 2016).

The End of the Beginning...

All the pieces fit together, as we model resilient strategies for communities anxious to move beyond hate. However, as communities struggle towards resilience, we can also see the dark side of the Web hardening. Reports of trolling that places it at the centre of a vision for a post-liberal world abound—Wired has reported on the extent of organised trolling of liberal writers, doxing (providing real-world information for real life harassment) their targets and driving them from the public world (Greenberg 2016). The desire to see racist neo-Nazi narratives as the new normal is happily voiced by their proponents all over the world, tying the post-Stormfront activists together in new networks of anger and hate. Before Brexit and Trump, these contributors to the Internet were marginal, criminal and brutal. Rather too quickly, they are moving towards centre stage (Cadwalladr 2017).

Resilience strategies in the face of cyber racism require serious social investment, innovative and creative leadership, serious skill development and integration of public and private sector initiatives with civil society. If there are troll farms on the other side of the Dark Web trying to erode and corrode social cohesion, empathy and collaboration, we need to plant many more cyber parks where civility, cohesion and respect can be nurtured and grow. These will not happen without sustained thinking, reflection, planning and resourcing. If spaces of race hate speech and the cyber racism it feeds remains barely attended and poorly contested, then we will be facing a corrupted public sphere, fragmented, fearful and inward-looking.

How Can the World Move Forward on Cyber Racism?

We have demonstrated how racism on the Internet has been produced and is sustained through the interrelations between the many levels of structure that give the Internet its power. The freedom that it accords offers advantages to those who advance racially charged agendas. At the global level, the political economy of the Internet has been developed to ensure economic benefits can accrue to the large corporations with minimum interference with their freedom to trade. In the USA, the conflict over net neutrality (the right of consumers to have equal access to any source on the Internet) remains a continuing point of contention as the Trump administration moves to give ownership of data and access to services to the service corporations (Public Knowledge 2017). The "freedom" of the Internet currently permits but does not ensure or enhance equality of access and freedom to navigate. Even so, freedom also facilitates the expansion of racism to travel widely and become "normalised" as part of the Web.

The UN ICERD groups have long been warning of these dangers, but have found little leverage to bring about change in the fundamental structures that underpin the edifice. Very recently, as the tide has risen, both the International Telecommunications Union (ITU) as the UN-linked planner of the infrastructure and technical standards, and the W3C foundation as the inheritor of the dreams of the designers of the World Wide Web, have recognised that the Internet and the Web they have supported for nearly two decades, are in dire trouble over racism and hate speech. Each has committed itself to exploring how it can redress the structural challenges that enable the Web to be used to undermine social cohesion and fragment cultural relations across the world.

This fundamental tension over freedom versus safety places pressure on corporations, which seek both to ensure their financial success and to defend themselves from governments which are more worried about social cohesion and communal solidarity than they are about the profitability of particular (often foreign) corporations. We have argued that the question of racism cannot be reduced to issues of freedom of speech, and left to different viewpoints battling it out for influence with the jury of public opinion. When French President Nicolas Sarkozy stated in 2011 that the Internet was not "a parallel universe which is free of rules of law or ethics or of any of the fundamental principles that must govern and do govern the social lives of our democratic states," he focused on the need for the users of cyberspace to "show tolerance and respect for diversity of language, culture and ideas" (Lasar 2011). Moreover, in the context of the case by a French anti-racist group LICRA against Yahoo over the (illegal in France) advertising for sale of Nazi memorabilia, French courts had ruled that the national government did have jurisdiction over the Internet, and that it was both feasible and practical for corporations to act in accordance with national laws (Goldsmith and Wu 2006). Despite the oft-repeated claim that the Internet exists beyond government, this is not true, though the variability of law across states makes any simple process of regulation or even shared common practice extraordinarily challenging.

In this context, as the rising tide of racism, xenophobia and antisemitism confronts societies in which social cohesion had been the norm, states and combinations of states (e.g., the Council of Europe) have devised strategies to modify the license claimed for the Internet in the past. Even so, the realities of social civility are being tested as technologies grow and change, interweaving with and transforming what had appeared to be stable just a few years before. Yet, the variability of the Internet and the distinctive qualities of the different space regimes that operate across its landscape often leave abandoned or bypassed secretive spaces where regulation does not intrude, control has been let slide or resisted and license operates with abandon. Writing just before the advent of social media in the early days of Web 2.0, Goldsmith and Wu (2006) proposed that the tradition of modernity, of being associated with territorial states, met the priorities most people had for "a human history that accommodates humanity in its diversity and allows it to flourish" (Goldsmith and Wu 2006). However, as social media have spread as a technology, so too has the erosion of nation-states and the uprising of transnational ideologies of hatred and exclusion, from White power at one extreme, to radical totalitarian armed insurrections claiming origins in Islamic beliefs at the other. As we have shown, the transnational spread of these ideologies and their associated violent practices would not have been feasible in terms of immediacy, scale and extent before the advent of Web 2.0.

Coordinated action across the globe that could push back against the spread of irrational hateful ideologies depends on a coming together of the three fields of power that operate in cyberspace—government, economy and civil society. We outline them in Fig 9.1. The conjoining of these fields needs to occur at the global, regional and national levels. We put forward some potential pathways that may meet the criteria widely identified by multitudes of reports, studies and advocacies. Broadly speaking, these criteria require the development of robust technological means of identifying the appearance of and penetration of hate speech,

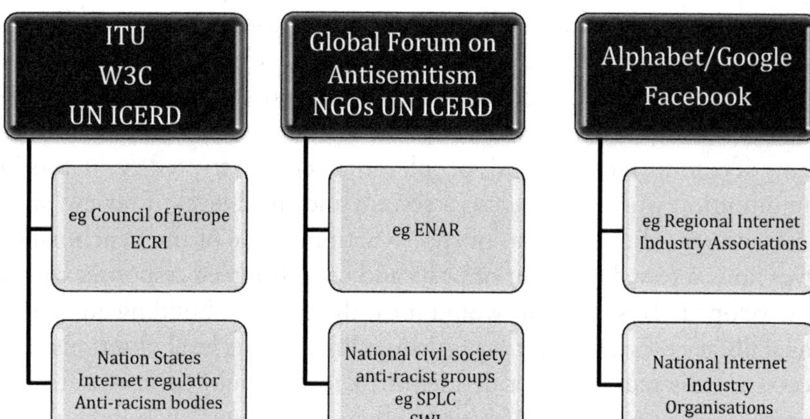

Fig. 9.1 Parallel stakeholder examples in responding to cyber racism—global, regional and national

while balancing the values of freedom to believe, speak and publish enshrined in the UN Human Rights conventions.

At the global level, the already existing interactions between governments, corporations and civil society could be more formalised, with the UN and its associated agencies establishing a standing group that included, at the very least, the major platforms and providers, to ensure transparency and reliability in action by infrastructure and content regulators, and systematic responsiveness to civil society concerns and priorities.

The Web 2.0 has increasingly perfected crowd sourcing technologies to test and respond to the relationship between peoples' aspirations and their choices (Matteo et al. 2014). The work of the Online Hate Prevention Institute with its FightAgainstHate software provides an example of how crowdsourcing to identify and name hate speech could be achieved, with initial identification then crowd-tested for robustness, bias and manipulation. The Australian test-bed could provide the basis for a scaled-up strategy. When allied with queries that ask what participants would like to see occur to identified hate posts and similar material, the software creates a social choice mechanism that allows machine learning of the hierarchy of responses that would appropriate in a range of situations. Thus, at one end it might simply trigger an automated reply such as "this post could cause some people distress," through to "this post should be removed, the account closed and the poster referred to authorities for investigation under criminal law." The platform would have its teams of community standards people, connected to trusted evaluators in communities, who would act as a second and third level of review, with local laws regarding freedom of speech acting as one of the benchmarks. Over time, a tested reservoir of posts and crowdsourced responses would be developed. This reservoir would form the basis for building an educational library of what counts as racism online, and its level of seriousness. Access to this material in a de-identified form would be crucial to the awareness-raising among Web users about what would count as online racism, what their rights would be to respond and the importance of applauding those who stand up against hate. Furthermore, such a library would provide access to resources that could help develop counter- and alternative speech, videos, memes and animations.

As we have seen, national laws and situations vary significantly, shaped by particular philosophies and histories. Nevertheless, the rise of racisms across the globe points to two elements that are shared—the challenge of ensuring a culture of civility permeates the online world, and that the havens for uncontrolled racisms are identified and isolated by the international community. We turn now to the Australian situation in order to identify what a national response within an international milieu could involve.

Australia: A National Direction

Australia has its unique national history and circumstances: a colonial settler society, an unresolved situation with its Indigenous peoples and a diverse immigrant society, which has yet to fully deal with the "end of White Australia" in the mid-1970s. The residues of these events have shaped national action and inaction on cyber racism (see Fig 9.2). While Australia has moved significantly to engage with cyberbullying, racism against groups does not get captured by the current focus on the protection of individual young people. In addition, while counter-radicalisation has been accorded high national priority, the associated racisms that are

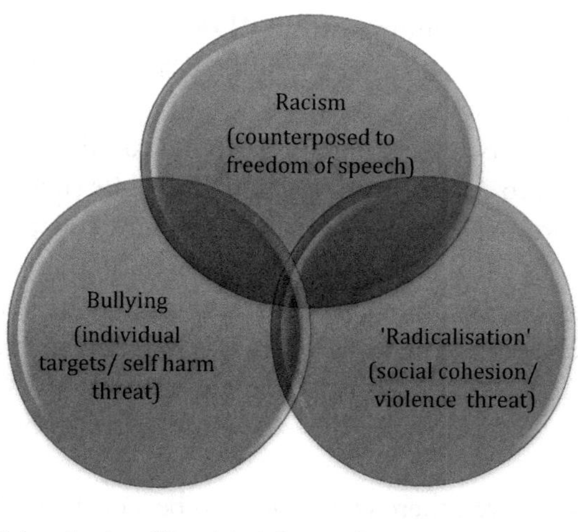

Fig. 9.2 How Cyber Racism fits with Cyber Bullying and Online 'radicalisation'

used by radical groups as a rationale for their actions and as a framework for defining their foes have not been well identified or addressed.

When we apply the same tripartite model that was used to describe the global situation, government, the economy and civil society all have key roles to play in combatting cyber racism (see Fig 9.3). National sovereignty clearly does give Australia the power to act against platforms that refuse to protect Australians from the marauding impact of cyber racism. However, Australia, as we have seen, has been reluctant to treat racism as a serious cyber problem, leaving it to the targets of racism to use the low-level systems that exist. Indeed, arguments over the use of the civil sanctions written into the RDA have been a prime target of alt-right and conservative groups, anxious not to have even the current limited restrictions on their activities in the sphere of racism.

In the Australian context, we suggest that the following layered structure could be more effective. The nation needs to be a player in the global sphere, so that its concerns have some bite on the international debate and in contributing to the international agenda. In recent years, the Australian government has backed away from any engagement with global management of cyber racism—it has refused to accept the UN Convention in this regard despite repeated invitations to do so, has briefly considered and then bypassed the European Convention protocol and has internally sought to reduce civil protection against cyber racism through constant undermining of the RDA. On the other hand, it has maintained a low level of support for the #RacismItStopswithMe campaign promoted through the AHRC.

In terms of its global involvement, Australia could withdraw its reservation in relation to Article 4 of the ICERD. While such a move would require some criminalisation of race hate speech, the scope already exists in the Commonwealth Crimes Act to prohibit advocacy of views that could lead people to commit acts of criminal violence against people on the basis of their race, ethnicity or religion. Once this segment of the law was clarified, Australia could then move to sign onto the Optional protocol on cyber racism of the Council of Europe Cybercrime convention.

At the national level, there would need to be a civil society forum that could meet the government and industry bodies as an equal stakeholder

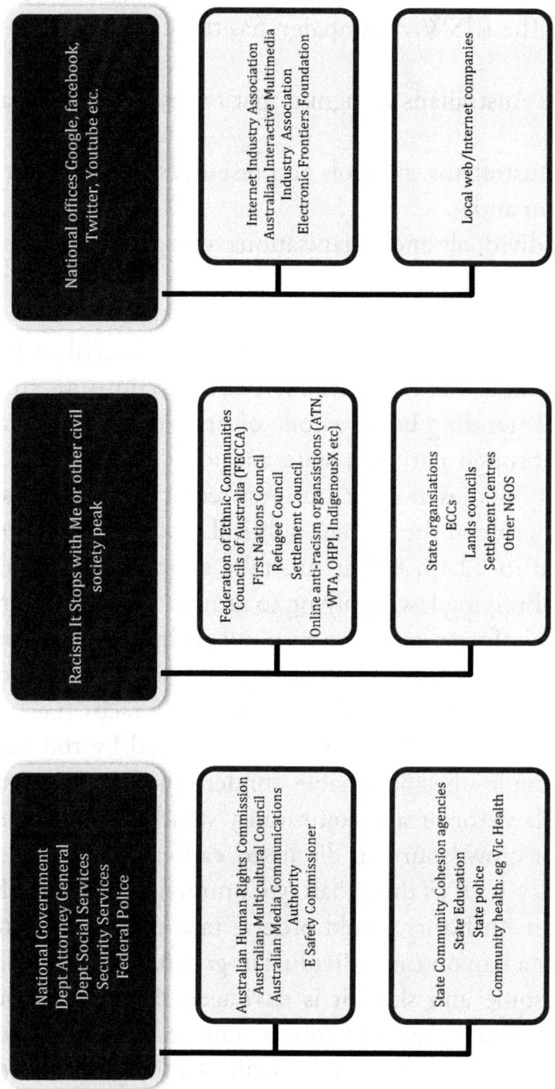

Fig. 9.3 Australian parallel stakeholders in combating cyber-racism

in the resolution of race hate issues. Such a potential body already exists in the group established through the AHRC, "Racism It Stops with Me" (RISWM), a stakeholder and supporters group for the national anti-racism policy[1]. The RISWM campaign has three goals:

1. Ensure more Australians recognise that racism is unacceptable in our community;
2. Give more Australians the tools and resources to take practical action against racism and
3. Empower individuals and organisations to prevent and respond effectively to racism.

There are no other national anti-racism bodies in civil society, though a number of national sector stakeholders already contribute to the space. With a national standing body, a code of practice could be created that works its way through national, state and local governments, peak and local civil society bodies and the business sector organisations that have an interest and a role in the current political economy of racism online.

We suggested in Chap. 8 that Australia adopt a version of the New Zealand Cyber Bullying law, applying to groups and covering racism and adults. Where platforms refuse to respond to individuals, governments could authorise and register civil society organisations, or government agencies, and take down material that is seriously racist (i.e., would trigger Section 18C of the RDA, but not be covered by the Section 18D defences, or could be actionable under the relevant state anti-discrimination laws, or breach community standards as defined through the processes of crowdsourcing discussed earlier). In such situations, a regulatory agency such as the E-Safety Commissioner, or the Media and Communication Authority, could order a take-down of the material or ultimately order a ban on the individual or group. While this process may appear cumbersome and slow, it is envisaged that once good working relationships are established and robust standards are clearly delineated, then action by platforms would be swift. Unlike the threats that the German government had to offer to Facebook in order to get the platform to perform as agreed, the Australian situation should be far smoother. In particular, because of the forum model, debates over the importance of free speech and the right to robust debate would be incorporated into setting and amending of community standards.

Community resilience will be helped by the creation of a satisfactory tripartite framework for setting community standards. However, the promotion of ultra-nationalist, alt-right, interethnic and interreligious disputes will continue. The identification of what counts as racist and hurtful material should become part of the education of young people using the Internet. While young people should be exposed to differences of opinion over cultural, religious and ethnic issues, the importance of learning how to engage, with civility, over points of disagreement has clearly been flagged as an element in reducing the heat in many online conflicts. Learning about how to create alternative material to contest the space that hateful material has colonised online opens up opportunities for creativity, empathy and the interaction that comes from cross-cultural communication. Flexibility and strength, fundamental components of a resilient community, come from well-resourced, carefully nurtured learning and practice. Racism online has been as successful as it has because too often the space it occupies has been vacated by those who do not relish any engagement with hateful and painful encounters. After a short while, it appears everyone there shares the racist views; and another step in the "normalisation" of hate has been taken.

The alternative, then, must lie in part in the creation of positive spaces where interaction supports learning and growth, where identities of difference are appreciated and explored and both commonalities and uniqueness are accorded appropriate room. In addition, sustained action is required by groups whose members mutually support each other in the stressful business at hand, to enter key sites where racism flourishes, offering calm rebuttals and alternative affirmative claims about difference. While initially such engagements can be painful and threatening, with trolls enjoying the chance to harass their opponents, participants learn to accept that these spaces are digital, not immediately material, recognising that sustained pushback has become crucial to their own well-being given the pervasive spread of hate material.

One of the longest established US anti-racist groups, the SPLC, published a 10-point guide to fighting racism in 2010, updated in 2017, that dealt primarily with "real-world" action (Southern Poverty Law Center

(SPLC) 2010). Many of these activities are transferable to the Internet, and may well be enhanced in that context. The SPLC asserted that apathy or inaction is taken by the wider society, racists "and worse the victims" as acceptance, and it argued that action on race hate can help reduce its acceptability. Building coalitions that unite people across different groups to share the work involved in pushing back against race hate provides opportunities for dialogues that the fragmentation of social relations caused by race hate online always seeks to undermine.

Support for the victims of race hate helps overcome their feelings of isolation and vulnerability. On the Internet, the isolation that people experience when they are targeted can be detrimental to their identity, self-worth and mental health; expressions of support provide a critical element in their capacity to "spring forward," and in the process such support builds wider community resilience and helps others learn how to be supportive and sustain the support. The Center urges victims to report every attack, express their experiences to local media and get a really clear sense of their legal rights—from reporting online intimidation to the platform managers, through to taking legal action. Legal circumstances vary by jurisdiction, so clarifying what to do and the costs in time and energy that different courses of action may involve need to be identified.

The Center noted that most attacks are not by organised groups, even though they may provide the ideology and narratives of justification that lone players utilise. While hate crimes can be pursued in the legal sphere, bias events fall outside the legal sphere—though again this varies by constituency, state and country. Nevertheless, such incidents "don't just victimize individuals; they torment entire communities." In the face of hate events, the Center proposed that people should create alternative events rather than seek confrontation. On the Internet, this requires a very creative sense of what might be possible, the discussion of which, in itself, provides an arena for building community resilience.

The denunciation of hate speech online can provide both symbolic and practical demonstrations of the consequences for perpetrators. The Center advises anti-hate groups not to debate perpetrators in media where the two positions are deemed equivalent, as this legitimatises the haters; even so, it remains important to work with the media to reveal how race hate works.

While the media offer one avenue of resistance, so too does lobbying political leaders to take race hate seriously. In Australia, the long drawn out struggle over Section 18C of the RDA drew tens of thousands of people into a lobbying coalition that resulted in the defeat of legislation intended to reduce protections under the Act. A loose group known as the "United Nations" saw Indigenous, Jewish, Arab, Chinese, European and other Australians (Baxendale 2017) lobby politicians and address the parliamentary inquiry that remained deadlock at its conclusion (Parliament of Australia 2017; Jakubowicz et al. 2017). Proponents for changes to the law had included the two coalition government conservative political parties and most of their parliamentarians, the leading conservative think tank Institute for Public Affairs and the whole of the News Limited stable of media—from *The Australian* broadsheet newspaper to the major popular newspapers in every major city, and the Sky cable TV network (Albrechtsen 2017).

The recognition that hate and prejudice have a long-term life and require a long-term strategy prompted the Center to propose building tolerant and united communities. In a context where corrosive fragmentation of communities emerges from unrestrained hate online, the building of communities that reject hate and promote civility continues to be the most effective pathway to community resilience. People promoting race hate online endeavour to compose communities attached to their ideology and programme, and thereby promote racism-based social and political action. As we have demonstrated, there are a significant number of these groups and individuals competing with one another to build their following, by feeding lurkers and likers with constant dribbles of attractively packaged material that reinforce the emotions of righteous anger and viciousness they seek to stimulate. Building communities of resistance and anti-racism remains a more challenging task, though one with a much wider sphere of potential influence and deeper roots in society. The Center describes a number of initiatives—"Not in my town," which rejects racism at the local level; Coloradans United Against Hate at the state level; and tolerance networks that link together existing coalitions and peace and justice groups in regions across the USA. Facing the early development of racist ideas among children offers a means to reduce the likelihood of later racist beliefs and actions as children group up. The learning of racial categories as the basis of social hierarchy can intrude deeply into a child's capacity to understand people from cultures different to their own. Thus, family environments that develop positive attitudes and call out prejudice when it occurs informally can play a role (Priest et al. 2014). Schools clearly play a critical role in developing cultures of tolerance, acceptance and civility (Australian Human Rights Commission 2016)—part of the armoury already well-rehearsed in countering cyberbullying.

The final step identified by the Center focused on the individual and their choices about the "racialised ecology" of their lives—pointing to the need to reflect on their own values, how they behaved with family and friends, their engagement with their community and their wider commitment to social action at a national level. As we focus on developing parameters for societal action in combatting cyber racism, we return to our matrix and the individuals who populate it.

Nine Elements in a Plan for Resilience

Our cyber racism matrix (Table 9.1) offers nine cells in which racism is generated and has impact, including triggering pushback, avoidance or capitulation.

In this final part of the book we suggest what might be needed to prompt each cell to move towards a non-racist civility on the Internet. We argue that there needs to be an ecology of resilience that mirrors the ecology that generates racism and reinforces it online (see Fig. 9.4).

Our concept of resilience encompasses the range of ecological spaces from the transnational to the individual. It identifies online users who range from anti-racist to racist, and from perpetrators to victims to bystanders. It proposes engagements that require, as many others have argued, close cooperation between civil society, the state and the industry. Without such cooperation, the spaces for racism to flourish will remain and expand.

By bringing the social ecology of individual users and groups of users fully into the analysis of cyber racism, we can more clearly identify the layers of processes and the interaction between them that produce contexts

Table 9.1 Attitudes to racism/ experiences of racism: nine fields of power

Racism/Online	Target	Perpetrator	Bystander
Opponent/not prejudiced	1 Looks for alternative speech opportunities; opposes racism online through activity; defends self from attack	2 Asserts own group superiority, while decrying racism	3 Once alerted becomes increasingly aware and active; joins supportive spaces
Unconcerned/not salient/mildly prejudiced	4 Alerted to racism by being targeted; withdraws from exposure	5 Unaware amplifier; goes with the flow; often joke teller; normalisation of racist discourse	6 Doesn't recognise racism online or doesn't engage; avoids exposure
Proponent/ strongly prejudiced	7 Identifies enemies and builds racial fight back	8 Uses Internet to build support, advocate position, harass targets	9 Lurks to like; aware amplifier

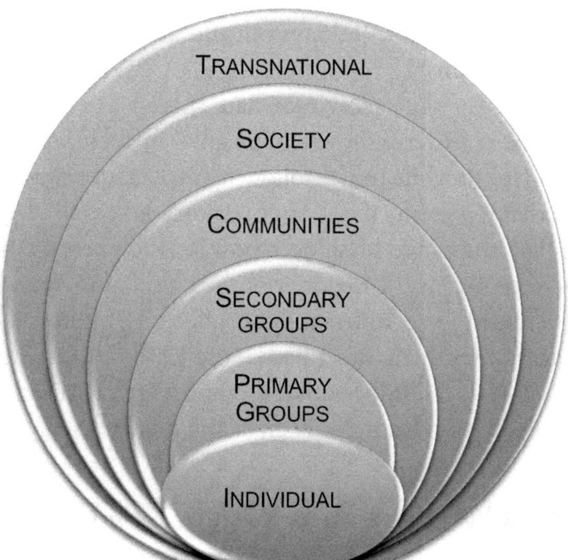

Fig. 9.4 Ecological rings of influences on attitudes to and experiences of racially-inflected difference

where cyber racism flourishes or declines. Ungar's review of the social ecology of resilience (Ungar 2011), while addressing child development as his focus, also provides valuable insights of use in our analysis. His model of resilience refers to the capacity of individuals to navigate their way to the resources they need, and negotiate for these resources to be provided and experienced in culturally meaningful ways (Ungar 2011, p. 10). Importantly, he refers to resilience as a process that "reflects the influence of culture on the naming of protective processes" (p. 10) where change occurs not because of what individuals do but rather because of how their environments change. In fact, he notes that a "subject-centered approach means that responsibility for resilience is wrongly placed on the victim of toxic environments" (p. 5). His "social ecological conceptualisation of resilience" offers as one of its principles "cultural relativity" in which culture operates as a productive force for psychosocial well-being.

By using some of the attitudinal questions discussed in Chap. 3 designed to reveal the parameters of racist thought, while expanding them into segments reflecting the type of encounters people admit to online, it is possible to get some sense of the changing weight of the cells identified in Table 9.1. Table 9.2 lays out the seven questions used covering old and new racisms.

The fields of power that allow race hate speech to emerge can be displayed as a matrix, as we have shown in Chap. 4. In Table 9.1, we numbered each cell, from 1 to 9, while in the following explanation we include a percentage figure that represents the average response across all seven questions. Cells 1–3 display the range of subject positions for people who are anti-racist. Cells 4–6 represent the majority of the population, for whom racism is not usually a salient issue, unless they are affected by it in some encounter that hurts or aggravates them, described by the 1998 Eureka study as "mildly prejudiced." Cells 7–9 include people who have very racialised world views, and seek satisfaction or advantage by engaging in race hate.

Cell 1: Targets Who Are Anti-Racist (2.2%)

As targets, Cell 1 contains the most likely people to push back against racist hate, looking for alternative speech opportunities. They are the

Table 9.2 Composite of six attitudes to race by position of subject in relation to online racism experience

		Target	Perpetrator	Bystander	No encounter
1: I am prejudiced					
I am prejudiced against other cultures	Strongly disagree	27 26.7%	8 21.6%	201 27.4%	311 23.2%
	Agree, neutral, disagree	72 71.3%	27 73%	527 71.9%	1016 75.8%
	Strongly agree	2 2%	2 5.4%	5 0.7%	13 1%
	Total	101 100%	37 100%	733 100%	1340 100%
2: I am uncomfortable					
I feel uncomfortable around people from other racial, ethnic or religious backgrounds	Strongly disagree	29 28.7%	8 21.6%	209 28.5%	282 21.0%
	Agree, neutral, disagree	65 64.4%	25 67.6%	497 67.8%	1032 77.0%
	Strongly agree	7 6.9%	4 10.8%	26 3.5%	26 1.9%
	Total	101 100%	37 100%	733 100%	1340 100%
3: Marrying in					
I think it is better if people marry within their own cultural group	Strongly disagree	17 16.8%	6 16.2%	170 23.2%	190 14.2%
	Agree, neutral, disagree	78 77.2%	28 75.7%	545 74.4%	1108 82.7%
	Strongly agree	6 5.9%	3 8.1%	18 2.5%	42 31.%
	Total	101 100%	37 100%	733 100%	1340 100%
4: Minorities weaken us					
The Australian way of life is weakened by people from minority racial, ethnic or religious backgrounds maintaining their cultural beliefs and values	Strongly disagree	13 12.9%	5 13.5%	127 17.3%	103 7.7%
	Agree, neutral, disagree	72 71.3%	26 70.3%	550 75.0%	1092 81.5%
	Strongly agree	16 15.8%	6 16.2%	56 7.6%	145 10.8%
	Total	101 100%	37 100%	733 100%	1340 100%

(continued)

Table 9.2 (continued)

		Target	Perpetrator	Bystander	No encounter
5: Minorities should be like us					
People from racial, ethnic, cultural and religious minority groups should behave more like mainstream Australians	Strongly disagree	9 8.9%	3 8.1%	45 61.%	22 1.6%
	Agree, neutral, disagree	69 68.3%	29 78.4%	599 81.7%	1063 79.3%
	Strongly agree	2 22.8%	5 13.5%	89 12.1%	255 19.0%
	Total	101 100%	37 100%	733 100%	1340 100%
6: Aborigines find it hard to succeed					
Being Aboriginal makes it harder to succeed in Australia today	Strongly agree	27 26.7%	10 27.0%	103 14.1%	89 6.6%
	Agree, neutral, disagree	63 62.4%	24 64.9%	566 77.2%	1130 84.3%
	Strongly disagree	11 10.9%	3 8.1%	64 8.7%	121 9.0%
	Total	101 100%	37 100%	733 100%	1340 100%
7: Aborigines get too much					
Aboriginal people get more government money than they should	Strongly disagree	13 12.9%	3 8.1%	75 10.2%	51 3.8%
	Agree, neutral, disagree	64 63.4%	24 64.9%	566 77.2%	1098 81.9%
	Strongly agree	24 23.8%	10 27.0%	92 12.6%	191 14.3%
	Total	101 100%	37 100%	733 100%	1340 100%

Source: CRaCR survey by fields of power Attitudes and Encounters

most likely to identify an encounter as being racist, recognising the more nuanced and possibly ambiguous hate speech and racist behaviour. They are also the most likely to identify their own behaviour or attitudes as potentially racist, though their values will be anti-racist. They also form the group that feels it has to bear the greatest burden in facing racism. They constantly make choices between avoiding its impact, responding to it as individuals or responding institutionally and as part of a larger civil society group.

This segment of Internet users, often thought of by government agencies as safely on the side of anti-racism, is also vulnerable, especially to sustained and purposeful attacks, including doxing. As we have seen in Chap. 5, they are also the segment most likely to be emotionally affected by exposure to race hate, and therefore the group most in need of support and encouragement. They are highly sensitive to the emotional impact of race hate attacks, though some will have hardened under the constant barrage they have experienced. They will be most likely recruits for sustained anti-racist work on the Internet in social media, as they can and often do provide some of the leaders and many role models. However, while these roles are important, for many bystanders who do not come from the targeted groups, advocates from targeted groups may not be fully trusted to reflect the "truth," but rather be seen in some way as self-interested and untruthful about the extent or impact of their claims to victimisation (Nelson et al. 2011). When supported, victims can gain value from the processes involved in confronting their harassers (Sanchez et al. 2016).

Strategies of value for Cell 1 Internet users can include the creation of networks of support, through civil society and social media. These might include tactics adopted by some Muslim women (Sargeant 2016) in Australia who have sought pledges and donations to charities from supporters for every hate-filled tweet or post they receive—thus the action of the hate posters increases resources going to the communities they want to harass (Di Stefano 2015). Supporters gain satisfaction from seeing how their contributions have helped defeat the impact of racism, while potentially the harassers realise that every time they troll they are in fact helping build the pushback against them.

Cell 2: Anti-Racists Who Use Racist Strategies (0.7%)

Cell 2 covers those opponents of racism who react against people they see as instigators of racism, using racist ideas and ideas of their own superiority in racial terms. By using racial categories to advance their political opposition to expressions of racism, they may well be reinforcing the categories they claim to decry. In multiracial societies, this cell may involve what some commentators describe as "anti-white" racism (Anti-Defamation League (ADL) 2017). This subject position can be highly counterproductive, as it allows their racist opponents to discard their arguments as self-serving and simply replicating that which they pretend to critique. In the Australian context, there was an upsurge of such criticisms of those who had defended 18C by those who had attacked it. For example, Andrew Bolt, gravely disappointed that his campaign to restrict the protections of 18C had failed, editorialised on his News Limited Sky News program "The Bolt Report," republished in the News Limited Tabloid *The Herald Sun*, that it was "beyond crazy the way we are being divided by race—we are being turned into a country of tribes" (Bolt 2017). He drew on News Limited commentator Miranda Devine who claimed "reverse racism is now acceptable in Australia." This rise in the discourse around anti-White racism has been discovered in the USA to now overwhelm concern about anti-Black bias among American Whites (Norton and Sommers 2011).

The concerns are based on the belief among many Whites that racism operates as a zero-sum game, that if anti-Black bias declines then anti-White bias rises. These perspectives also feed into the debate over whether Blacks in a society dominated by Whites can in fact be racist towards Whites. Given White privilege and the benefits that flow to all Whites from the exploitation and suppression of Blacks in historical time, while non-Whites can be biased or prejudiced, some people argue that they cannot be racist (Blay 2015). However, other commentators have indicated that for them racism is a universal problem based on widely shared attitudes rather than politico-economic domination, in that a racist social system will cause all intergroup conflict to be racially inflected. "Racism is not confined to one race—all communities need to join hands to end it. The sad reality is racism can be found in any group in Australia. No group is exempt. The culture of racism in Australia needs to be addressed without targeting one race or community" (Raja 2014).

The debate over anti-White and interethnic racism, its extent, its impact and its effects, can be sustained in a civil form online in order to test out such propositions without rancour, cynicism or insult while upholding principles of anti-racism. The distinction between racism and prejudice contains far more than a technical concern for semantic accuracy—rather it goes to the heart of conceptions of ethno-cultural diversity and the nature of the power relations that produce the differences between groups.

Cell 3: Anti-Racists Who Are Not Targets (15.3%)

The anti-racist bystanders have the greatest potential to contribute to building civil and tolerant Internet cultures. Because they are already aware of racism and the prejudices it subsumes, once they are alerted to its presence, they are able to engage with the perpetrators and other bystanders to shift the tone and focus of the conversation. Bystanders' anti-racism has been described as "action taken by 'ordinary' people in response to incidents of interpersonal or systemic racism" (Nelson et al. 2011). Where bystanders who hold anti-racist views are present, they can influence other bystanders, while offering support to victims, and call out perpetrators, by demonstrating opposition from more than the targeted group or individual (Kahn et al. 2016; Sanchez et al. 2016).

Anti-racism strategies need therefore to identify, recruit, apply and sustain the interest and commitment of the Cell 3 users. Discussing the process of online moderation across many platforms, *The Verge* quoted a former Facebook content policy chief, who summarised how moderation worked and the role of users in that process.

> Platforms rely on users in three profound ways that alter that linear relationship: One, users produce content—our stories, photos, videos, comments, and interactions are core assets for the platforms we use; two, user behaviors—amassed, analyzed, packed, and repackaged—are the source of advertising revenue; and three, users play a critical role in moderation, since almost every content moderation system depends on users flagging content and filing complaints, shaping the norms that support a platform's brand. In other words, users are not so much customers as uncompensated digital laborers who play dynamic and indispensable functions (despite being largely uninformed about the ways in which their labor is being used and capitalized). (Buni and Chemaly 2016)

Activated bystanders therefore have a very powerful latent resource, their capacity to unite in a sustained push to move moderation towards the goal of civility, against the corporate desire to optimise opportunities for revenue. As Roanna Carleton-Taylor, founder of UK online group "Resisting Hate," has argued, the importance of resisting online hate speech cannot be underestimated as "words help to shape and define the ideals and the standards within our communities. Words, put simply, change how people think" (Carleton-Taylor 2017). A key resource that can help organisations and alert activists to grasp the broad range of strategies available can be found through Civilination (Weckerle 2017), a US-based group that provides strategies and contact lists for over 70 civility organisations. Its founder Andrea Weckerle published a useful overview of how to deal with online conflict (Weckerle 2013).

While over 30% of Internet users may have had a recognised encounter with racism, most were bystanders rather than targets. Of that 30%, perhaps 15% were anti-racist in their core values and attitudes. The majority of encounters actually occurred for the large middle group of people unconcerned by or unaware of racism prior to the encounter, and who saw themselves as barely touched by it after the encounter. These form the population of the second line, Cells 4–6.

Cell 4: Mildly Prejudiced Targets of Racism (7.9%)

The majority of the Australian population was described by the Eureka study in 1998 as "mildly prejudiced" (Eureka Research 1998). Even so, as racism was not generally salient to their life experience, they were either unaware of their encounters with it, or did not perceive it as being of importance when they did encounter it. However, some individuals discovered that they were the targets of harassment or hate speech, either because they belonged to a category of people who were more generally insulted, or because they had been targeted personally because of their race, ethnicity or religion.

Cell 4 Internet users are vulnerable to a rapid escalation in the salience of racism to their user experience, and can move in one of two opposing directions. Alerted to racist speech, they may adopt it in return, using racist hate speech to push back against their attackers. In this case, they can amplify racist hate speech as an acceptable mode of discourse, thus normalising it further and reinforcing the world view of those proponents of racism (that we will discuss in Cell 7). On the other hand, they may withdraw from any spaces where they might encounter racism, thus allowing the hate speech to continue, feeding the sense among racists and their followers that race hate speech both works to intimidate its targets and leaves the field clearer for the elaboration of racist discourses to flourish.

This group can be brought into an ongoing civil pushback against racism by ensuring there are safe places on the Internet to report and discuss their experiences, which helps to ensure that they become aware of how widespread the phenomenon of racism is. They can also be supported to understand how racism works and the important part they can play in not being intimidated into withdrawal. Of course, they should not simply be exposed to harassment without recourse.

Cell 5: Mildly Prejudiced Normalising Perpetrators of Racism (2.8%)

The users in Cell 5 are possibly the most critical players in facilitating the spread of racism online. Where they see "normalised" racism, they already believe such views (at the milder scale not advocating violence) are acceptable, particularly if they are meme-based or are in some way humorous. We describe them as "unaware amplifiers," as they comprise the majority of the swarms that form around racist posts on Facebook or retweeters on Twitter. They will be regular social media users and have a network of friends who share their social values, their mildly racist attitudes and their sense of humour such as it is. They will typically regard humour about race issues as not racist, or minimise their sense of its impact (Grigg and Lenore 2015; Matamoros-Fernandez 2016). However, it is their willingness to like, re-post and retweet "normal" racism, especially against Indigenous and Muslim targets, that gives them the influence that they have.

As noted in Chap. 3, our findings suggest that a greater awareness of the consequences and operation of racism can have a limiting effect on peoples' willingness to amplify racism through repetition and transmission. Thus, strategies that emphasise the hurt that racist hate speech can cause may have an effect on this group, who are not the trolls and activist racists of Cell 7. Key social skills that would be needed to reduce the likelihood of racism include learning empathy and civility in communication. They also learn about how racism operates by having others call out behaviour that may be racist, but which they have not yet recognised as such.

Cell 6: Mildly Prejudiced Bystanders (63.1%)

Cell 6 encompasses the single largest segment of Internet users, including those who have encountered racism and have disregarded it, or those who have not recognised they have encountered it at all. Whatever their personal attitudes on race, they do not feel implicated in online debates over racism, often believing racism occurs as a normal part of social life, but not one of interest to them. For the most part where they do recognise racist speech they let it "wash over them," and feel no responsibility for its presence. While they may not intend to amplify they racism they encounter, they may often do so unintentionally.

Overall, they prefer to operate in spaces where they will not be exposed, and they will often block, but rarely report, any disturbing material they encounter. Where a friend posts material that discomforts them, they are most likely to de-friend rather than protest, push back or argue.

While they comprise the largest group, potentially providing an important mass of action around reducing racism online, these users are hard to reach, as they avoid calls to social responsibility. However, they can be approached via their friend networks, using engaging imagery that raises their awareness through demonstrating the salience for them of opposing, rather than acquiescing, to racism. In their social networks, opinion leaders remain the point of first entry, as their values can influence the orientations of other network participants. They respond to calls from well-regarded authority figures, who reward them with recognition and approval if they behave in appropriate ways (as, for example, in buying advertised products promoted by cultural role models) (Curran 2012; Rieder 2013; Miller 2016). In some cases, their opinion leaders may have stronger views than their network friends, allowing them to offer a way into raising and sustaining these issues, advocating for bystander action, through offering role models of e-leadership in confronting racism.

If offended, they move away. This group also generates the greatest value for providers and advertisers, demonstrated by the advertiser boycott when Facebook and Google both proved unable to prevent the appearance of hate sites when people went to product advertisements during early 2017. Even though Facebook and the other platforms now have to employ many thousands of checkers for reported objectionable sites or pages (Buni and Chemaly 2016), human effort was unable to deal with the millions of events on Facebook, Google and YouTube revealed by The London Times and the Wall Street Journal in their investigations of this phenomenon (Mostrous 2017; Alba 2017; Watson 2017).

Cell 7: Racist Targets of Racism (1%)

Strongly prejudiced targets see the world in highly racialised terms, promoting racially charged discourses in response to attacks they experience (and often precipitate) (Gaynor 2014). As we discussed in Chap. 3, Cells 7 and 8 may overlap, with some perpetrators also identifying as victims. They may see themselves as engaged in a race war on the Internet, hunting out their opponents and striking back as strongly as they can. Their perception of victimhood frames their discursive strategies, displacing all responsibility for their behaviour onto their "enemies."

They share with Cell 8 a vested interest in a racialised Internet, through which they can promote their vision of resistance to what they see as the hegemonic presence of their racially or religiously defined oppressors. With such Manichaen perspectives, these users are more likely to fulfil the other characteristics of trolls, though they may also have a sense of their own wounded dignity in the interchanges they generate (Pasquale 2015).

Individuals in this cell may be open to strategies for de-escalating conflict online, such as those discussed by Civilination (Weckerle 2013, 2017). As Jimmy Wales, the founder of Wikipedia, notes on the

Civilination website, the interaction between opponents online can become toxic very quickly, with the most astringent voices battling it out and more reserved participants in a discussion standing back or dissipating as the vitriol accelerates (Wales 2016). So if the cycle of acceleration can be broken, then it may be possible to develop tactical manoeuvres to remove heat from a situation and introduce civility, thereby allowing the conversation to proceed, and a wider group of participants to enter and extend the range of meanings and perspectives on offer. The US organisation Trollbusters offers advice to women on such strategies (as well as defensive moves for better protection from trolls) on their Facebook page and twitter feeds. Their approach focuses on the targets of sexist harassment, in order to "counter hate with love by sending positive, affirming messaging to the point of attack. We provide a hedge of protection around the individual, helping to provide emotional support and reputation management during cyber crises" (Trollbusters 2017).

Cell 8: Racist Perpetrators: Troll Haven (0.5%)

Our research suggests Cell 8 contains about half of 1% of Internet users. Nevertheless, they produce a significant part of the intentional race hate speech on the Internet, and from the literature they would probably all fit the definition of a troll. Most White power proponents are aligned with or denizens of well-known race hate organisations, either running their own Facebook groups or websites, or using their Twitter accounts to harass their targets while building their following. They tend to be highly competitive with each other, while seeking to define their specific brand. While historically many were associated with Stormfront, the rapid spread of sites and pages has transformed the Internet and generated more strident and sophisticated White power and other racist sites.

Some Cell 8 members are strong ethno-religious nationalists who target political enemies of their own claims to nationhood (for instance, groups that pit Hindu Indians (Poonia 2017) or Burmese Buddhists against Pakistani or Rohingya Muslims (Rajagopalan 2017)). They may stress fundamentalist religious beliefs with hatred of non-believers (ultra-Christian or radical Jihadist Muslim) (Muehlenberg 2016), or they might advance a racial superiority agenda tied to national expansionism such as some of the Huaren (Han Chinese) online communities (Dikotter 2001; Barabantseva 2011; Hornby 2017).

The proponents of racial superiority drift easily into racial harassment of those who are from different origins. While the border zone may be grey, with justifiable differences of opinion, the more dramatic and hostile trolling that occurs can be more readily identified. Indeed, many trolls perform the role of moderator on their own Facebook or Reddit pages, and thus could be more closely surveilled by the cyber safety teams on these platforms.

As the major platforms have refined their protection strategies, the overarching advice from many of the online safety groups suggests a mixture of reporting and blocking posts, with the reporting practices now increasingly allowing platforms to identify and ban individuals whatever their front registration name might be. Moreover, both Twitter and Reddit have decided that even hate speech that is not illegal nevertheless intimidates and prevents the full expression of freedom of speech by those harassed, and thus should be banned. With the press towards "real names" only being used on platforms, it is possible for anti-racist groups to "out" posters of race hate—through the creation of "rogues galleries," and the reporting of harassing posters to their employers and civil society groups to which they belong, a form of reverse "doxing" (The Anti-Bogan 2017).

Cell 9: Standing by with Prejudice (5.7%)

The prejudiced bystander, people we describe as "lurking to like" on race hate sites, also perform the role of amplification (Gorski 2003). While they tend to form the swarms rather than provide the leadership roles, they are more open to being "peeled" away from their associations. Given they visit hate sites in order to feel they are part of a shared community of values, thereby reinforcing community by attending, some have been successfully addressed through efforts such as All Together Now's Exit White Power project (All Together Now 2017; Young et al. 2016).

Such approaches would be part of a wider strategy; the online community provides a central identity. Such bystanders need to be given another way of thinking about themselves that offers community without depending on hate.

Conclusion

The application of carefully assessed evidence to the complex reality of online racism demonstrates that pushing back against the spread of hate speech is both possible and realisable, once the underlying dynamic has been exposed to analysis. Given its birth in political and economic systems of exploitation, ultimately racism cannot be defeated without massive changes in these systems. However, strategic advances based on alliances between governments, industry and civil society are quite feasible, especially when the chimera of freedom of speech has been tested against the reality of the erosion of democracy that comes from the flight to private spaces of illusionary safety. If the Internet and the social media it affords are to remain a true digital commons, then ultimately the rules of civility and the possibility of trust have to be promoted.

Our matrix of online racism and resistance points to the layered strategies required to build resilience and facilitate the spread of civility. Such an approach necessarily starts in the domestic sphere where adults and teenagers can model non-racist practice for children, while demonstrating civil engagement between themselves. As with the strategies developed to resist cyberbullying, so too cyber racism can be addressed through schools where an ethics of online civility and a practice of empathetic awareness allow the slogan of "empathy trumps hate" to be internalised. Civil engagement over difficult issues links with learning how to de-escalate conflict and seek shared understanding, even in the context of communal histories riddled with violence and hate. The crucial role of empowered and courageous bystanders needs to be recognised and supported, and active pushback respected, admired and rewarded with support.

This book has travelled the very messy and often painful terrain in which racism permeates the Internet, corroding the social capital that ultimately underpins the possibility of multicultural societies. Given that it is easier to trigger violence and anger than to pursue peace and reconciliation, the tasks involved remain challenging. However, a composite strategy can be proposed that ultimately links global partnerships between governments, industry and civil society to national and local ones. Such alliances for civility offer the possibility of realising a more respectful world, in which riacially-inflected differences are increasingly de-politicised, and the diversity of cultures and peoples who have to co-exist discover the means to do so in the new and borderless digital commons in which we are all involved.

Notes

1. https://itstopswithme.humanrights.gov.au/resources-1

References

Alba, D. (2017, March 25). YouTube's Ad Problems Finally Blow Up in Google's Face. *Wired*. https://www.wired.com/2017/03/youtubes-ad-problems-finally-blow-googles-face/. Accessed 13 Apr 2017.

Albrechtsen, J. (2017). There Is No Room for Section 18c in a Truly Liberal Australia. *The Australian*, News Ltd. http://www.theaustralian.com.au/opinion/columnists/janet-albrechtsen/there-is-no-room-for-section-18c-in-a-truly-liberal-n-australia/news-story/4a3edf4f69e7e3eb42928caad537950e. Accessed 22 Feb 2017.

Aldrich, D. P., & Meyer, M. A. (2015). Social Capital and Community Resilience. *American Behavioral Scientist, 59*(2), 254–269.

All Together Now. (2017). *Extremism*. http://alltogethernow.org.au/extremism/

Anti-Defamation League (ADL). (2017). Anti-Racist Is a Code for Anti-White. *Hate Slogans/Slang Terms*. https://www.adl.org/education/references/hate-symbols/anti-racist-is-a-code-for-anti-white

Australian Human Rights Commission (AHRC). (2013). Human Rights in Cyberspace (Background Paper). AHRC. https://www.humanrights.gov.au/our-work/rights-and-freedoms/publications/background-paper-human-rights-cyberspace. Accessed 1 Mar 2017.

Australian Human Rights Commission. (2016). *Building Belonging: A Toolkit for Early Childhood Educators on Cultural Diversity and Responding to Prejudice – Educator Guide*, Australian Human Rights Commission.

Back, L. (2002). Aryans Reading Adorno: Cyber-Culture and Twenty-First-Century Racism. *Ethnic and Racial Studies, 25*(4), 628–651.

Barabantseva, E. (2011). Who Are "Overseas Chinese Ethnic Minorities"? China's Search for Transnational Ethnic Unity. *Modern China*. http://journals.sagepub.com/doi/abs/10.1177/0097700411424565. Accessed 15 Mar 2017.

Bartlett, J., & Krasodomski-Jones, A. (2015). *Counter-Speech*. London: Demos: Centre for Analysis of Social Media. https://www.demos.co.uk/project/counter-speech/

Baxendale, R. (2017, March 22). Ethnic Leaders United Against Substantive Changes to Section 18C. *The Australian*. Available at: http://www.theaustralian.com.au/national-affairs/ethnic-leaders-united-against-substantive-changes-to-section-18c/news-story/534963bdb7dc4346bc75dbafad7fb688

BBC News. (2016). Twitter Suspends Alt-Right Figureheads. *BBC News*. http://www.bbc.com/news/technology-37997864

Berman, G., & Paradies, Y. (2010). Racism, Disadvantage and Multiculturalism: Towards Effective Anti-Racist Praxis. *Ethnic and Racial Studies, 33*(2), 214–232.

Blay, Z. (2015). 4 'Reverse Racism' Myths That Need to Stop: "Why Isn't There a White History Month?!". *Huffington Post*. http://www.huffingtonpost.com.au/entry/reverse-racism-isnt-a-thing_us_55d60a91e4b07addcb45da97

Bleich, E. (2011). *The Freedom to Be Racist? How the United States and Europe Struggle to Preserve Freedom and Combat Racism*. Oxford: Oxford University Press.

Bolt, A. (2017, April 5). Why This Anti-White Racism?. *The Bolt Report*, Herald Sun. http://www.heraldsun.com.au/blogs/andrew-bolt/why-this-antiwhite-racism/news-story/bb36667e0294a427db9954b57af9bb81. Accessed 11 Apr 2017.

Briskman, L. (2015). The Creeping Blight of Islamophobia in Australia. *International Journal for Crime, Justice and Social Democracy, 4*(3), 112–121.

Brookings, E., & Singer, P. (2016, November). War Goes Viral How Social Media Is Being Weaponized Across the World. *The Atlantic*.

Buni, C., & Chemaly, S. (2016). The Secret Rules of the Internet. *The Verge*. http://www.theverge.com/2016/4/13/11387934/internet-moderator-history-youtube-facebook-reddit-censorship-free-speech

Cadwalladr, C. (2017). Robert Mercer: The Big Data Billionaire Waging War on Mainstream Media. *The Guardian*. https://www.theguardian.com/politics/2017/feb/26/robert-mercer-breitbart-war-on-media-steve-bannon-donald-trump-nigel-farage. Accessed 14 Mar 2017.

Calmes, J. (2017, August 12). Trump Responds to Charlottesville Violence with Vague Statements Blaming 'Many Sides'. *Los Angeles Times*. http://www.latimes.com/politics/la-pol-updates-trump-tweets-charlottesville-violence-htmlstory.html. Accessed 17 Aug 2017.

Capitanchik, D., & Whine, M. (1996). *The Governance of Cyberspace: Racism on the Internet*. London: Institute for Jewish Policy Research.

Carleton-Taylor, R. (2017). Online Hate Speech—Why Words Do Matter. *Huffington Post (UK)*. http://www.huffingtonpost.co.uk/roanna-carletontaylor/online-hate-speech-why-wo_1_b_14245300.html. Accessed 23 Mar 2017.

Citron, D. (2014). *Hate Crimes in Cyberspace*. Cambridge: Harvard University Press.

Curran, J. (2012). Reinterpreting the Internet. In J. Curran, N. Fenton, & D. Freedman (Eds.), *Misunderstanding the Internet* (pp. 3–33). London: Routledge.

Di Stefano, M. (2015, November 13). This Muslim Woman Accidentally Started a Charity Using Hateful Tweets. *BuxxFeed News*. https://www.buzzfeed.com/markdistefano/su-su-susan-carland?utm_term=.gbe3MPreO-.ygzE4JBVk

Dikotter, F. (2001). Nationalist Myth-Making: The Construction of the Chinese Race. *Human Rights in China (HRIC)*. http://www.hrichina.org/en/content/4573. Accessed 12 Apr 2017.

DW. (2015, September 15). Is There a Way to Tackle Hate Speech and Harmful Content on Facebook? *Deutsche Welle News*.

DW. (2016, December 15). International Auschwitz Committee Accuses Facebook of Arrogant Approach to Hate Speech. *DW—Deutsche Welle News*. http://www.dw.com/en/international-auschwitz-committee-accuses-facebook-of-arrogant-approach-to-hate-speech/a-36787344

England, C. (2017). Pewdiepie Dropped by Disney Over YouTube Star's Anti-Semitic Videos. *Independent*.

Eureka Research. (1998). *The Anti-Racism Campaign: Quantitative Market Research to Guide Campaign Development*. Canberra: Multicultural Affairs Branch, Anti-Racism Campaign Unit, Department of Immigration and Multicultural Affairs. https://andrewjakubowicz.files.wordpress.com/2011/11/dimiaantiracismreportquant1998.pdf. Accessed 28 Nov 2016.

Fleishman, C., & Smith, A. (2016). (((Echoes))), Exposed: The Secret Symbol Neo-Nazis Use to Target Jews Online. *Tech.Mic*. https://mic.com/articles/144228/echoes-exposed-the-secret-symbol-neo-nazis-use-to-target-jews-online-.pQLaSlPIC. Accessed 23 Feb 2017.

Gagliardone, I., Gal, D., Alves, T., & Martinez, G. (2015). *Countering Online Hate Speech*. Paris: UNESCO. http://unesdoc.unesco.org/images/0023/002332/233231e.pdf. Accessed 21 Feb 2017.

Gaynor, B. (2014). The Anti-Bogan, Islam and Rape. *BernardGaynor.com.au*. http://bernardgaynor.com.au/the-anti-bogan-islam-and-rape/
Global Forum for Combating Antisemitism. (2015). *GFCA 2015 Final Statement on Combating Cyberhate and Antisemitism on the Internet* accessed 12 February 2017 at http://mfa.gov.il/mfa/abouttheministry/conferences-seminars/gfca2013/documents/gfca2015booklet.pdf.
Goldsmith, J., & Wu, T. (2006). *Who Controls the Internet? Illusions of a Borderless World*. New York: Oxford University Press.
Gorski, P. C. (2003). White Privilege and White Power: Online Resources on White Racism and Anti-Racism. *Multicultural Perspectives, 5*(4), 28–30.
Greenberg, A. (2016, October). Inside Google's Internet Justice League and Its AI-Powered War on Trolls. *Wired*.
Griffiths, B., & Pedersen, A. (2009). Prejudice and the Function of Attitudes Relating to Muslim Australians and Indigenous Australians. *Australian Journal of Psychology, 61*(4), 228–238.
Grigg, K., & Lenore, M. (2015). "Just a Joke": Young Australian Understandings of Racism. *International Journal of Intercultural Relations, 47*, 195–208. doi:10.1016/j.ijintrel.2015.06.006. Accessed 17 Aug 2016.
Grimshaw, P., & McGregor, R. (2006). *Collisions of Cultures and Identities: Settlers and Indigenous Peoples*. Melbourne: University of Melbourne, Department of History.
Group of Eminent Persons of the Council of Europe. (2011). *Living Together: Combining Diversity and Freedom in 21st-Century Europe*. Council of Europe, Strasbourg. Accessed 12 June 2016 at https://www.coe.int/t/dg4/highereducation/2011/KYIV%20WEBSITE/Report%20on%20diversity.pdf.
Healy, S. (2006). *Cultural Resilience, Identity and the Restructuring of Political Power in Bolivia*. Paper Submitted for the 11th Biennial Conference of the International Association for the Study of Common Property, Bali, Indonesia, 19 June.
Hoffman, D. (1996). *The Web of Hate*. New York: Anti-Defamation League.
Hoffman, D. (1997). *High-Tech Hate : Extremist Use of the Internet*. New York: Anti-Defamation League.
Hornby, L. (2017, January 9). China Battles to Control Growing Online Nationalism. *The Financial Times*. https://www.ft.com/content/5ae7b358-ce3c-11e6-864f-20dcb35cede2. Accessed 23 Jan 2017.
Houston, J. B., Spialek, M. L., Cox, J., Greenwood, M. M., & First, J. (2015). The Centrality of Communication and Media in Fostering Community Resilience: A Framework for Assessment and Intervention. *American Behavioral Scientist, 59*(2), 270–283. http://abs.sagepub.com/content/59/2/270.abstract. Accessed 1 Feb 2015.

Hunter, A., & Chandler, G. (1999). Adolescent Resilience. *Image—The Journal of Nursing Scholarship, 31*(3), 243–247.

Jakubowicz, A. (2017). What Did Galaxy's Poll Tell Us About Freedom of Speech and 18C? Not What the IPA Said It Did. *The Conversation*. https://theconversation.com/what-did-galaxys-poll-tell-us-about-freedom-of-speech-and-18c-not-what-the-ipa-said-it-did-72197. Accessed 3 Feb.

Jakubowicz, A., Dunn, K., & Sharples, R. (2017). Australians Believe 18C Protections Should Stay. *The Conversation*. https://theconversation.com/australians-believe-18c-protections-should-stay-73049. Accessed 17 Feb.

Jewish Telegraphic Agency. (2017). *Wiesenthal Center Calls for Task Force to Catch Callers Making JCC Bomb Threats*. http://www.jta.org/2017/02/21/news-opinion/united-states/wiesenthal-center-calls-for-task-force-to-catch-callers-making-jcc-bomb-threats

Kahn, K. B., Barreto, M., Kaiser, C. R., & Rego, M. S. (2016). When Do High and Low Status Group Members Support Confrontation? The Role of Perceived Pervasiveness of Prejudice. *British Journal of Social Psychology, 55*(1), 27–43.

Keller, L. (2009). Experts Discuss the Role of Race Propaganda After White Massachusetts Man Kills Two African Immigrants. *Fighting Hate*, Southern Poverty Law Center. https://www.splcenter.org/fighting-hate/intelligence-report/2009/experts-discuss-role-race-propaganda-after-white-massachusetts-man-kills-two-african

Lasar, M. (2011). Nazi Hunting: How France First "Civilized" the Internet. *Ars Technica*. https://arstechnica.com/tech-policy/2011/06/how-france-proved-that-the-internet-is-not-global/. Accessed 25 Feb 2017.

Ledogar, R. J., & Fleming, J. (2008). Resilience, an Evolving Concept: A Review of Literature Relevant to Aboriginal Research. *Pimatisiwim, 6*(2), 7–24.

Levine, M. (2013). *Controversial, Harmful and Hateful Speech on Facebook*. https://www.facebook.com/notes/facebook-safety/controversial-harmful-and-hateful-speech-on-facebook/574430655911054/

Luthar, S. (2006). Resilience in Development: A Synthesis of Research Across Five Decades. In D. Cicchetti & D. Cohen (Eds.), *Developmental Psychopathology: Risk, Disorder, and Adaptation* (pp. 740–795). New York: Wiley.

Maier, L. (2016). Germany Investigates Mark Zuckerberg and Facebook Over Slow Removal of Hate Speech. *Forward*. http://forward.com/news/world/353625/germany-investigates-mark-zuckerberg-and-facebook-over-slow-removal-of-hate/. Accessed 23 Mar 2017.

Matamoros-Fernandez, A. (2016). Platformed Racism: The Mediation and Circulation of an Australian Racebased Controversy on Twitter, Facebook and YouTube. *Association of Internet Researchers Annual Conference*. http://eprints.qut.edu.au/101370/1/AoIR2016_platformed_racism.pdf. Accessed 14 Mar 2017.

Matteo, V., Guiver, J., Kazai, G., Kohli, P., & Shokouhi, M. (2014). Community-Based Bayesian Aggregation Models for Crowdsourcing. *International World Wide Web Conference Committee (IW3C2)*, Seoul, Korea, April 7–11: IW3C2.

McFarlane, L., & Bocji, P. (2003). An Exploration of Predatory Behaviour in Cyberspace: Towards a Typology of Cyberstalkers. *First Monday, 8*(9). http://firstmonday.org/issues/issue8_9/mcfarlane/index.html

Miller, T. (2016). Cybertarian Flexibility—When Prosumers Join the Cognitariat, All That Is Scholarship Melts into Air. In M. Curtin & K. Sanson (Eds.), *Precarious Creativity: Global Media, Local Labor*. Oakland: University of California Press.

Mostrous, A. (2017, March 17). YouTube Hate Preachers Share Screens with Household Names. *The Times*. http://www.thetimes.co.uk/article/youtube-hate-preachers-share-screens-with-household-names-kdmpmkkjk. Accessed 25 Mar 2017.

Muehlenberg, B. (2016). Hate-Preaching Muslims Acceptable, Love-Preaching Christians Unacceptable. *CultureWatch*. https://billmuehlenberg.com/2016/06/14/hate-preaching-muslims-acceptable-love-preaching-christians-unacceptable/

Nelson, J., Dunn, K., & Paradies, Y. (2011). Bystander Anti-Racism: A Review of the Literature. *Analyses of Social Issues and Public Policy, 11*(1), 263–284.

Norton, M., & Sommers, S. (2011). Whites See Racism as a Zero-Sum Game That They Are Now Losing. *Perspectives on Psychological Science, 6*(3), 215–218. doi:10.1177/1745691611406922.

Notar, C., Padgett, S., & Roden, J. (2013a). Cyberbullying: A Review of the Literature. *Universal Journal of Educational Research, 1*(1), 1–9.

Notar, C., Padgett, S., & Roden, J. (2013b). Cyberbullying: Resources for Intervention and Prevention. *Universal Journal of Educational Research, 1*(3), 133–145.

Oboler, A. (2008, April). Online Antisemitism 2.0. "Social Antisemitism on the Social Web". *Post-Holocaust and Antisemitism Series*, (67). Accessed 12 Nov 2016. http://jcpa.org/article/online-antisemitism-2-0-social-antisemitism-on-the-social-web/.

Oboler, A. (2012). Aboriginal Memes and Online Hate. *The Online Hate Prevention Institute*, OHPI. http://ohpi.org.au/aboriginal-memes-and-online-hate/

Oboler, A. (2014). Legal Doctrines Applied to Online Hate Speech. *Computers and Law*, 87 (July), 9–15.

Oboler, A. (2016). Measuring the Hate: The State of Antisemitism in Social Media. *The Online Hate Prevention Institute*, OHPI. http://ohpi.org.au/measuring-antisemitism/

Paradies, Y. (2017). Racism and Health. In *International Encyclopedia of Public Health* (pp. 249–259) [Online]. Version. http://www.sciencedirect.com/science/article/pii/B9780128036785003726. Accessed 2 Mar 2017.

Parliament of Australia. (2017). *Freedom of Speech in Australia: Inquiry into the Operation of Part IIA of the Racial Discrimination Act 1975 (Cth) and Related Procedures Under the Australian Human Rights Commission Act 1986 (Cth).* Canberra: Parliament of Australia. http://www.aph.gov.au/Parliamentary_Business/Committees/Joint/Human_Rights_inquiries/FreedomspeechAustralia/Report. Accessed 3 Apr 2017.

Pasquale, F. (2015). How to Tame an Internet Troll. *The Chronicle of Higher Education*. http://www.chronicle.com/article/How-to-Tame-an-Internet-Troll/233357

Pedersen, A. (2010). Attitudes Toward Out-Groups and the Perception of Consensus: All Feet Do Not Wear One Shoe. *Journal of Community and Applied Social Psychology, 18*, 543–554.

Poonia, N. (2017). Opinion | Zaira Wasim Troll Episode Reveals Misogyny Online. *News18.com*. http://www.news18.com/news/india/opinion-zaira-wasim-troll-episode-reveals-misogyny-online-1337511.html. Accessed 9 Feb 2017.

Priest, N., et al. (2014). Understanding the Complexities of Ethnic-Racial Socialization Processes for Both Minority and Majority Groups: A 30-Year Review. *International Journal of Intercultural Relations, 43*(B), 139–155.

Public Knowledge. (2017). *Net Neutrality*. Public Knoweldge. https://www.publicknowledge.org/issues/net-neutrality. Accessed 15 Apr 2017.

Raja, C. (2014). We Must Recognise Racism in Order to End It. *The Drum (ABC News).*

Rajagopalan, M. (2017, March 19). Internet Trolls Are Using Facebook to Target Myanmar's Muslims. *Buzzfeed News*. https://www.buzzfeed.com/meghara/how-fake-news-and-online-hate-are-making-life-hell-for?utm_term=.meJZxkdY1-.lekQqGmE0. Accessed 23 Mar 2017.

Reynolds, L., & Scott, R. (2016). *Digital Citizens: Countering Extremism Online.* London: Demos. Available at: https://www.demos.co.uk/wp-content/uploads/2016/12/Digital-Citizenship-web-1.pdf.

Richardson, S. (1998). Net Spreads Message, Draws in Recruits. *Montgomery Advertiser*. https://www.stormfront.org/dblack/press062898.htm. Accessed 28 June 1998.

Rieder, B. (2013, May 2–4). Studying Facebook Via Data Extraction: The Netvizz Application. *WebSci'13*.

Rutter, M. (2000). Resilience Re-considered: Conceptual Considerations, Empirical Findings, and Policy Implications. In J. Shonkoff & S. Meisels (Eds.), *Handbook of Early Childhood Intervention* (pp. 651–682). New York: Cambridge University Press.

Sanchez, D. T., Himmelstein, M. S., Young, D. M., Albuja, A. F., & Garcia, J. A. (2016). Confronting as Autonomy Promotion: Speaking Up Against Discrimination and Psychological Well-being in Racial Minorities. *Journal of Health Psychology, 21*(9), 1999–2007.

Sargeant, C. (2016). A Muslim Advocate Wrote About Pauline Hanson & Boy Did the Racists Go Wild. *Pedestrian.tv*. https://www.pedestrian.tv/news/arts-and-culture/a-muslim-advocate-wrote-about-pauline-hanson--boy-/9c1cb83d-27da-44ca-92fb-ffc8e44df0d1.htm

Sentencing Advisory Council. (2009). Sentencing for Offences Motivated by Hatred or Prejudice. *Sentencing Advisory Council, Advice*, July, Sentencing Advisory Council.

Silverman, T., Stewart, C. J., Birdwell, J., & Amanullah, Z. (2016). *The Impact of Counter-Narratives:Insights from a Year-Long Cross-Platform Pilot Study of Counter-Narrative Curation, Targeting, Evaluation and Impact*. Available at: http://www.strategicdialogue.org/wp-content/uploads/2016/08/Impact-of-Counter-Narratives_ONLINE_1.pdf

Simon, R., & Schaler, J. (2007). Anti-Semitism the World Over in the Twenty-first Century. *Current Psychology, 26*(3), 152–182.

Simon Wiesenthal Center. (2017). Simon Wiesenthal Center's 2017 Digital Terrorism & Hate Report Card: Social Media Giants Fail to Curb Online Extremism. *Simon Wiesenthal Center*. http://www.wiesenthal.com/site/apps/nlnet/content2.aspx?c=lsKWLbPJLnF&b=4441467&ct=14988437. Accessed 15 Apr 2017.

Simon Wiesenthal Center Snider Social Action Institute. (2016). *Digital Terrorism and Hate 2016*. http://digitalhate.net/index.php. Accessed 19 Dec 2016.

Simon Wiesenthal Institute. (2009). *Facebook, YouTube +: How Social Media Outlets Impact Digital Terrorism and Hate*. Simon Wiesenthal Institute. http://www.wiesenthal.com/site/apps/nlnet/content2.aspx?c=lsKWLbPJLnF&b=5711841&ct=7131713. Accessed 12 Oct 2016.

Southern Poverty Law Center (SPLC). (2010). Ten Ways to Fight Hate: A Community Response Guide. *Hatewatch*. https://www.splcenter.org/20100216/ten-ways-fight-hate-community-response-guide

Southern Poverty Law Center (SPLC). (2017). *Hate Groups 1990–2016*. https://www.splcenter.org/hate-map.

Stein, J. (2016, August 18/September 12). How Trolls Are Ruining the Internet. *Time*. Accessed 21 Feb 2017.

Stevens, T., & Neuman, P. (2009). *Countering Online Radicalisation: A Strategy for Action*. London: International Centre for the Study of Radicalisation and Political Violence (ICSR) Community Security Trust. Available at: http://icsr.info/wp-content/uploads/2012/10/1236768491ICSROnlineRadicalisationReport.pdf

The Anti-Bogan. (2017). People Must Stand Up Against These Things': Man Films Racial Tirade at Coburg Park. *The Anti-Bogan*. https://theantibogan.wordpress.com/2017/01/06/people-must-stand-up-against-these-things-man-films-racial-tirade-at-coburg-park/

The New Zealand Law Commission. (2012). Harmful Digital Communication: The Adequacy of the Current Sanction and Remedies. *Ministerial Briefing Paper*.

The YouTube Team. (2017). Community Guidelines. *YouTube/Policy and Safety*. https://www.youtube.com/yt/policyandsafety/communityguidelines.html

Toor, A. (2016, August 4). Facebook, Google, and Twitter Combat Online Extremism with Targeted Videos. *The Verge*. http://www.theverge.com/2016/8/4/12373978/facebook-google-twitter-extremism-counter-speech-isd-isis

Trollbusters. (2017). *TrollBusters: Fighting Cyberbullying with Love*. https://www.facebook.com/onlinetrollbusters

Ungar, M. (2011). The Social Ecology of Resilience: Addressing Contextual and Cultural Ambiguity of a Nascent Construct. *American Journal of Orthopsychiatry, 81*(1), 1–17.

Waldron, J. (2010). Dignity and Defamation: The Visibility of Hate. *Harvard Law Review, 123*(7), 1596–1601.

Wales, J. (2016). Jimmy Wales on How to Foster a Positive Online Culture. *Civilination, YouTube*. https://www.youtube.com/watch?time_continue=18&v=b1W0l5vBIH8. Accessed 17 Mar 2017.

Watson, L. (2017). A Bunch of New Companies Just Pulled Their Google Ads Over Racist Videos. *Gizmodo*. https://www.gizmodo.com.au/2017/03/a-bunch-of-new-companies-just-pulled-their-google-ads-over-racist-videos/-Ld2tOgwTFKhL5mek.99. Accessed 25 Mar 2017.

Weckerle, A. (2013). *Civility in the Digital Age: How Companies and People Can Triumph Over Haters, Trolls, Bullies and Other Jerks*. Indianapolis: Pearson Education.

Weckerle, A. (2017). *Civilination*. http://civilination.org/. Accessed 27 Mar 2017.

Williams, G. (2009, April 7). Racist Premise of Our Constitution Remains. *Sydney Morning Herald*. http://www.smh.com.au/federal-politics/racist-premise-of-our-constitution-remains-20090406-9utm.html. Accessed 5 Jan 2017.

Young, H., Rooze, M., Russell, J., Ebner, J., & Schulten, N. (2016). *Evidence-Based Policy Advice: Final Report. Prevention, De-radicalisation and Citizenship*. TerRa: Terrorism and Radicalisation. http://terratoolkit.eu/wp-content/uploads/2016/12/TERRA-Evidence-based-Policy-Advice_English_Final-Report.pdf. Accessed 4 Mar 2017.

Index

A. Jakubowicz et al., *Cyber Racism and Community Resilience*, Palgrave Hate Studies,
DOI 10.1007/978-3-319-64388-5

N